Lecture Notes in Mathematics

Edited by J.-M. Morel, F. Takens and B. Teissier

Editorial Policy
for the publication of monographs

1. Lecture Notes aim to report new developments in all areas of mathematics – quickly, informally and at a high level. Monograph manuscripts should be reasonably self-contained and rounded off. Thus they may, and often will, present not only results of the author but also related work by other people. They may be based on specialized lecture courses. Furthermore, the manuscripts should provide sufficient motivation, examples and applications. This clearly distinguishes Lecture Notes from journal articles or technical reports which normally are very concise. Articles intended for a journal but too long to be accepted by most journals, usually do not have this "lecture notes" character. For similar reasons it is unusual for doctoral theses to be accepted for the Lecture Notes series.

2. Manuscripts should be submitted (preferably in duplicate) either to one of the series editors or to Springer-Verlag, Heidelberg. In general, manuscripts will be sent out to 2 external referees for evaluation. If a decision cannot yet be reached on the basis of the first 2 reports, further referees may be contacted: the author will be informed of this. A final decision to publish can be made only on the basis of the complete manuscript, however a refereeing process leading to a preliminary decision can be based on a pre-final or incomplete manuscript. The strict minimum amount of material that will be considered should include a detailed outline describing the planned contents of each chapter, a bibliography and several sample chapters.
Authors should be aware that incomplete or insufficiently close to final manuscripts almost always result in longer refereeing times and nevertheless unclear referees' recommendations, making further refereeing of a final draft necessary.
Authors should also be aware that parallel submission of their manuscript to another publisher while under consideration for LNM will in general lead to immediate rejection.

3. Manuscripts should in general be submitted in English.
Final manuscripts should contain at least 100 pages of mathematical text and should include
- a table of contents;
- an informative introduction, with adequate motivation and perhaps some historical remarks: it should be accessible to a reader not intimately familiar with the topic treated;
- a subject index: as a rule this is genuinely helpful for the reader.

Continued on inside back-cover

Lecture Notes in Mathematics 1803

Springer
Berlin
Heidelberg
New York
Hong Kong
London
Milan
Paris
Tokyo

Georg Dolzmann

Variational Methods for Crystalline Microstructure - Analysis and Computation

Springer

Author

Georg Dolzmann

Department of Mathematics
University of Maryland
College Park
MD 20742
Maryland, USA

e-mail: dolzmann@math.umd.edu
http://www.math.umd.edu/~dolzmann/

Cataloging-in-Publication Data applied for

A catalog record for this book is available from the Library of Congress.

Bibliographic information published by Die Deutsche Bibliothek
Die Deutsche Bibliothek lists this publication in the Deutsche Nationalbibliografie;
detailed bibliographic data is available in the Internet at http://dnb.ddb.de

Mathematics Subject Classification (2000): 74B20, 74G15, 74G65, 74N15, 65M60

ISSN 0075-8434
ISBN 3-540-00114-X Springer-Verlag Berlin Heidelberg New York

Springer-Verlag Berlin Heidelberg New York a member of BertelsmannSpringer
Science + Business Media GmbH

http://www.springer.de

Typesetting: Camera-ready TEX output by the author

SPIN: 10899540 41/3142/ du - 543210 - Printed on acid-free paper

Preface

The mathematical modeling of microstructures in solids is a fascinating topic that combines ideas from different fields such as analysis, numerical simulation, and materials science. Beginning in the 80s, variational methods have been playing a prominent rôle in modern theories for microstructures, and surprising developments in the calculus of variations were stimulated by questions arising in this context.

This text grew out of my *Habilitationsschrift* at the University of Leizpig, and would not have been possible without the constant support and encouragement of all my friends during the past years. In particular I would like to thank S. Müller for having given me the privilege of being a member of his group during my years in Leipzig in which the bulk of the work was completed.

Finally, the financial support through the Max Planck Institute for Mathematics in the Sciences, Leipzig, my home institution, the University of Maryland at College Park, and the NSF through grant DMS0104118 is gratefully acknowledged.

College Park, August 2002 *Georg Dolzmann*

Contents

1. Introduction

Many material systems show fascinating microstructures on different length scales in response to applied strains, stresses, or electromagnetic fields. They are at the heart of often surprising mechanical properties of the materials and a lot of research has been directed towards the understanding of the underlying mechanisms. In this text, we focus on two particular systems, shape memory materials and nematic elastomers, which display similar microstructures, see Figure 1.1, despite being completely different in nature. The reason for this remarkable fact is that the oscillations in the state variables are triggered by the same principle: breaking of symmetry associated with solid to solid phase transitions. In the first system we find an austenite-martensite transition, while the second system possesses an isotropic to nematic transition.

An extraordinarily successful model for the analysis of phase transitions and microstructures in elastic materials was proposed by Ball&James and Chipot&Kinderlehrer based on nonlinear elasticity. They shifted the focus from the purely kinematic theory studied so far to a variational theory. The fundamental assumption in their approach is that the observed microstructures correspond to elements of minimizing sequences rather than minimizers for a suitable free energy functional with an energy density that reflects the breaking of the symmetry by the phase transition. This leads to a variational problem of the type: minimize

$$\mathcal{J}(\boldsymbol{u},T) = \int_{\Omega} W(D\boldsymbol{u}(\boldsymbol{x}),T)\mathrm{d}\boldsymbol{x},$$

where $\Omega \subset \mathbb{R}^3$ denotes the reference configuration of the elastic body, $\boldsymbol{x}$ the spatial variable, $\boldsymbol{u} : \Omega \to \mathbb{R}^3$ the deformation, T the temperature, and $W : \mathbb{M}^{3\times3} \times \mathbb{R}^+ \to \mathbb{R}^+$ the energy density. The precise form of W depends on a large number of material parameters and is often not explicitly known. However, the strength of the theory is that no analytical formula for the energy density is needed. The behavior of deformations with small energy should be driven by the structure of the set of minima of W, the so-called energy wells, which are entirely determined by the broken symmetry.

These considerations lead naturally to the following two requirements for the energy density W. First, the fundamental axiom in continuum mechanics that the material response be invariant under changes of observers, i.e.,

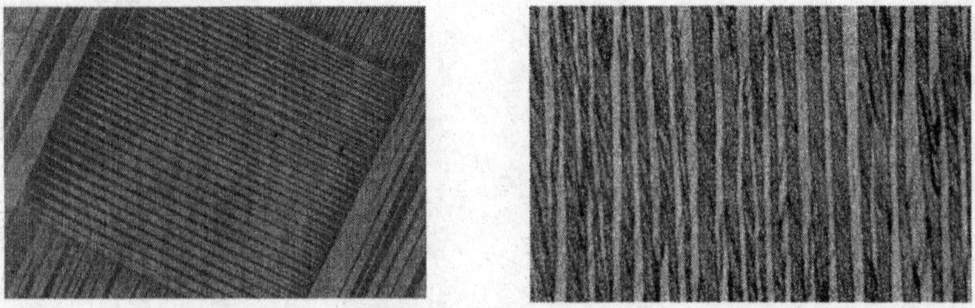

Fig. 1.1. Microstructures in a single crystal CuAlNi (courtesy of Chu&James, University of Minnesota, Minneapolis) and in a nematic elastomer (courtesy of Kundler&Finkelmann, University Freiburg).

$$W(RF,T) = W(F,T) \text{ for all } R \in \mathrm{SO}(3). \qquad (1.1)$$

Secondly, the invariance reflecting the symmetry of the high temperature phase, i.e.,

$$W(R^T FR,T) = W(F,T) \text{ for all } R \in \mathcal{P}_a, \qquad (1.2)$$

where $\mathcal{P}_a$ is the point group of the material in the high temperature phase. Here we restrict ourselves to invariance under the point group since the assumption that the energy be invariant under all bijections of the underlying crystalline lattice onto itself leads to a very degenerated situation with a fluid-like behavior of the material under dead-load boundary conditions. The two hypotheses (1.1) and (1.2) have far reaching consequences which we are now going to discuss briefly (see the Appendix for notation and terminology). We focus on isothermal situations, and we assume therefore that $W \geq 0$ and that the zero set K is not empty,

$$K(T) = \{X : W(X,T) = 0\} \neq \emptyset \text{ for all } T.$$

We deduce from (1.1) and (1.2) that

$$U \in K(T) \quad \Rightarrow \quad QUR \in K(T) \text{ for all } Q \in \mathrm{SO}(3),\ R \in \mathcal{P}_a. \qquad (1.3)$$

This implies that $K(T)$ is typically a finite union so-called energy wells,

$$K(T) = \mathrm{SO}(3)U_1 \cup \ldots \cup \mathrm{SO}(3)U_k. \qquad (1.4)$$

We refer to sets with such a structure often as multi-well sets. Here the matrices U_i describe the k different variants of the phases and k is determined from the point groups of the austenite and the martensite alone. A set of the form $\mathrm{SO}(3)U_i$ will in the sequel frequently be called energy well.

We now describe the framework for the mathematical analysis of martensitic transformations and its connection with quasiconvex hulls.

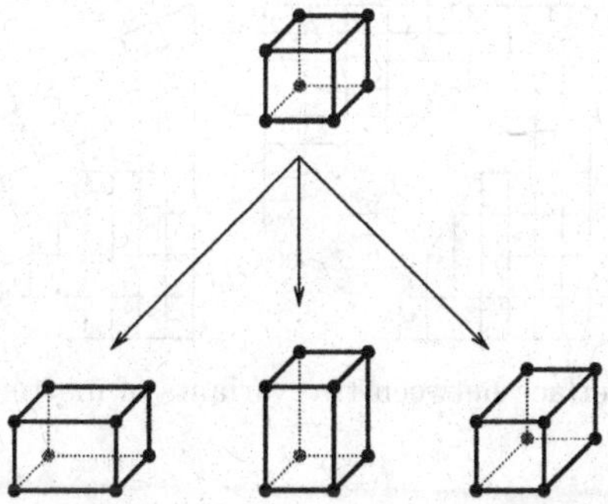

Fig. 1.2. The cubic to tetragonal phase transformation.

1.1 Martensitic Transformations and Quasiconvex Hulls

A fundamental example of an austenite-martensite transformation is the cubic to tetragonal transformation that is found in single crystals of certain Indium-Thallium alloys. The cubic symmetry of the austenitic or high temperature phase is broken upon cooling of the material below the transformation temperature. The three tetragonal variants that correspond to elongation of the cubic unit cell along one of the three cubic axes and contraction in the two perpendicular directions, are in the low temperature phase states of minimal energy, see Figure 1.2. If we use the undistorted austenitic phase as the reference configuration of the body under consideration, then the three tetragonal variants correspond to affine mappings described by the matrices

$$U_1 = \begin{pmatrix} \eta_2 & 0 & 0 \\ 0 & \eta_1 & 0 \\ 0 & 0 & \eta_1 \end{pmatrix}, \quad U_2 = \begin{pmatrix} \eta_1 & 0 & 0 \\ 0 & \eta_2 & 0 \\ 0 & 0 & \eta_1 \end{pmatrix}, \quad U_3 = \begin{pmatrix} \eta_1 & 0 & 0 \\ 0 & \eta_1 & 0 \\ 0 & 0 & \eta_2 \end{pmatrix}$$

with $\eta_2 > 1 > \eta_1 > 0$ (if the lattice parameter of the cubic unit cell is equal to one, then η_1 and η_2 are the lattice parameters of the tetragonal cell, i.e., are the lengths of the shorter and the longer sides of the tetragonal cell, respectively). In accordance with (1.3), the variants are related by

$$U_2 = R_2^T U_1 R_2, \quad U_3 = R_3^T U_1 R_3,$$

where R_2 and R_3 are elements in the cubic point group given by

$$R_2 = \begin{pmatrix} 0 & 1 & 0 \\ 1 & 0 & 0 \\ 0 & 0 & -1 \end{pmatrix}, \quad R_3 = \begin{pmatrix} 0 & 0 & 1 \\ 0 & -1 & 0 \\ 1 & 0 & 0 \end{pmatrix}.$$

A short calculation shows that no further variants can be generated by the action of the cubic point group.

The origin for the formation of microstructure lies in the fact that the different variants can coexist in one single crystal. They can be formed purely

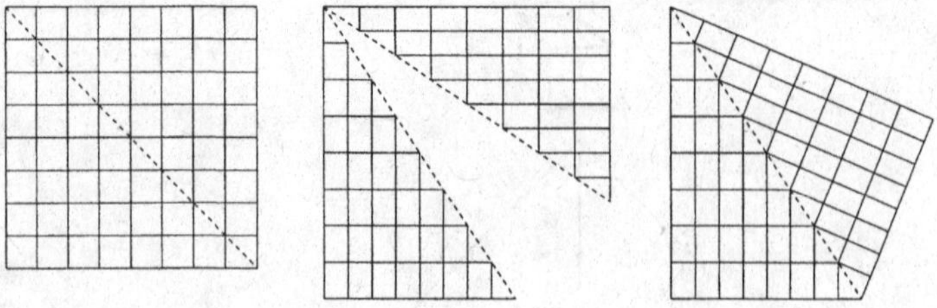

Fig. 1.3. Formation of an interface between two variants of martensite in a single crystal.

displacively, without diffusion of the atoms in the underlying lattice. This is illustrated in Figure 1.3. Consider a cut along a plane with normal $(1,1,0)$ The upper part is stretched in direction $(1,0,0)$ while the lower part is elongated in direction $(0,1,0)$. This corresponds to transforming the material into the phases described by U_1 and U_2, respectively. After a rigid rotation of the upper part, the pieces match exactly and the local neighborhood relations of the atoms have not been changed.

The austenite-martensite transition has an important consequence: the so-called shape memory effect, which leads to a number of interesting technological applications. A piece of material with a given shape for high temperatures can be easily deformed at low temperatures by rearranging the martensitic variants. Upon heating above the transformation temperature, the material returns to the uniquely determined high temperature shape, see Figure 1.4.

Mathematically, the existence of planar interfaces between two variants of martensite is equivalent to the existence of rank-one connections between the corresponding energy wells. Here we say that two wells $\mathrm{SO}(3)U_i$ and $\mathrm{SO}(3)U_j$, $i \neq j$, are rank-one connected if there exists a rotation $Q \in \mathrm{SO}(3)$ such that

$$QU_i - U_j = \boldsymbol{a} \otimes \boldsymbol{n} \quad \text{(Hadamard's jump condition)}, \tag{1.5}$$

where the matrix $\boldsymbol{a}\otimes\boldsymbol{n}$ is defined by $(\boldsymbol{a}\otimes\boldsymbol{n})_{kl} = (a_k n_l)$ for $\boldsymbol{a}, \boldsymbol{n} \in \mathbb{R}^3$. If (1.5) holds, then $\boldsymbol{n}$ is the normal to the interface. More importantly, the existence of rank-one connections together with the basic assumption that the energy density W be positive outside of $K(T)$ implies that W cannot be a convex function along rank-one lines. We conclude that W cannot be quasiconvex since rank-one convexity is a necessary condition for quasiconvexity. Recall that a function $W : \mathbb{M}^{m\times n} \to \mathbb{R}$ is said to be quasiconvex if the inequality

$$\int_{[0,1]^n} W(F + D\varphi)\mathrm{d}\boldsymbol{x} \geq \int_{[0,1]^n} W(F)\mathrm{d}\boldsymbol{x}$$

holds for all $F \in \mathbb{M}^{m\times n}$ and $\varphi \in C_0^\infty([0,1]^n;\mathbb{R}^m)$. Quasiconvexity of W is (under suitable growth and coercivity assumptions) equivalent to weak sequential lower semicontinuity of the corresponding energy functional and

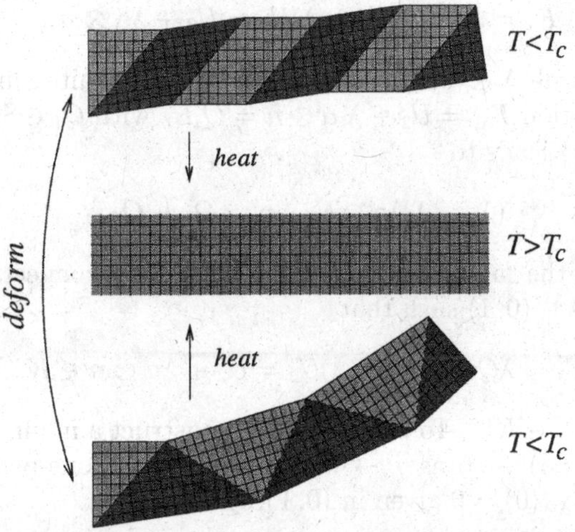

Fig. 1.4. The shape memory effect. The figure shows two macroscopically different deformations of the same lattice by using different arrangements of the martensitic variants. The upper configuration uses just two deformation gradients, the lower one six.

therefore the crucial notion of convexity in the vector valued calculus of variations (see Section A.1 for details about the relevant notions of convexity).

The mathematical interest in these problems lies exactly in this lack of quasiconvexity, that excludes a (naïve) application of the direct method in the calculus of variations. Many questions concerning existence, regularity, or uniqueness of minimizers remain open despite considerable progress in the past years. In this text we focus mainly on the question of for which affine boundary data the infimum of the energy is zero. This is not only a challenging mathematical problem, but also of considerable interest in applications since it characterizes for example all affine deformations of a shape memory material that can be recovered upon heating. From now on we drop the explicit dependence on the temperature T in the notation since we are restricting ourselves to isothermal processes. We therefore define the quasiconvex hull K^{qc} of the compact zero set K of W (at fixed temperature) as

$$K^{qc} = \{F : \inf_{\substack{u \in W^{1,\infty}(\Omega;\mathbb{R}^3) \\ u(x)=Fx \text{ on } \partial\Omega}} \int_\Omega W(Du)\mathrm{d}x = 0\}.$$

The following exemplary construction shows that K^{qc} is typically nontrivial, i.e., $K^{qc} \neq K$, and provides at the same time a link between the computation of K^{qc} and the fine-scale oscillations (or 'microstructure') between different variants observed in experiments, see Figure 1.1. Suppose that $QU_1 - U_2 = a \otimes n$ and let

$$F_\lambda = \lambda Q U_1 + (1-\lambda) U_2 = U_2 + \lambda \boldsymbol{a} \otimes \boldsymbol{n}. \tag{1.6}$$

Then $F_\lambda \notin K$ for all $\lambda \in (0,1)$, except for possibly a finite number of values since the assumption $F_{\lambda_i} = U_2 + \lambda_i \boldsymbol{a} \otimes \boldsymbol{n} = Q_i U_\ell$ with $Q_i \in \mathrm{SO}(3)$, $i = 1, 2$, and $\ell \in \{1, \ldots, k\}$ leads to

$$(\lambda_1 - \lambda_2) \boldsymbol{a} \otimes U_\ell^{-T} \boldsymbol{n} = Q_1 - Q_2.$$

This contradicts the fact that there are no rank-one connections in SO(3). We now choose $\lambda \in (0,1)$ such that

$$F_\lambda = \lambda Q U_1 + (1-\lambda) U_2 = U_2 + \lambda \boldsymbol{a} \otimes \boldsymbol{n} \notin K,$$

and assert that $F_\lambda \in K^{\mathrm{qc}}$. To prove this, we construct a minimizing sequence $\boldsymbol{u}_j$ such that $\mathcal{J}(\boldsymbol{u}_j) \to 0$ as $j \to \infty$. Let χ'_λ be the one-periodic function $\mathbb{R} \to \{0,1\}$ with $\chi_\lambda(0) = 0$ given in $(0,1]$ by

$$\chi'_\lambda(t) = \begin{cases} 0 & \text{if } t \in (0,\lambda], \\ 1 & \text{if } t \in (\lambda, 1], \end{cases}$$

and define

$$\chi_\lambda(t) = \int_0^t \chi'_\lambda(s) \mathrm{d}s, \quad \chi_{\lambda,j}(t) = \frac{1}{j} \chi_\lambda(jt) \text{ for } j \in \mathbb{N}.$$

Then

$$\widetilde{\boldsymbol{u}}_j(\boldsymbol{x}) = U_2 \boldsymbol{x} + \chi_{\lambda,j}(\langle \boldsymbol{x}, \boldsymbol{n} \rangle) \, \boldsymbol{a}$$

satisfies $D\widetilde{\boldsymbol{u}}_j(\boldsymbol{x}) \in \{QU_1, U_2\}$ a.e., $\widetilde{\boldsymbol{u}}_j$ converges to $F_\lambda \boldsymbol{x}$ strongly in L^∞ and weakly in $W^{1,\infty}$, and we only need to modify $\widetilde{\boldsymbol{u}}_j$ close to the boundary of Ω in order to correct the boundary data. This can be done for example by choosing a cut-off function $\varphi \in C^\infty([0,\infty))$ such that $\varphi \equiv 0$ in $[0, \frac{1}{2})$ and $\varphi \equiv 1$ in $[1, \infty)$. Then

$$\boldsymbol{u}_j(\boldsymbol{x}) = \big(1 - \varphi(j \operatorname{dist}(\boldsymbol{x}, \partial\Omega))\big) F_\lambda \boldsymbol{x} + \varphi(j \operatorname{dist}(\boldsymbol{x}, \partial\Omega)) \widetilde{\boldsymbol{u}}_j(\boldsymbol{x}) \tag{1.7}$$

has the desired properties, and a short calculation shows that the energy in the boundary layer converges to zero as $j \to \infty$. Hence $\mathcal{J}(\boldsymbol{u}_j) \to 0$, while $\mathcal{J}(F_\lambda \boldsymbol{x}) > 0$. Therefore it is energetically advantageous for the material to form fine microstructure, i.e., minimizing sequences develop increasingly rapid oscillations. This argument shows that $F_\lambda \in K^{\mathrm{qc}}$.

It is clear that this process for the construction of oscillating sequences and elements in K^{qc} can be iterated. In fact, if F_λ and G_μ are matrices with the foregoing properties that satisfy additionally $\operatorname{rank}(F_\lambda - G_\mu) = 1$, then $\gamma F_\lambda + (1-\gamma) G_\mu \in K^{\mathrm{qc}}$ for $\gamma \in [0,1]$.

We denote by K^{lc} the set of all matrices that can be generated in finitely many iterations, the so-called lamination convex hull of K. It yields an important lower bound for the quasiconvex hull K^{qc}:

$$K^{\mathrm{lc}} \subseteq K^{\mathrm{qc}}.$$

This is an extremely useful way to construct elements in K^{qc}, which we will refer to as lamination method, but it has also its limitations, mainly due to the fact that it requires to find explicitly rank-one connected matrices in the set K. Therefore the following equivalent definition of K^{qc}, which is in nice analogy with the dual definition of the convex hull of a compact set, is of fundamental importance

$$K^{\mathrm{qc}} = \big\{F : f(F) \leq \sup_{X \in K} f(X) \text{ for all } f : \mathbb{M}^{m\times n} \to \mathbb{R} \text{ quasiconvex}\big\}. \quad (1.8)$$

Thus K^{qc} is the set of all matrices that cannot be separated from K by quasiconvex functions. While this characterization seems to be at a first glance of little interest (the list of quasiconvex functions that is known in closed form is rather short), it allows us to relate K^{qc} to two more easily accessible hulls of K, the rank-one convex hull K^{rc} and the polyconvex hull K^{pc} which are defined analogously to (1.8) by replacing quasiconvexity with rank-one convexity and polyconvexity, respectively (see Section A.1 for further information). All these hulls will be referred to as 'semiconvex' hulls. The method for calculating the different semiconvex hulls based on this definition - separating points from a set by semiconvex functions - will be called the separation method in the sequel. Since rank-one convexity is a necessary condition for quasiconvexity and polyconvexity a sufficient one, we have the chain of inclusions

$$K^{\mathrm{lc}} \subseteq K^{\mathrm{rc}} \subseteq K^{\mathrm{qc}} \subseteq K^{\mathrm{pc}},$$

and frequently the most practicable way to obtain formulae for K^{qc} is to identify K^{lc} and K^{pc}.

There exists an equivalent characterization of the semiconvex hulls of K that turns out to be a suitable generalization of the representation of the convex hull of K as the set of all centers of mass of (nonnegative) probability measures supported on K. This formulation arises naturally by the search for a good description of the behavior of minimizing sequences for the energy functional. The sequence $\boldsymbol{u}_j$ constructed in (1.7) converges weakly in $W^{1,\infty}$ to the affine function $\boldsymbol{u}(\boldsymbol{x}) = F_\lambda \boldsymbol{x}$. This limit does not provide any information about the oscillations present in the sequence $\boldsymbol{u}_j$. The right limiting object, that encodes essential information about these oscillations, is the Young measure $\{\nu_{\boldsymbol{x}}\}_{\boldsymbol{x}\in\Omega}$ generated by the sequence of deformation gradients $D\boldsymbol{u}_j$. This approach was developed by L. C. Young in the context of optimal control problems, and introduced to the analysis of oscillations in partial differential equations by Tartar. By the fundamental theorem on Young measures (see Section A.1 for a statement) we may choose a subsequence (not relabeled) of the sequence $\boldsymbol{u}_j$ such that the sequence $D\boldsymbol{u}_j$ generates a gradient Young measure $\{\nu_{\boldsymbol{x}}\}_{\boldsymbol{x}\in\Omega}$ that allows us to calculate the limiting energy along the subsequence via the formula

$$\lim_{j\to\infty} \mathcal{J}(\boldsymbol{u}_j) = \lim_{j\to\infty} \int_\Omega W(D\boldsymbol{u}_j)\mathrm{d}\boldsymbol{x} = \int_\Omega \int_{\mathbb{M}^{3\times 3}} W(A)\mathrm{d}\nu_{\boldsymbol{x}}(A)\mathrm{d}\boldsymbol{x}.$$

We conclude that $\operatorname{supp}\nu_{\boldsymbol{x}} \subseteq K$ a.e. since $\mathcal{J}(\boldsymbol{u}_j) \to 0$ as $j \to \infty$. The averaging technique for gradient Young measures ensures the existence of a homogeneous gradient Young measure $\bar{\nu}$ with $\langle \bar{\nu}, id \rangle = F_\lambda$ and $\operatorname{supp}\bar{\nu} \subseteq K$. Here we say that the gradient Young measure $\{\nu_{\boldsymbol{x}}\}_{\boldsymbol{x}\in\Omega}$ is homogeneous if there exists a probability measure ν such that $\nu_{\boldsymbol{x}} = \nu$ for a.e. $\boldsymbol{x}$, see Section A.1 for details. For example, the sequence $\boldsymbol{u}_j$ in (1.7) generates the homogeneous gradient Young measure $\nu = \lambda\delta_{QU_1} + (1-\lambda)\delta_{U_2}$ which is usually referred to as a simple laminate. It turns out that K^{qc} is exactly the set of centers of mass of homogeneous gradient Young measures supported on K. Since this special class of probability measures is by the work of Kinderlehrer and Pedregal characterized by the validity of Jensen's inequality for quasiconvex functions we obtain

$$K^{\mathrm{qc}} = \{F = \langle \nu, id \rangle : \nu \in \mathcal{P}(K), f(\langle \nu, id \rangle) \le \langle \nu, f \rangle \text{ for all } f : \mathbb{M}^{m\times n} \to \mathbb{R} \text{ quasiconvex}\},$$

where $\mathcal{P}(K)$ denotes the set of all probability measures supported on K. This formula gives the convex hull of K if we replace in the definition quasiconvexity by convexity, since Jensen's inequality holds for all probability measures. The obvious generalizations to rank-one convexity and polyconvexity provide equivalent definitions for the other semiconvex hulls which are extremely useful in the analysis of properties of generic elements in these hulls. In particular, since the minors M of a matrix F are polyaffine functions (i.e., both M and $-M$ are polyconvex) we conclude that the polyconvex hull is determined from measures $\nu \in \mathcal{P}$ that satisfy the so-called minor relations

$$\langle \nu, M \rangle = M(\langle \nu, id \rangle) \text{ for all minors } M.$$

In the three-dimensional situation this reduces to

$$K^{\mathrm{pc}} = \{F = \langle \nu, id \rangle : \nu \in \mathcal{P}(K), \operatorname{cof} F = \langle \nu, \operatorname{cof} \rangle, \det F = \langle \nu, \det \rangle\}. \quad (1.9)$$

1.2 Outline of the Text

With the definitions of the foregoing section at hand, we now briefly describe the topics covered in this text. A more detailed description can be found at the beginning of the each chapter.

In Chapter 2 we focus on the question of how to find closed formulae for semiconvex hulls of compact sets. There does not yet exist a universal method for the resolution of this problem, but three different approaches are emerging as very powerful tools to which we refer to as the separation method, the lamination method, and the splitting method, respectively. As a very instructive example for the separation and the splitting method, we analyze a discrete set of eight points. Then we characterize the semiconvex hulls for compact sets in 2×2 matrices with fixed determinant that are invariant under multiplication form the left by SO(2). We thus find a closed formula for all sets arising in two-dimensional models for martensitic phase transformations. The results are then extended to sets invariant under O(2, 3) which are relevant for the description of thin film models proposed by Bhattacharya and James. As a preparation for the relaxation results in Chapter 3 we conclude with the analysis of sets defined by singular values.

Chapter 3 is inspired by the experimental pictures of striped domain patterns in nematic elastomers, see Figure 1.1, which arise in connection with a nematic to isotropic phase transformation. For this material, Bladon, Terentjev and Warner derived a closed formula W_{BTW} for the free energy density which depends on the deformation gradient F and the nematic director $\boldsymbol{n}$, but not on derivatives of $\boldsymbol{n}$. From the point of view of energy minimization, one can first minimize in the director field $\boldsymbol{n}$ and one obtains a new energy W that depends only on the singular values of the deformation gradient. This is a consequence of the isotropy of the high temperature phase which has in contrast to crystalline materials no distinguished directions. We derive an explicit formula for the macroscopic energy W^{qc} of the system which takes into account the energy reduction by (asymptotically) optimal fine structures in the material.

We begin the discussion of aspects related to the numerical analysis of microstructures in Chapter 4. The standard finite element method seeks a minimizer of the nonconvex energy in a finite dimensional space. Assume for example that Ω_h is a quasiuniform triangulation of Ω and that $\mathcal{S}_h(\Omega_h)$ is a finite element space on Ω_h (a typical choice being the space of all continuous functions that are affine on the triangles in Ω_h). Suitable growth conditions on the energy density imply the existence of a minimizer $\boldsymbol{u}_h \in \mathcal{S}_h(\Omega_h)$ that satisfies $\mathcal{J}(\boldsymbol{u}_h) \leq \mathcal{J}(\boldsymbol{v}_h)$ for all $\boldsymbol{v}_h \in \mathcal{S}_h(\Omega_h)$. To be more specific, let us assume that we minimize the energy subject to affine boundary conditions of the form (1.6). If one chooses for $\boldsymbol{v}_h$ an interpolation of the functions $\boldsymbol{u}_j$ in (1.7) with j carefully chosen depending on h, then one obtains easily that the energy converges to zero at a certain rate for $h \to 0$,

$$\mathcal{J}(\boldsymbol{u}_h) \leq ch^{\alpha}, \quad \alpha, c > 0. \tag{1.10}$$

The fundamental question is now what this information about the energy implies about the finite element minimizer. Recent existence results for nonconvex problems indicate that minimizers of $\mathcal{J}$ are not unique (if they exist), and in this case the bound (1.10) is rather weak. It merely shows that the infimum can be well approximated in the finite element space (absence of the Lavrentiev or gap phenomenon). On the other hand, if $\mathcal{J}$ fails to have a minimizer, then it is interesting to investigate whether for a suitable set of boundary conditions the minimizing microstructure (the Young measure ν)

is unique and what (1.10) implies for u_h as h tends to zero. In particular, if ν is unique, then the sequence Du_h should display very specific oscillations, namely those recorded in ν. This is the motivation behind Luskin's stability theory for microstructures, and we present a general framework that allows one to give a precise meaning to this intuitive idea. Our approach is inspired by the idea that stability should be a natural consequence of uniqueness, and we verify this philosophy for affine boundary conditions $F \in K^{qc}$ based on an algebraic condition, called condition (C_b), on the set K. The new feature in our analysis is to base all estimates on inequalities for polyconvex measures. This method turns out to be very flexible and we include extensions to thin film theories and more general boundary conditions that correspond to higher order laminates.

We apply the general theory developed in Chapter 4 to examples of martensitic phase transformations in Chapter 5. Our focus is to analyze the uniqueness of simple laminates ν based on our algebraic condition (C_b). It turns out that typically simple laminates are uniquely determined from their center of mass unless the lattice parameters in the definition of the set K satisfy a certain algebraic condition. In theses exceptional case, we provide explicit characterizations for the possible microstructures underlying the affine deformation $F = \langle \nu, id \rangle$.

Algorithmic aspects in the numerical analysis and computation of microstructures are addressed in Chapter 6. Nonconvex variational problems typically fail to have a classical solution or they have solutions with intrinsically complicated geometries that cannot be approximate numerically. A natural remedy here is either to replace the original functional by its relaxation or to broaden the class of solutions, i.e., to pass from minimizers in the classical sense to minimizing Young measures. The latter approach requires a discretization of the space of all probability measures, and it is an open problem to find an efficient way to accomplish this. However, (finite) laminates are accessible to computations and the explicitly known examples indicate that this subclass is in fact sufficient in many cases (the Tartar square being a notable exception, see Section 6.2). In this chapter we first discuss an algorithm for the computation of the rank-one convex envelope of a given energy density which is an upper bound for the relaxed or quasiconvexified energy density in the relaxed energy functional. We then modify this algorithm to find minimizing laminates of finite order and we prove rigorously convergence of the proposed algorithms under reasonable assumptions.

In Chapter 7 we present detailed references to literature closely related to the text. Additional references can be found in the appendices in which we summarize some background material that is not included in the text. Appendix A contains information about notions of convexity and classical criteria for the existence of rank-one connections between matrices. Basic nomenclature in crystallography is summarized in Appendix B, and the mathematical notation used throughout the text is collected in Appendix C.

2. Semiconvex Hulls of Compact Sets

Quasiconvex hulls of sets and envelopes of functions are at the heart of the analysis of phase transformations by variational techniques. In this chapter we address the question of how to find for a given compact set K its quasiconvex hull K^{qc}. In general, this is an open problem since all characterizations of K^{qc} are intimately connected to the notion of quasiconvexity of functions in the sense of Morrey, and the understanding of quasiconvexity is one of the great challenges in the calculus of variations. However, three conceptual approaches can be identified that allow one to resolve the problem for important classes of sets K.

The separation method. By definition, K^{qc} is the set of all matrices F that cannot be separated by quasiconvex functions from K. If an inner bound for K^{qc} is known, i.e., a set $\mathcal{A}$ with $\mathcal{A} \subseteq K^{\mathrm{qc}}$, for example $\mathcal{A} = K^{\mathrm{lc}}$ or $\mathcal{A} = K^{\mathrm{rc}}$, then it suffices to find for all $F \in \mathbb{M}^{m\times n} \setminus \mathcal{A}$ a quasiconvex function f with $f \leq 0$ in $\mathcal{A}$ and $f(F) > 0$. An example of this approach is our proof for the formulae of the semiconvex hulls of the set K with eight points in Theorem 2.1.1 based on Šverák's examples of quasiconvex functions on symmetric matrices.

The lamination method. Since quasiconvexity implies rank-one convexity, the segments $\lambda F_1 + (1 - \lambda)F_2$, $\lambda \in [0, 1]$ are contained in K^{qc} if the end points F_1 and F_2 belong to K^{qc} and satisfy $\operatorname{rank}(F_1 - F_2) = 1$. This fact motivated the definition of the lamination convex hull K^{lc} which is one of the fundamental inner bounds for K^{qc}. The lamination method tries to identify K^{lc} and then to use K^{lc} as an inner bound for the separation method.

The splitting method. This method is well adapted to situations where a good outer bound $\mathcal{A}$ for K^{qc} is known, i.e., $K^{\mathrm{qc}} \subseteq \mathcal{A}$. It can be interpreted as a reversion of the lamination method. Assume for example that $\mathcal{A}$ is given by a finite number of inequalities, as for example in Theorems 2.1.1 or 2.2.3 below. We call a point F an unconstrained point, if we have strict inequality in all inequalities defining $\mathcal{A}$, and a constrained point else. Assume that F is an unconstrained point. By continuity, $F_t = F(\mathbb{I} + t\boldsymbol{a} \otimes \boldsymbol{b})$ belongs to $\mathcal{A}$ for all $\boldsymbol{a} \in \mathbb{R}^m$, $\boldsymbol{b} \in \mathbb{R}^n$ and t sufficiently small. Since K is compact, all the hulls are compact and we may suppose that $\mathcal{A}$ is compact. We now define $t^{\pm}$ to be the smallest positive (largest negative) parameter t for which the matrix F_t satisfies equality in at least one of the inequalities in the definition of $\mathcal{A}$, i.e., is a constrained point. Therefore it suffices to prove that the constrained points belong to K^{lc} in order to show that $\mathcal{A} \subseteq K^{\mathrm{lc}}$. Then

$$K^{\mathrm{qc}} \subseteq \mathcal{A} \subseteq K^{\mathrm{lc}} \subseteq K^{\mathrm{qc}}$$

and thus $\mathcal{A} = K^{\mathrm{qc}}$. Having equality in one of the inequalities defining $\mathcal{A}$ provides additional information and often simplifies the proof that certain matrices belong to K^{lc}. Moreover, this procedure can be iterated and is also applicable to sets with a determinant constraint, since $\det F_t = \det F$ if $\langle a, b\rangle = 0$. A convenient choice of $\mathcal{A}$ is often K^{pc}. We use this general strategy, which we call the splitting method, extensively in the following sections.

A natural question that arises in this context is whether the polyconvex hull K^{pc} coincides with the rank-one convex hull K^{rc} (or even the lamination convex hull K^{lc}), since in these cases a characterization of K^{qc} is automatically obtained. More generally, one can ask whether $\mathcal{M}^{\mathrm{pc}}(K) = \mathcal{M}^{\mathrm{rc}}(K)$, i.e., whether the set of all polyconvex measures satisfying the minors relations is equal to the set of all laminates characterized by Jensen's inequality for all rank-one convex functions. It turns out that this is typically not true, and to illustrate this point we consider the two-well problem where K is given by

$$K = \mathrm{SO}(2)U_1 \cup \mathrm{SO}(2)U_2, \quad U_1 = \begin{pmatrix} \alpha & 0 \\ 0 & 1/\alpha \end{pmatrix}, \quad U_2 = \begin{pmatrix} 1/\alpha & 0 \\ 0 & \alpha \end{pmatrix}$$

with $\alpha > 1$. Then $\mathcal{M}^{\mathrm{pc}}$ is the set of all probability measures that satisfy the minors relation

$$\det \langle \nu, id\rangle = \langle \nu, \det \rangle. \tag{2.1}$$

Consider now the special class of all probability measures $\nu \in \mathcal{M}^{\mathrm{pc}}(K)$ that are supported on three points,

$$\nu = \lambda_1\delta_{X_1} + \lambda_2\delta_{X_2} + \lambda_3\delta_{X_3}, \quad X_i \in K,\ \lambda_i \in (0,1),\ \lambda_1 + \lambda_2 + \lambda_3 = 1.$$

We assert that every $\nu \in \mathcal{M}^{\mathrm{qc}} \subseteq \mathcal{M}^{\mathrm{pc}}$ of this form is in fact a laminate. Indeed, it follows from Šverák's results that at least two of the three matrices X_i must be rank-one connected. We may thus assume that $X_i \in \mathrm{SO}(2)U_i$ for $i = 1, 2$ with $\mathrm{rank}(X_1 - X_2) = 1$. The minors relation and the expansion

$$\begin{aligned} \det F = \sum_{i=1}^{3} \lambda_i \det X_i - \frac{\lambda_1\lambda_2}{1-\lambda_3}\det(X_2 - X_1) \\ - \lambda_3(1-\lambda_3)\det\Big(\big(\frac{\lambda_1}{1-\lambda_3}X_1 + \frac{\lambda_2}{1-\lambda_3}X_2\big) - X_3\Big) \end{aligned}$$

now imply that

$$\mathrm{rank}\Big(X_3 - \big(\frac{\lambda_1}{\lambda_1+\lambda_2}X_1 + \frac{\lambda_2}{\lambda_1+\lambda_2}X_2\big)\Big) = 1,$$

i.e., that $\nu \in \mathcal{M}^{\mathrm{rc}}(K)$ is a second order laminate. In order to prove that $\mathcal{M}^{\mathrm{pc}}(K) \neq \mathcal{M}^{\mathrm{qc}}(K)$ it suffices to construct a solution of (2.1) with matrices X_i that are not rank-one connected. For a specific example, we choose $\alpha > 1$ to be the solution of $(\alpha + 1/\alpha)^2 = 8$ and we define

$$X_1 = U_1, \quad X_2 = JU_1 = \begin{pmatrix} 0 & -1/\alpha \\ \alpha & 0 \end{pmatrix}, \quad X_3 = JU_2 = \begin{pmatrix} 0 & -\alpha \\ 1/\alpha & 0 \end{pmatrix},$$

where J denotes the counter-clockwise rotation by $\pi/2$, and we fix $\lambda = 1/3$. Then

$$\det \sum_{i=1}^{3} \lambda_i X_i = \frac{1}{9} \begin{pmatrix} \alpha & -(\alpha + 1/\alpha) \\ \alpha + 1/\alpha & 1/\alpha \end{pmatrix} = \frac{1}{9}(1 + (\alpha + \frac{1}{\alpha})^2) = 1,$$

and consequently $\nu \in \mathcal{M}^{\mathrm{pc}}(K) \setminus \mathcal{M}^{\mathrm{qc}}(K)$. Note, however, that $\langle \nu, id \rangle \in K^{\mathrm{qc}}$. We use the idea to find elements in K^{pc} by solving the minors relations for example in Theorem 2.2.6 to find an SO(2) invariant set K with $K^{\mathrm{pc}} \neq K^{\mathrm{rc}}$.

This chapter is divided into several sections in which we discuss quasiconvex hulls for various classes of sets K. We begin with a very illustrative example of a discrete set in symmetric 2×2 matrices first analyzed by Dacorogna and Tanteri. For this set, the separation method provides one with an outer bound for K^{qc} and the splitting method allows one to show that $K^{\mathrm{qc}} = K^{\mathrm{lc}}$ for a large range of parameters. We then turn towards sets with constant determinant that are invariant under SO(2). This is the class of sets relevant in two-dimensional models for phase transformations. Theorem 2.2.5 gives an explicit characterization of the semiconvex hulls. These results are then extended to O(2, 3) invariant sets related to the modeling of thin films. Finally, we study sets defined by singular values and the results obtained here are an important ingredient in the derivation of the quasiconvex envelope of a model energy for nematic elastomers in Chapter 3.

2.1 The Eight Point Example

We begin the analysis of semiconvex hulls with the following example of a discrete set with eight points in symmetric 2×2 matrices that was introduced by Dacorogna and Tanteri. Theorem 2.1.1 extends their results and the proof illustrates the power of the separation and the splitting method. Before we state the theorem, we discuss briefly quasiconvexity in symmetric matrices.

Let $\mathcal{S}^n \subset \mathbb{M}^{n \times n}$ denote the subspace of all symmetric matrices. A function $f : \mathcal{S}^n \to \mathbb{R}$ is said to be quasiconvex if for all matrices $F \in \mathcal{S}^n$ and all $\varphi \in W_0^{2,\infty}(\Omega)$ the inequality

$$\int_\Omega f(F + D^2\varphi) dx \geq \int_\Omega f(F) dx$$

holds. The proof of our characterization of the quasiconvex hull relies on Šverák's result that the function

$$g_0(F) = \begin{cases} \det F & \text{if } F \text{ is positive definite,} \\ 0 & \text{else,} \end{cases}$$

is quasiconvex on symmetric matrices. This function gives new restrictions on gradient Young measures ν supported on symmetric matrices since they have to satisfy the inequality

$$g_0(\langle \nu, id \rangle) \le \langle \nu, g_0 \rangle. \tag{2.2}$$

It is an open problem whether these functions can be extended to all $n \times n$ matrices and whether they can be used directly for separation on $\mathbb{M}^{n\times n}$.

The statement of the following theorem emphasizes the fact that the description of K^{pc} (K^{qc}, K^{rc}) involves additional conditions compared to the formulae for $\text{conv}(K)$ (K^{pc}, K^{qc}).

Theorem 2.1.1. *Let*

$$K = \left\{ F = \begin{pmatrix} x & y \\ y & z \end{pmatrix} : |x| = a, \ |y| = b, \ |z| = c \right\}$$

with constants a, b, $c > 0$. *Then*

$$\text{conv}(K) = \left\{ F = \begin{pmatrix} x & y \\ y & z \end{pmatrix} : |x| \le a, \ |y| \le b, \ |z| \le c \right\}$$

and

$$K^{\text{pc}} = \left\{ F \in \text{conv}(K) : (x-a)(z+c) \le y^2 - b^2, \ (x+a)(z-c) \le y^2 - b^2 \right\}.$$

Moreover, the following assertions hold:
i) If $ac - b^2 < 0$ *then*

$$K^{(2)} = K^{\text{lc}} = K^{\text{rc}} = \left\{ F \in \text{conv}(K) : |y| = b \right\}.$$

ii) If $ac - b^2 \ge 0$ *then* $K^{(4)} = K^{\text{lc}} = K^{\text{rc}} = K^{\text{qc}}$ *and*

$$\begin{aligned} K^{\text{qc}} = \{ F \in K^{\text{pc}} : \ & (x-a)(z-c) \ge (|y| - b)^2, \\ & (x+a)(z+c) \ge (|y|-b)^2 \}. \end{aligned}$$

Remark 2.1.2. Note that K^{qc} is quasiconvex as a set in all 2×2 matrices, not only as a subset of the symmetric matrices.

Remark 2.1.3. It is an open problem to find a formula for the quasiconvex hull of K in the case $ac - b^2 < 0$.

Remark 2.1.4. A short calculation shows that the additional inequalities in the definition of $K^{\rm lc}$ are true for all $F \in K^{\rm pc}$ if $ac - b^2 = 0$ and that consequently $K^{\rm lc} = K^{\rm pc}$. This was already shown by Dacorogna and Tanteri. The authors also obtained the formula for $K^{\rm lc}$ in the case $ac - b^2 < 0$ and observed that $K^{\rm lc}$ is always contained in the intersection of the convex hull of K with the exterior of the two hyperboloids $(x-a)(z+c) = y^2 - b^2$ and $(x+a)(z-c) = y^2 - b^2$. However, they did not identify the latter set as $K^{\rm pc}$.

We now turn towards the proof of the theorem which we split into several steps. First we derive the formula for the polyconvex hull of K. Then the formulae for the lamination convex hulls in statements i) and ii) in the theorem are established. Finally we present the proof for the representation of the quasiconvex hull for $ac - b^2 \geq 0$.

The Polyconvex Hull of *K*. Among the different notions of convexity, polyconvexity has the most similarities with classical convexity. One instance is the following representation for the polyconvex hull $K^{\rm pc}$,

$$K^{\rm pc} = \big\{F \in \mathbb{M}^{2\times 2} : (F, \det F) \in \operatorname{conv}(\widetilde{K})\big\}, \tag{2.3}$$

where

$$\widetilde{K} = \big\{(F, \det F) : F \in K\big\} \subset \mathbb{R}^5 .$$

By definition, K consists of symmetric matrices, and therefore $\widetilde{K}$ and $\operatorname{conv}(\widetilde{K})$ are contained in a four-dimensional subspace of $\mathbb{R}^5$. We restrict our calculations to this subspace by the identifications

$$K = \left\{ \begin{pmatrix} a \\ c \\ b \end{pmatrix}, \begin{pmatrix} -a \\ c \\ b \end{pmatrix}, \begin{pmatrix} a \\ -c \\ b \end{pmatrix}, \begin{pmatrix} -a \\ -c \\ b \end{pmatrix}, \begin{pmatrix} a \\ c \\ -b \end{pmatrix}, \begin{pmatrix} -a \\ c \\ -b \end{pmatrix}, \begin{pmatrix} a \\ -c \\ -b \end{pmatrix}, \begin{pmatrix} -a \\ -c \\ -b \end{pmatrix} \right\}$$

and

$$\widetilde{K} = \big\{(x, z, y, xz - y^2) : (x, z, y) \in K\big\}.$$

We denote the eight points in $\widetilde{K}$ by $\widetilde{\boldsymbol{f}}_1, \dots, \widetilde{\boldsymbol{f}}_8$.

Since K is a finite set, $\operatorname{conv}(\widetilde{K})$ is a polyhedron in $\mathbb{R}^4$, which is the intersection of a finite number of half spaces. Moreover, on each face of $\operatorname{conv}(\widetilde{K})$ we must have at least four points in $\widetilde{K}$ that span a three-dimensional hyperplane in $\mathbb{R}^4$. A short calculation shows that the following list of six normals completely describes the convex hull of $\widetilde{K}$:

$$\begin{aligned} \boldsymbol{n}_1 &= (c, a, 0, -1), & \boldsymbol{n}_2 &= (-c, a, 0, 1,), \\ \boldsymbol{n}_3 &= (c, -a, 0, 1), & \boldsymbol{n}_4 &= (-c, -a, 0, -1), \\ \boldsymbol{n}_5 &= (0, 0, 1, 0), & \boldsymbol{n}_6 &= (0, 0, -1, 0). \end{aligned}$$

It turns out that the hyperplanes defined by $\boldsymbol{n}_1, \ldots, \boldsymbol{n}_4$ contain six points in K,

$$\begin{aligned}
\langle \widetilde{\boldsymbol{f}}_4, \boldsymbol{n}_1 \rangle = \langle \widetilde{\boldsymbol{f}}_8, \boldsymbol{n}_1 \rangle &= -3ac + b^2 < ac + b^2 = \langle \widetilde{\boldsymbol{f}}_i, \boldsymbol{n}_1 \rangle, \quad i \notin \{4,8\},\\
\langle \widetilde{\boldsymbol{f}}_3, \boldsymbol{n}_2 \rangle = \langle \widetilde{\boldsymbol{f}}_7, \boldsymbol{n}_2 \rangle &= -3ac - b^2 < ac - b^2 = \langle \widetilde{\boldsymbol{f}}_i, \boldsymbol{n}_2 \rangle, \quad i \notin \{3,7\},\\
\langle \widetilde{\boldsymbol{f}}_2, \boldsymbol{n}_3 \rangle = \langle \widetilde{\boldsymbol{f}}_6, \boldsymbol{n}_3 \rangle &= -3ac - b^2 < ac - b^2 = \langle \widetilde{\boldsymbol{f}}_i, \boldsymbol{n}_3 \rangle, \quad i \notin \{2,6\},\\
\langle \widetilde{\boldsymbol{f}}_1, \boldsymbol{n}_4 \rangle = \langle \widetilde{\boldsymbol{f}}_5, \boldsymbol{n}_4 \rangle &= -3ac + b^2 < ac + b^2 = \langle \widetilde{\boldsymbol{f}}_i, \boldsymbol{n}_4 \rangle, \quad i \notin \{1,5\},
\end{aligned}$$

and that the faces of the polyhedron defined by $\boldsymbol{n}_5$ and $\boldsymbol{n}_6$ contain four points,

$$\begin{aligned}
\langle \widetilde{\boldsymbol{f}}_j, \boldsymbol{n}_5 \rangle &= -b < b = \langle \widetilde{\boldsymbol{f}}_i, \boldsymbol{n}_5 \rangle, \qquad i = 1,2,3,4,\ j = 5,6,7,8,\\
\langle \widetilde{\boldsymbol{f}}_j, \boldsymbol{n}_6 \rangle &= -b < b = \langle \widetilde{\boldsymbol{f}}_i, \boldsymbol{n}_6 \rangle, \qquad i = 5,6,7,8,\ j = 1,2,3,4.
\end{aligned}$$

We include a few details of the argument that leads to this characterization. The general idea is to choose (at least) four of the eight points $\boldsymbol{f}_i$ in K and to check whether the corresponding points $\widetilde{\boldsymbol{f}}_i$ span a three-dimensional plane in $\widetilde{K}$. If this plane constitutes a separating plane (a face), then its normal is added to the description of $\operatorname{conv}(\widetilde{K})$.

As an example we choose $\boldsymbol{f}_i$, $i = 1,3,5,7$. We need to check whether these points define a plane in $\mathbb{R}^4$, i.e., whether the three vectors $\widetilde{\boldsymbol{f}}_1 - \widetilde{\boldsymbol{f}}_7$, $\widetilde{\boldsymbol{f}}_3 - \widetilde{\boldsymbol{f}}_7$, and $\widetilde{\boldsymbol{f}}_5 - \widetilde{\boldsymbol{f}}_7$ are linearly independent. It turns out that these vectors are linearly dependent, and thus it is possible to add a further vector to the list of vectors, say $\boldsymbol{f}_2$. Now the vectors $\widetilde{\boldsymbol{f}}_j - \widetilde{\boldsymbol{f}}_7$, $j = 1, 2, 3, 5$, are linearly independent since the rank of the matrix

$$A = \begin{pmatrix} 0 & -2a & 0 & 0 \\ 2c & 2c & 0 & 2c \\ 2b & 2b & 2b & 0 \\ 2ac & 0 & 0 & 2ac \end{pmatrix}$$

is three (if one deletes the second column which corresponds to $\widetilde{\boldsymbol{f}}_2 - \widetilde{\boldsymbol{f}}_7$, then the rank is only two). The corresponding normal vector has to satisfy $A^T\boldsymbol{n} = 0$ and this leads to the linear system

$$\begin{pmatrix} 0 & c & b & ac \\ -a & c & b & 0 \\ 0 & 0 & b & 0 \\ 0 & c & 0 & ac \end{pmatrix} \boldsymbol{n} = \boldsymbol{0}$$

that has the solution $\boldsymbol{n}_1 = (c, a, 0, -1)$. The corresponding equation in the description of K^{pc} is

$$\langle \widetilde{\boldsymbol{f}}, \boldsymbol{n}_1 \rangle = cx + az - (xz - y^2) \le ac + b^2.$$

Geometrically this is a hyperboloid which contains in fact six of the eight points in K. It turns out that there are four hyperboloids that are important in the description of K^{pc}. In the following diagrams we sketch the set K in $\mathbb{R}^3$ (with the x-axis to the left, the y-axis out of the paper plane, and the z-axis oriented upwards) and we circle the points in K that define the corresponding separating planes or hyperboloids.

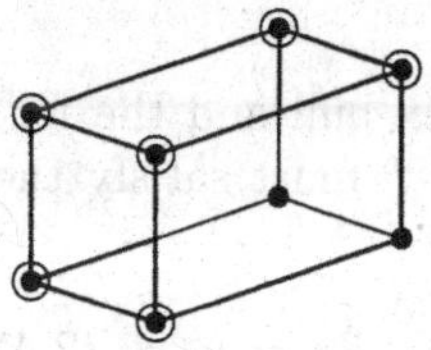

Separating hyperboloid 1: We choose $j = 1, 2, 3, 5$. The corresponding system for the normal is given by

$$\begin{pmatrix} 0 & 0 & 1 & 0 \\ 0 & 1 & 0 & a \\ 1 & 0 & 0 & c \end{pmatrix} \boldsymbol{n} = \boldsymbol{0}$$

and the solution gives $\boldsymbol{n}_1 = (c, a, 0, -1)$.

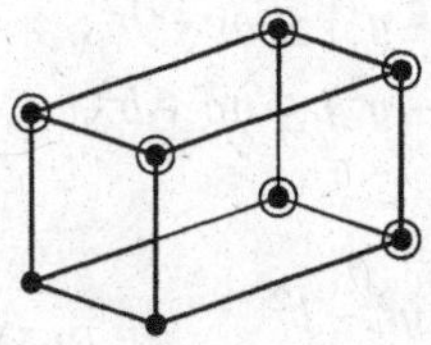

Separating hyperboloid 2: We choose $j = 1, 2, 4, 6$. The corresponding system for the normal is given by

$$\begin{pmatrix} 1 & 0 & 0 & c \\ 0 & 0 & 1 & 0 \\ 0 & -1 & 0 & a \end{pmatrix} \boldsymbol{n} = \boldsymbol{0}$$

and the solution gives $\boldsymbol{n}_2 = (-c, a, 0, 1)$.

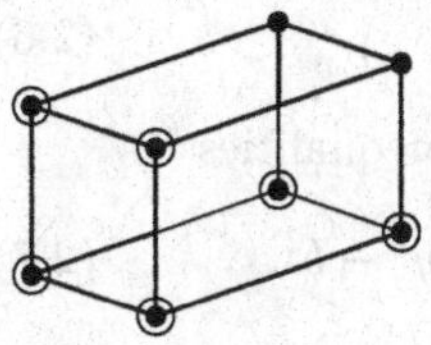

Separating hyperboloid 3: We choose $j = 1, 3, 4, 7$. The corresponding system for the normal is given by

$$\begin{pmatrix} 0 & 0 & 1 & 0 \\ 0 & 1 & 0 & a \\ 1 & 0 & 0 & -c \end{pmatrix} \boldsymbol{n} = \boldsymbol{0}$$

and the solution gives $\boldsymbol{n}_3 = (c, -a, 0, 1)$.

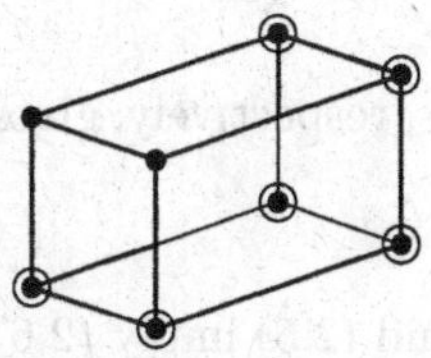

Separating hyperboloid 4: We choose $j = 2, 3, 4, 8$. The corresponding system for the normal is given by

$$\begin{pmatrix} 1 & 0 & 0 & -c \\ 0 & 0 & 1 & 0 \\ 0 & 1 & 0 & -a \end{pmatrix} \boldsymbol{n} = \boldsymbol{0}$$

and the solution gives $\boldsymbol{n}_4 = (c, a, 0, 1)$.

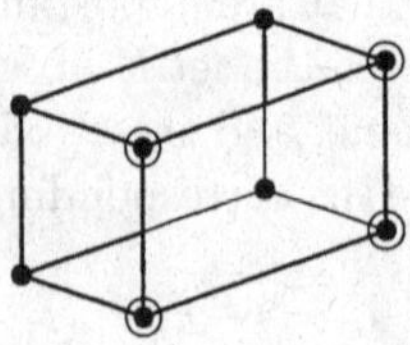

Four points on a b-face of the cube. The corresponding system for the normal is given by

$$\begin{pmatrix} 1\,0\,0\,c \\ 0\,1\,0\,a \\ a\,c\,0\,0 \end{pmatrix} \boldsymbol{n} = \boldsymbol{0}$$

and the solution gives $\boldsymbol{n}_5 = (0,0,1,0)$. Replacing b by $-b$ leads to $\boldsymbol{n}_6 = (0,0,-1,0)$.

In view of the representation (2.3) for the polyconvex hull and the formulae for the normals, this implies that all points in K^{pc} must satisfy the convex inequality

$$|y| \le b \tag{2.4}$$

as well as the additional inequalities

$$\begin{aligned} cx + az - (xz - y^2) &\le ac + b^2, & -cx + az + (xz - y^2) &\le ac - b^2, \\ cx - az + (xz - y^2) &\le ac - b^2, & -cx - az - (xz - y^2) &\le ac + b^2, \end{aligned}$$

which we can rewrite as

$$\begin{array}{rlrl} -(x-a)(z-c) \le -y^2 + b^2, & (x+a)(z-c) \le \;\; y^2 - b^2, \\ (x-a)(z+c) \le \;\; y^2 - b^2, & -(x+a)(z+c) \le -y^2 + b^2. \end{array} \tag{2.5}$$

We now assert that this system of inequalities is equivalent to the conditions

$$|x| \le a, \quad |z| \le c, \quad |y| \le b \tag{2.6}$$

describing the convex hull of K and the two additional inequalities

$$(x+a)(z-c) \le y^2 - b^2, \quad (x-a)(z+c) \le y^2 - b^2. \tag{2.7}$$

This proves the formula for the polyconvex hull of K. In fact, the sum of the two upper and the two lower inequalities in (2.5) implies

$$az \le ac \quad \text{and} \quad -az \le ac,$$

and the sum of the two left and the two right inequalities, respectively, gives

$$cx \le ac \quad \text{and} \quad -cx \le ac.$$

Therefore $|z| \le c$ and $|x| \le a$ and this proves that (2.4) and (2.5) imply (2.6) and (2.7). Conversely, if the convex inequalities $|x| \le a$, $|z| \le c$, and $|y| \le b$ in (2.6) hold, then $x - a \le 0$, $z - c \le 0$ and $-y^2 + b^2 \ge 0$. Consequently $-(x-a)(z-c) \le -y^2 + b^2$. Similarly, we have $x + a \ge 0$, $z + c \ge 0$ and thus $-(x+a)(z+c) \le -y^2 + b^2$, as asserted. This concludes the proof of the formula for K^{pc} for all parameters a, b, $c > 0$.

The Lamination Convex Hull of K for $ac - b^2 < 0$. We now turn towards proving the formula for K^{lc} and we assume first that $ac - b^2 < 0$. We let

$$\mathcal{A} = \{F \in \mathrm{conv}(K) : |y| = b\}.$$

In this case, none of the matrices in $\mathcal{A}$ with $y = b$ is rank-one connected to any of the matrices in $\mathcal{A}$ with $y = -b$, and the assertion follows essentially from the following well-known locality property of the rank-one convex hull.

Proposition 2.1.5. *Assume that K is compact and that K^{rc} consists of two compact components C_1 and C_2 with $C_1 \cap C_2 = \emptyset$. Then*

$$K^{\mathrm{rc}} = (K \cap C_1)^{\mathrm{rc}} \cup (K \cap C_2)^{\mathrm{rc}}. \tag{2.8}$$

Clearly, all elements in $\mathcal{A}$ can be constructed using the rank-one connections between the four matrices in K with $y = b$ and $y = -b$, respectively. The observation is now that the polyconvex hull is not connected, since $K^{\mathrm{pc}} \cap \{F : |y| \le \varepsilon\} = \emptyset$ for $\varepsilon > 0$ so small that $\varepsilon^2 < b^2 - ac$. Indeed, summation of the two inequalities in the definition of K^{pc} implies $ac - xz \ge b^2 - y^2$ or, equivalently, $0 > ac - b^2 + y^2 \ge xz$. Thus necessarily either $x > 0$ and $z < 0$ or $x < 0$ and $z > 0$. In the former case the first inequality cannot hold since

$$(z - a)(z + c) \le y^2 - b^2 \quad \Leftrightarrow \quad 0 \le x(z + c) - az \le ac - b^2 + y^2 < 0.$$

In the latter case the second inequality is violated. We may now apply Proposition 2.1.5 and we conclude that $K^{\mathrm{lc}} = K^{\mathrm{rc}} = \mathcal{A}$.

The Lamination Convex Hull of K for $ac - b^2 \ge 0$. Assume now that $ac - b^2 \ge 0$, and let $\mathcal{A}$ be given by

$$\mathcal{A} = \{F \in K^{\mathrm{pc}} : (x - a)(z - c) \ge (|y| - b)^2,\ (x + a)(z + c) \ge (|y| - b)^2\}.$$

By symmetry, we may suppose in the following arguments that $y \ge 0$. Then this set is described by three types of inequalities, namely the *stripes*

$$|x| \le a, \quad |z| \le c, \quad |y| \le b \tag{2.9}$$

defining the convex hull of K, the *hyperboloids*

$$(x - a)(z + c) \le y^2 - b^2, \quad (x + a)(z - c) \le y^2 - b^2, \tag{2.10}$$

in the definition of K^{pc}, and the *cones*

$$(x - a)(z - c) \ge (y - b)^2, \quad (x + a)(z + c) \ge (y - b)^2. \tag{2.11}$$

To simplify notation, we write

$$\bar{F} = \begin{pmatrix} \xi & \eta \\ \eta & \zeta \end{pmatrix}.$$

Since $\mathcal{A}$ is compact, it suffices to prove that all points $\bar{F} \in \mathcal{A}$ that satisfy equality in at least one of the inequalities in the definition of $\mathcal{A}$ can be constructed as laminates. To see this, assume that $\bar{F}$ lies in the interior of $\mathcal{A}$. The idea is to split $\bar{F}$ along a rank-one line into two rank-one connected matrices $\bar{F}^{\pm}$ that satisfy equality in at least one of the inequalities in the definition of $\mathcal{A}$. We set

$$\begin{aligned} t^- &= \sup\{t < 0 : \bar{F} + tw \otimes w \text{ satisfies at least one equality in } \mathcal{A}\}, \\ t^+ &= \inf\,\{t > 0 : \bar{F} + tw \otimes w \text{ satisfies at least one equality in } \mathcal{A}\}. \end{aligned}$$

By assumption, $t^- < 0 < t^+$ and we define $\bar{F}^{\pm} = \bar{F} + t^{\pm} w \otimes w$. Then $\bar{F} = (t^- \bar{F}^+ - t^- F^+)/(t^+ - t^-)$ and it suffices to show that $\bar{F}^{\pm}$ are contained in K^{lc}.

Assume thus that $\bar{F} \in \mathcal{A}$ satisfies equality in at least one inequality in the definition of $\mathcal{A}$. We have to prove that this implies $\bar{F} \in K^{\mathrm{lc}}$. This is immediate for the convex inequalities $|x| \le a$, $|y| \le b$, and $|z| \le c$. For example, if $\xi = a$, then by (2.10) $|\eta| = b$ and by symmetry we may assume that $\eta = b$. Then (2.9) implies that $\zeta = \lambda c + (1-\lambda)(-c)$ for some $\lambda \in [0,1]$ and thus

$$\bar{F} = \lambda \begin{pmatrix} a & b \\ b & c \end{pmatrix} + (1-\lambda) \begin{pmatrix} a & b \\ b & -c \end{pmatrix}, \quad \begin{pmatrix} a & b \\ b & c \end{pmatrix} - \begin{pmatrix} a & b \\ b & -c \end{pmatrix} = 2c e_2 \otimes e_2.$$

The argument is similar for $|\zeta| = c$. Finally, if $|\eta| = b$ and $\eta \ge 0$, then

$$(\xi, \eta) \in \operatorname{conv}\big\{(a,c), (-a,c), (a,-c), (-a,-c)\big\},$$

and therefore $\bar{F} \in K^{(2)}$.

Assume next that $\bar{F}$ lies on the surface of one of the cones

$$(x-a)(z-c) \ge (y-b)^2, \quad (x+a)(z+c) \ge (y-b)^2.$$

These cones are the rank-one cones centered at points in K, and we may suppose that $\bar{F}$ is contained in the rank-one cone C_1 given by

$$C_1 = \Big\{F : \det\Big[F - \begin{pmatrix} a & b \\ b & c \end{pmatrix}\Big] = (x-a)(z-c) - (y-b)^2 = 0\Big\};$$

the argument is similar in the other case. The cone C_1 intersects the part of the boundary of the convex hull of K that is contained in the plane $\{y = -b\}$, which by the foregoing arguments is contained in $K^{(2)}$. We will show that $\bar{F}$ belongs to a rank-one segment between a point G in this intersection and the point $F_1 \in K$, where F_1 and G are given by

$$F_1 = \begin{pmatrix} a & b \\ b & c \end{pmatrix} \quad \text{and} \quad G = \begin{pmatrix} \overline{x} & -b \\ -b & \overline{z} \end{pmatrix}, \qquad |\overline{x}| \le a,\ |\overline{z}| \le c.$$

This implies $\bar{F} \in K^{(3)} \subseteq K^{\mathrm{lc}}$. In order to prove this fact, let

$$R = F_1 - \bar{F} = \begin{pmatrix} a-\xi & b-\eta \\ b-\eta & c-\zeta \end{pmatrix}.$$

By assumption, $\det R = 0$, and we seek a $t \in \mathbb{R}$ such that

$$F_1 + tR = \begin{pmatrix} a+t(a-\xi) & b+t(b-\eta) \\ b+t(b-\eta) & c+t(c-\zeta) \end{pmatrix} = G.$$

This implies

$$t = -\frac{2b}{b-\eta}$$

and thus

$$\overline{x} = a - \frac{2b(a-\xi)}{b-\eta}, \quad \overline{z} = c - \frac{2b(c-\xi)}{b-\eta}.$$

Clearly $\overline{x} \le a$ and we only have to check that $\overline{x} \ge -a$, or equivalently

$$\frac{a}{b} \ge \frac{a-\xi}{b-\eta}.$$

To establish this inequality, we subtract the equality $(x-a)(z-c) = (y-b)^2$ in the definition of C_1 from the inequality $(x+a)(z-c) \le y^2 - b^2$ in the definition of K^{pc}, and we obtain that $\bar{F}$ satisfies

$$2a(\zeta - c) \le (-2b)(b-\eta).$$

Therefore, again in view of the definition of C_1,

$$\frac{a}{b} \ge \frac{b-\eta}{c-\zeta} = \frac{a-\xi}{b-\eta},$$

and this proves the bounds for $\overline{x}$; the arguments for $\overline{z}$ are similar. Since $G \in K^{(2)}$ we conclude

$$\bar{F} = \frac{1+t}{t} F_1 - \frac{1}{t} G = \frac{b+\eta}{2b} F_1 + \frac{b-\eta}{2b} G \in K^{(3)}.$$

It remains to consider the case that $\bar{F} \in \mathcal{A}$ satisfies equality in one of the inequalities defining the one-sheeted hyperboloids. Assume thus that

$$(\xi + a)(\zeta - c) = \eta^2 - b^2.$$

The idea is to use the geometric property of one-sheeted hyperboloids H already observed by Šverák, namely that for each point F on H there exist two straight lines intersecting at F that are contained in H, and that correspond to rank-one lines in the space of symmetric matrices. More precisely, we seek solutions $\boldsymbol{w} = (u, v) \in \mathbb{S}^1$ of

$$\bar{F} + t\boldsymbol{w} \otimes \boldsymbol{w} \in H \quad \text{or} \quad (\xi + tu^2 + a)(\zeta + tv^2 - c) = (\eta + tuv)^2 - b^2.$$

This is equivalent to the quadratic equation

$$u^2(\zeta - c) + v^2(\xi + a) = 2uv\eta.$$

Since $u = 0$ and $v = 0$ are only solutions for $\xi = -a$ and $\zeta = c$, respectively, we may assume that $u,\, v \neq 0$. In this case there are two solutions for the ratio $\tau = u/v$, given by

$$\tau_{1,2} = \frac{\eta \pm b}{\zeta - c}.$$

The strategy is now to split $\bar{F}$ into two points $\bar{F}^{\pm}$ along one of these rank-one lines that satisfy equality in at least two of the inequalities in the definition of $\mathcal{A}$. Let

$$t^- = \sup\{t < 0 : \bar{F} + t\boldsymbol{w} \otimes \boldsymbol{w} \text{ realizes two equalities in } \mathcal{A}\},$$
$$t^+ = \inf\{t > 0 : \bar{F} + t\boldsymbol{w} \otimes \boldsymbol{w} \text{ realizes two equalities in } \mathcal{A}\}.$$

By assumption, $t^- < 0 < t^+$ and we define $\bar{F}^{\pm} = \bar{F} + t^{\pm}\boldsymbol{w} \otimes \boldsymbol{w}$. In view of the foregoing arguments, the matrices $\bar{F}^{\pm}$ belong either to $K^{(3)}$ or to the intersection $\widetilde{H}$ of the two hyperboloids,

$$\widetilde{H} = \{F : (x + a)(z - c) = y^2 - b^2,\, (x - a)(z + c) = y^2 - b^2\}.$$

The formula for the lamination convex hull is therefore established if we show that $\widetilde{H} \subseteq K^{\mathrm{lc}}$. By symmetry it suffices again to prove this for all $F \in \widetilde{H}$ with $y \geq 0$. Now, if $F \in \widetilde{H}$, then

$$az = cx, \quad \text{and} \quad xz - ac = y^2 - b^2.$$

Thus the intersection of the two hyperboloids can be parameterized for $y \geq 0$ by

$$t \mapsto \left(\sigma\sqrt{\frac{a}{c}}\sqrt{t^2 + ac - b^2}, t, \sigma\sqrt{\frac{c}{a}}\sqrt{t^2 + ac - b^2}\right), \quad \sigma \in \{\pm 1\}, \quad t \geq 0.$$

We may assume that $\sigma = 1$. In this case the inequality $(x-a)(z-c) \geq (y-b)^2$ in the definition of $\mathcal{A}$ is equivalent to $(ac - b^2)(b - t)^2 \leq 0$ and this implies $t = b$, and thus $F \in K$, if $ac - b^2 > 0$. If $ac - b^2 = 0$, then the intersection of the hyperboloids coincides with the rank-one line between

$$\begin{pmatrix} a & b \\ b & c \end{pmatrix} \text{ and } \begin{pmatrix} -a & -b \\ -b & -c \end{pmatrix}, \quad \text{or} \quad \begin{pmatrix} -a & b \\ b & -c \end{pmatrix} \text{ and } \begin{pmatrix} a & -b \\ -b & c \end{pmatrix},$$

and consequently $F \in K^{(1)}$. This proves the formula for the lamination convex hull.

The Quasiconvex Hull of K for $ac - b^2 \geq 0$. It remains to prove that for $ac - b^2 \geq 0$ all points in $K^{\text{pc}} \setminus K^{\text{lc}}$ can be separated from K (or equivalently from K^{lc}) with quasiconvex functions. Recall that by Remark 2.1.4 the quasiconvex and the polyconvex hull coincide for $ac - b^2 = 0$. We may therefore assume in the following that $ac - b^2 > 0$. We divide the proof of this assertion into three steps. First we show that the additional inequalities in the definition of K^{lc} are only active for $x,\ z \geq 0$ or $x,\ z \leq 0$. Then we construct a sufficiently rich family of quasiconvex functions that separates points from K, and finally we prove the theorem.

Reduction to the Case $x,\ y,\ z \geq 0$. By symmetry we may always assume that $y \geq 0$. In this case the formula for K^{lc} contains the additional inequalities

$$(x + a)(z + c) \geq (y - b)^2, \quad (x - a)(z - c) \geq (y - b)^2. \tag{2.12}$$

Assume, for example, that $F \in K^{\text{pc}}$ with $x \leq 0$ and $z \geq 0$. The inequalities in (2.12) can be rewritten as

$$(x \pm a)(z \pm c) \geq b^2 - y^2 + 2y^2 - 2by.$$

It follows from $F \in K^{\text{pc}}$ that $-(x + a)(z - c) \geq b^2 - y^2$. The foregoing inequalities are thus true if

$$(x \pm a)(z \pm c) \geq -(x + a)(z - c) + 2y^2 - 2by$$

is satisfied. The equation with the minus and the plus sign are equivalent to

$$2x(z - c) + 2y(b - y) \geq 0 \quad \text{and} \quad 2z(x + a) + 2y(b - y) \geq 0, \tag{2.13}$$

respectively. Since by assumption $x \leq 0$, $z \leq c$, and $y \in [0, b]$, the first inequality in (2.13) holds and this implies the first inequality (2.12). Similarly, the second inequality in (2.13) is true in view of $z \geq 0$ and $x \geq -a$, and consequently the second inequality in (2.12) follows.

Construction of Quasiconvex Functions. From now on we assume that $x,\ y,\ z \geq 0$ and that $x \neq a$, $z \neq c$ and $y \neq b$. We need to show that all points in K^{pc} with $(x - a)(z - c) < (y - b)^2$ can be separated from K by quasiconvex functions. This will be done using the Šverák's remarkable result that the functions

$$g_\ell(F) = \begin{cases} |\det F| & \text{if the index of } F \text{ is } \ell, \\ 0 & \text{otherwise,} \end{cases}$$

are quasiconvex on symmetric matrices. Here the index of the symmetric matrix F is the number of its negative eigenvalues.

We begin by calculating the intersection of the boundary of the cone $(x - a)(z - c) \geq (y - b)^2$ with K^{pc} for fixed $y \in [0, b)$. This intersection can be parameterized by

$$t \mapsto \begin{pmatrix} t & y \\ y & c+(y-b)^2/(t-a) \end{pmatrix}, \quad t \in I_y = \left[\frac{ay}{b}, a + \frac{b(y-b)}{c}\right],$$

and we write $t \mapsto F(y,t)$ or $t \mapsto F_{y,t}$ for simplicity. A short calculation shows that $|I_y| = (ac - b^2)(b-y)/(bc) > 0$. We define quasiconvex functions $f_{y,t}$ on the space of all symmetric matrices by

$$f_{y,t}(F) = g_0(F - F_{y,t}), \quad y \in [0,b),\, t \in I_y,$$

and show first that $f_{y,t} = 0$ on K. In order to do this, it suffices to prove that all the matrices of the form $F - F(y,t)$ with $F \in K$ are not positive definite. In fact,

$$\det\left[\begin{pmatrix} a & \pm b \\ \pm b & \pm c \end{pmatrix} - F_{y,t}\right] = (a-t)(\pm c - c) + (y-b)^2 - (\pm b - y)^2 \leq 0,$$

and thus all matrices of the form $F - F_{y,t}$, with $F \in K$ and $F_{11} = a$ are not positive definite. Moreover,

$$\left[\begin{pmatrix} -a & \pm b \\ \pm b & \pm c \end{pmatrix} - F_{y,t}\right] = \begin{pmatrix} -a-t & \pm b - y \\ \pm b - y & \pm c - c - \frac{(y-b)^2}{t-a} \end{pmatrix},$$

and consequently all the matrices $X = F - F_{y,t}$ with $F \in K$ and $F_{11} = -a$ satisfy $X_{11} \leq 0$ and are therefore not positive definite.

Separation of Points from K^{lc} with Quasiconvex Functions. Recall that we assume that

$$\bar{F} = \begin{pmatrix} \xi & \eta \\ \eta & \zeta \end{pmatrix} \quad \text{with } \xi, \eta, \zeta \geq 0 \text{ and } \xi \neq a,\, \zeta \neq c,\, \eta \neq b.$$

We have to show that all matrices $\bar{F} \in K^{\mathrm{pc}}$ with

$$(\xi - a)(\zeta - c) < (\eta - b)^2 \tag{2.14}$$

can be separated from K by a quasiconvex function. We will achieve this by analyzing different regions for ξ which are related to the intersection of K^{qc} with the plane $y = \eta$. In this plane, the intersection of K^{qc} with the quadrant $x \geq 0$ and $z \geq 0$ is bounded by the three hyperbolic arcs

$$(x-a)(z-c) = (\eta - b)^2 \quad \text{and} \quad (x \pm a)(z \mp c) = \eta^2 - b^2.$$

In the following we consider four different regions for $\xi \geq 0$ which are defined by the points where two of these hyperbolic arcs intersect (see Figure 2.1). More precisely, the hyperbola $(x-a)(z-c) = (\eta-b)^2$ intersects the hyperbola $(x+a)(z-c) = \eta^2 - b^2$ for $x_1 = a\eta/b$ and the hyperbola $(x-a)(z+c) = \eta^2 - b^2$ for $x_2 = a + b(\eta - b)/c$. The four cases now correspond to $\xi \in [0, x_1]$, $\xi \in (x_1, x_2)$, $\xi = x_2$, and $\xi \in (x_2, a)$, respectively. We begin with the last case first.

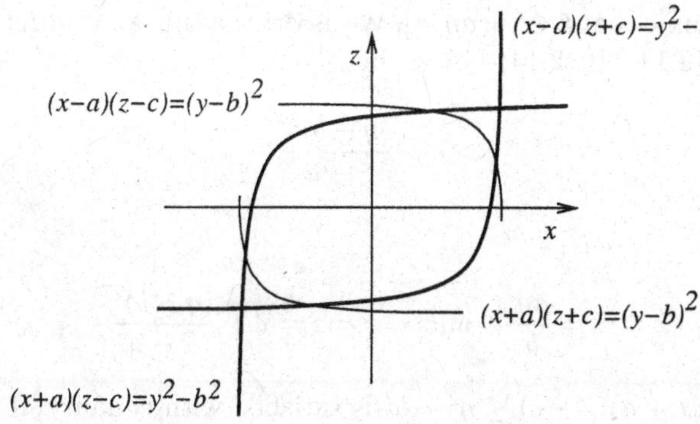

Fig. 2.1. *The polyconvex hull (bounded by the thick solid lines) and the quasiconvex hull (the intersection of the four hyperbolic arcs) of K in the plane $\{y = \eta > 0\}$.*

Case a) Assume that $\xi > a + b(\eta - b)/c$. If $(\xi - a)(\zeta + c) \leq \eta^2 - b^2$, then

$$\zeta \geq -c + \frac{b^2 - \eta^2}{a - \xi} > -c - \frac{c(b^2 - \eta^2)}{b(\eta - b)} = \frac{c\eta}{b}.$$

We define

$$G_\eta = F\big(\eta, a + \frac{b(\eta - b)}{c}\big), \; Z = \bar{F} - G_\eta = \begin{pmatrix} \xi - a - b(\eta - b)/c & 0 \\ 0 & \zeta - c\eta/b \end{pmatrix}.$$

Then Z is positive definite and in view of Section 2.1 the function $g_0(F - G_\eta)$ separates $\bar{F}$ from K^{lc}. On the other hand, if $(\xi - a)(\zeta + c) > \eta^2 - b^2$, then $\bar{F}$ does not belong to K^{pc}.

Case b) Assume that $\xi = a + b(\eta - b)/c$. We assert that in view of (2.14) we may find an $\widetilde{x} \in I_\eta = (a\eta/b, \xi)$ such that

$$Z = \bar{F} - F(\eta, \widetilde{x}) = \begin{pmatrix} \xi - \widetilde{x} & 0 \\ 0 & \zeta - c - (\eta - b)^2/(\widetilde{x} - a) \end{pmatrix}$$

is positive definite. This follows easily since $\bar{F}$ is positive definite if and only if $\xi > \widetilde{x}$ and

$$\zeta - c - \frac{(\eta - b)^2}{\widetilde{x} - a} > 0 \quad \text{or} \quad (\widetilde{x} - a)(\zeta - c) - (\eta - b)^2 < 0.$$

In view of (2.14) we can choose $\widetilde{x} < \xi$ close enough to x such that the latter inequality holds. Therefore we can separate $\bar{F}$ from K^{lc} with the function $g_0(F - F(\eta, \widetilde{x}))$.

Case c) Assume that $\xi \in \big(a\eta/b, a + b(\eta - b)/c\big)$. The conclusion follows as in case b), since we can choose by continuity $\widetilde{x} \in \big(a\eta/b, \xi\big)$ such that $\bar{F} - F(\eta, \widetilde{x})$ is positive definite.

Case d) Assume that $\xi \in [0, a\eta/b]$. We assert that no point in K^{pc} satisfies the inequality (2.14). If (2.14) holds, then

$$\zeta > c + \frac{(\eta - b)^2}{\xi - a}.$$

However, for

$$x = \widetilde{x} = \frac{a\eta}{b} \quad \text{and} \quad z = \widetilde{z} = c + \frac{(\eta - b)^2}{\widetilde{x} - a}$$

the inequality $(x + a)(z - c) \leq \eta^2 - b^2$ is satisfied with equality. If

$$\xi \leq \frac{a\eta}{b} \quad \text{and} \quad \zeta > c + \frac{(\eta - b)^2}{\widetilde{x} - a},$$

then $(\xi + a)(\zeta - c) > \eta^2 - b^2$, a contradiction. The same argument applied to gradient Young measures ν supported on K shows in connection with the inequality (2.2) that K is quasiconvex as a subset in the space of all 2×2 matrices. This concludes the proof of the theorem.

2.2 Sets Invariant Under SO(2)

We now turn to the case of SO(2) invariant sets K which is relevant for two-dimensional models in elasticity. After a discussion of k-well problems with finite k, we present the general result for arbitrary compact sets K with equal determinant. The section is concluded by the surprising example of the set K given by

$$K = \mathrm{SO}(2)\begin{pmatrix} \alpha & 0 \\ 0 & \beta \end{pmatrix} \cup \mathrm{SO}(2)\begin{pmatrix} \beta & 0 \\ 0 & \alpha \end{pmatrix} \cup \mathrm{SO}(2), \quad \alpha > \beta > 1,$$

for which K^{rc} and K^{pc} do not coincide. This example shows that the general result for sets with constant determinant cannot be extended to sets without this constraint.

The One-well Problem. We begin with the case of one well, $K = \mathrm{SO}(2)$. In view of the definition of the polyconvex hull we must have $\det F = 1$ for all $F \in K^{\mathrm{pc}}$. However, the only elements F in the *convex* hull of SO(2) with $\det F = 1$ are proper rotations and therefore

$$K^{\mathrm{rc}} = K^{\mathrm{qc}} = K^{\mathrm{pc}} = \mathrm{SO}(2).$$

Incompatible Wells. The result for the one-well problem has the following generalization. Assume that $K = \mathrm{SO}(2)U_1 \cup \ldots \cup \mathrm{SO}(2)U_k$, $k \geq 1$, where the wells $\mathrm{SO}(2)U_i$ and $\mathrm{SO}(2)U_j$ are incompatible for $i \neq j$, i.e., $\det(QU_i - U_j) \neq 0$ for $i \neq j$ and $Q \in \mathrm{SO}(2)$. In this situation, Šverák proved that the hulls are still trivial,

$$K^{\mathrm{rc}} = K^{\mathrm{qc}} = K^{\mathrm{pc}} = \mathrm{SO}(2)U_1 \cup \ldots \cup \mathrm{SO}(2)U_k.$$

Motivated by applications to phase transformations we analyze next the so-called k-well problem in two dimensions where K is given by

$$F = \mathrm{SO}(2)U_1 \cup \ldots \cup \mathrm{SO}(2)U_k, \quad k \geq 2,$$

with positive definite matrices U_i that satisfy $\det U_i = \Delta > 0$ for $i = 1, \ldots, k$.

The Two-well Problem. This problem has attracted a lot of attention since it is the basic model for martensitic transformations in two dimensions. Here K is given by $K = \mathrm{SO}(2)U_1 \cup \mathrm{SO}(2)U_2$ with $\det U_1, \det U_2 > 0$. We may assume that $U_1 = \mathbb{I}$ and that U_2 is diagonal, i.e., that K is of the form

$$K = \mathrm{SO}(2) \cup \mathrm{SO}(2)H, \quad H = \mathrm{diag}(\lambda, \mu),\ 0 < \mu \leq 1 \leq \lambda. \tag{2.15}$$

In view of Proposition A.2.5, the two wells are compatible. A short calculation shows that there exists only one rank-one connection between the wells if one of the two eigenvalues of H is equal to one (in fact, $\mathrm{rank}(\mathbb{I} - H) = 1$), and that there exist two rank-one connections if both eigenvalues are different from one.

Remark 2.2.1. Assume that $\lambda > 1 = \mu$, (the case $\mu < 1 = \lambda$ is similar) and let

$$K = \mathrm{SO}(2) \cup \mathrm{SO}(2)U, \quad U = \begin{pmatrix} \lambda & 0 \\ 0 & 1 \end{pmatrix}.$$

Then $K^{\mathrm{pc}} = K^{\mathrm{qc}} = K^{\mathrm{rc}} = K^{\mathrm{lc}} = K^{(1)}$ and

$$K^{\mathrm{qc}} = \Big\{F : F = QU_s,\ Q \in \mathrm{SO}(2),\ U_s = \begin{pmatrix} s & 0 \\ 0 & 1 \end{pmatrix},\ s \in [1, \lambda]\Big\}.$$

We include a short proof of this fact which is a nice application of the minors relation.

Proof. Assume that $\nu \in \mathcal{M}^{\mathrm{pc}}(K)$ is a polyconvex measure represented by $\nu = t\varrho + (1-t)\sigma U$, where ϱ and σ are probability measures supported on $\mathrm{SO}(2)$ and $t \in [0,1]$. Let

$$R = \langle \varrho, id \rangle = \begin{pmatrix} r_1 & -r_2 \\ r_2 & r_1 \end{pmatrix}, \quad S = \langle \sigma, id \rangle = \begin{pmatrix} s_1 & -s_2 \\ s_2 & s_1 \end{pmatrix} \in \mathrm{conv}(\mathrm{SO}(2)).$$

Then the minors relations imply

$$F = tR + (1-t)SU, \quad \det F = t + (1-t)\lambda.$$

It follows from

$$R : \operatorname{cof}(SU) = (r_1 s_1 + r_2 s_2)(1+\lambda),$$

the expansion (C.3) for the determinant, and Young's inequality that

$$\begin{aligned}\det F &= t^2 \det R + t(1-t)R : \operatorname{cof}(SU) + (1-t)^2 \det S \det U \\ &\le t^2 + t(1-t)(\lambda+1) + (1-t)^2\lambda \\ &= t + (1-t)\lambda\end{aligned}$$

with strict inequality unless

$$r_1 = s_1,\ r_2 = s_2,\ \det R = 1,\ \det S = 1.$$

It follows that $R = S \in \mathrm{SO}(2)$ and that ν is given by

$$\nu = t\delta_R + (1-t)\delta_{RU},$$

i.e., $F \in K^{(1)}$. This proves the assertion. □

The general two-well problem was solved by Šverák. The crucial observation is that any element in the *convex* hull of K is of the form $F = Y + ZH$ where Y and Z are conformal matrices. Therefore it is natural to introduce new coordinates in $\mathbb{M}^{2\times 2}$ by

$$F = \begin{pmatrix} y_1 & -y_2 \\ y_2 & y_1 \end{pmatrix} + \begin{pmatrix} z_1 & -z_2 \\ z_2 & z_1 \end{pmatrix}\begin{pmatrix} \lambda & 0 \\ 0 & \mu \end{pmatrix}$$

and to write $F = (\boldsymbol{y}, \boldsymbol{z})$ with $\boldsymbol{y}, \boldsymbol{z} \in \mathbb{R}^2$.

Theorem 2.2.2. *Assume that K is given by (2.15) with $0 < \mu < 1 < \lambda$. Then $K^{\mathrm{pc}} = K^{\mathrm{qc}} = K^{\mathrm{rc}} = K^{(3)}$. Moreover, for* $\det H = 1$,

$$K^{\mathrm{pc}} = \{X = (\boldsymbol{y}, \boldsymbol{z}) : |\boldsymbol{y}| + |\boldsymbol{z}| \le 1 \text{ and } \det F = 1\},$$

while for $\det H > 1$

$$K^{\mathrm{pc}} = \Big\{X = (\boldsymbol{y}, \boldsymbol{z}) : |\boldsymbol{y}| \le 1 - \frac{\det X - 1}{\det H - 1} \text{ and } |\boldsymbol{z}| \le \frac{\det X - 1}{\det H - 1}\Big\}.$$

A different formula follows for the case of equal determinant ($\lambda\mu = 1$) from the results below.

The k-well Problem with Equal Determinant. In the following we assume that

$$K = \mathrm{SO}(2)U_1 \cup \ldots \cup \mathrm{SO}(2)U_k, \quad \det U_i = \Delta > 0,\ i = 1, \ldots, k.$$

The characterization of the semiconvex hulls in this case relies on the fact that every pair of wells $\mathrm{SO}(2)U_i$ and $\mathrm{SO}(2)U_j$, $i \neq j$, is rank-one connected, see Proposition A.2.5. In order to describe the semiconvex hulls of K, we consider the set $\widetilde{K} = \{C = F^TF : F \in K\}$ as a subset of all positive definite and symmetric matrices. We define $\widetilde{K}^{\mathrm{pc}}$, $\widetilde{K}^{\mathrm{qc}}$, and $\widetilde{K}^{\mathrm{rc}}$ analogously. It is an immediate consequence of the representations of the semiconvex hulls as centers of mass of probability measures or as zero sets of semiconvex functions that all matrices $F \in K^{\mathrm{pc}}$ satisfy

$$|F\boldsymbol{e}|^2 \leq \max\{|U_1\boldsymbol{e}|^2, \ldots, |U_k\boldsymbol{e}|^2\} \quad \text{for all } \boldsymbol{e} \in \mathbb{S}^1.$$

This implies that, for all $\boldsymbol{e} \in \mathbb{S}^1$, the set $\widetilde{K}^{\mathrm{pc}}$ is contained in the half space

$$H = \{C \text{ symmetric with } C : (\boldsymbol{e} \otimes \boldsymbol{e}) \leq \max\{|U_1\boldsymbol{e}|^2, \ldots, |U_k\boldsymbol{e}|^2\}\}$$

in the three-dimensional space of all symmetric matrices. Moreover, if QU_i and U_j are rank-one connected, $QU_i - U_j = \boldsymbol{a} \otimes \boldsymbol{n}$, then $|U_i\boldsymbol{n}^\perp|^2 = |U_j\boldsymbol{n}^\perp|^2$, and all the matrices $F_\lambda = \lambda QU_i + (1-\lambda)U_j$ on the rank-one segment between QU_i and U_j satisfy $|F_\lambda \boldsymbol{n}^\perp|^2 = |U_i\boldsymbol{n}^\perp|^2 = |U_j\boldsymbol{n}^\perp|^2$. Consequently the matrices

$$F_\lambda^T F_\lambda = U_j^T U_j + \lambda U_j^T \boldsymbol{a} \otimes \boldsymbol{n} + \boldsymbol{n} \otimes U_j^T \boldsymbol{a} + \lambda^2 |\boldsymbol{a}|^2 \boldsymbol{n} \otimes \boldsymbol{n}$$

lie on the hyperplane $\{C : (\boldsymbol{n}^\perp \otimes \boldsymbol{n}^\perp) = |U_i\boldsymbol{n}^\perp|^2 = |U_j\boldsymbol{n}^\perp|^2\}$ and satisfy $\det(F_\lambda^T F_\lambda) = \Delta^2$ since the determinant is affine on rank-one lines (and hence constant on the entire rank-one segment since $\det QU_j = \det U_i = \Delta$). This hyperplane intersects the surface $\mathcal{S}$ of constant determinant Δ^2 in the space of positive definite symmetric matrices in a one-dimensional curve. This suggests that the rank-one convex hull of K corresponds to a polygon on $\mathcal{S}$ which is bounded by curves generated from rank-one connections between the wells.

The next theorem asserts that these curves form indeed the boundary of the semiconvex hulls if the matrices U_i and U_j have a certain maximality property. This property states that *all* matrices U_ℓ lie on one side of the curve generated by (one of) the two rank-one connections between the wells $\mathrm{SO}(2)U_i$ and $\mathrm{SO}(2)U_j$. This implies that we may relabel the matrices U_i such that

$$|U_1\boldsymbol{n}^\perp|^2 = \ldots = |U_\ell \boldsymbol{n}^\perp| > \max_{j \geq \ell+1} |U_j\boldsymbol{n}^\perp|^2. \tag{2.16}$$

The second part of the theorem asserts that there exists a finite description of the hulls of K, in the sense that they are described by a finite number $\ell \leq k$ of inequalities. These inequalities are related to matrices U_i that correspond to a 'corner' of the hulls, i.e., to matrices for which there exists a vector $\boldsymbol{v} \in \mathbb{S}^1$ such that

$$|U_i \boldsymbol{v}|^2 > \max_{j \neq i} |U_j \boldsymbol{v}|^2.$$

Theorem 2.2.3. *Assume that* $U_1, \ldots, U_k \in \mathbb{M}^{2\times 2}$, $k \geq 2$, *are symmetric and positive definite matrices with* $\det U_i = \Delta > 0$ *for* $i = 1, \ldots, k$, *and that* $\mathrm{SO}(2)U_i \neq \mathrm{SO}(2)U_j$ *for* $i \neq j$. *Let* $\mathcal{U} = \{U_1, \ldots, U_k\}$,

$$K = \mathrm{SO}(2)U_1 \cup \ldots \cup \mathrm{SO}(2)U_k,$$

and define the set of corners $\mathcal{C}$ *by*

$$\mathcal{C} = \{U \in \mathcal{U} : \exists \boldsymbol{v} \in \mathbb{S}^1 \text{ with } |U\boldsymbol{v}|^2 > \max_{V \in \mathcal{U}\setminus\{U\}} |V\boldsymbol{v}|^2\}. \tag{2.17}$$

Suppose that $\mathcal{C}$ *contains* ℓ *elements,* $\ell \leq k$. *Then*

$$K^{(2)} = K^{\mathrm{rc}} = K^{\mathrm{qc}} = K^{\mathrm{pc}}$$

and there exists a set $\mathcal{E}_\ell = \{\boldsymbol{e}_1, \ldots, \boldsymbol{e}_\ell\} \subset \mathbb{S}^1$ *such that*

$$K^{\mathrm{qc}} = \{F : \det F = \Delta, \, |F\boldsymbol{e}_i|^2 \leq \max_{U \in \mathcal{C}} |U\boldsymbol{e}_i|^2, \, i = 1, \ldots, \ell\}.$$

Proof. Let

$$\mathcal{A} = \{F : \det F = \Delta, \, |F\boldsymbol{e}_i|^2 \leq \max_{U \in \mathcal{C}} |U\boldsymbol{e}_i|^2, \, i = 1, \ldots, \ell\}.$$

We first show that the set $\mathcal{A}$ is defined by polyconvex conditions, that is, $K^{\mathrm{pc}} \subseteq \mathcal{A}$, and then we apply the splitting method to prove that $\mathcal{A} \subseteq K^{(2)}$. Therefore $K^{\mathrm{pc}} \subseteq \mathcal{A} \subseteq K^{(2)} \subseteq K^{\mathrm{pc}}$ and all inclusions are equalities. This establishes the theorem.

In order to demonstrate the inclusion $K^{\mathrm{pc}} \subseteq \mathcal{A}$ we define functions g and $h_i : \mathbb{M}^{2\times 2} \to \mathbb{R}$ by

$$g(X) = (\det X - \Delta)^2, \quad h_i(X) = \big(|X\boldsymbol{e}_i| - \max_{U \in \mathcal{C}} |U\boldsymbol{e}_i|\big)^+, \, i = 1, \ldots, \ell,$$

where $(t)^+ = \max\{t, 0\}$. Then $\mathcal{A}$ is equivalently given by

$$\mathcal{A} = \{F \in \mathbb{M}^{2\times 2} : g(F) \leq 0, \, h_i(F) \leq 0, \, i = 1, \ldots, \ell\}.$$

We conclude that $K^{\mathrm{pc}} \subseteq \mathcal{A}$ since the functions g and h_i are polyconvex with $g(F) \leq 0$ and $h_i(F) \leq 0$ for $F \in K$.

It remains to prove that $\mathcal{A} \subseteq K^{(2)}$. This is done by showing that $\mathcal{A}$ is compact and that the boundary of $\mathcal{A}$ consists of ℓ so-called maximal arcs defined in the following way. Suppose that U_i, $U_j \in \mathcal{C}$, $i \neq j$, and that there exists an $\boldsymbol{e} \in \mathbb{S}^1$ such that

$$|U_i e| = |U_j e| \geq \max_{U \in \mathcal{C} \setminus \{U_i, U_j\}} |Ue|.$$

Proposition A.2.5 implies the existence of $Q_i \in \mathrm{SO}(2)$ and $s_i \in \mathbb{R}$ such that $Q_i U_i - U_j = s_i U_j e \otimes e^\perp$. Then we call the set

$$\Gamma_{ij}(e) = \{F = Q(U_j + ts_i U_j e \otimes e^\perp) : Q \in \mathrm{SO}(2),\, t \in [0,1]\}$$

a maximal arc with endpoints U_i and U_j. By construction,

$$U_j + ts_j U_j e \otimes e^\perp = (1-t)U_j + tQ_i U_i \in K^{(1)},$$

and hence $\Gamma_{ij}(e) \subset K^{(1)}$. We now divide the proof into a series of steps in which we first construct the set $\mathcal{E}$ in the assertion of the theorem (Step 4) and show that the boundary of $\mathcal{A}$ consists of ℓ maximal arcs (Step 6). Then the assertion follows by an application of the splitting method.

Step 1: Suppose that $i, j \in \{1, \ldots, k\}$. If there exists a $v \in \mathbb{S}^1$ and an $\varepsilon > 0$ such that

$$|U_i w| = |U_j w| \quad \text{for all } w \text{ with } |w - v| < \varepsilon,$$

then $\mathrm{SO}(2)U_i = \mathrm{SO}(2)U_j$.

Indeed, if this equality holds, then $|U_i(v + tv^\perp)| = |U_j(v + tv^\perp)|$ for t small enough and thus

$$|U_i v|^2 = |U_j v|^2, \quad \langle U_i v, U_i v^\perp \rangle = \langle U_j v, U_j v^\perp \rangle, \quad |U_i v^\perp|^2 = |U_j v^\perp|^2,$$

and hence $U_i^T U_i = U_j^T U_j$. Since $\det U_i = \det U_j$, this implies by the polar decomposition theorem that there exists a $Q \in \mathrm{SO}(2)$ with $U_i = QU_j$.

Step 2: The set $\mathcal{C}$ is well-defined and contains at least two matrices, that is, $\ell \geq 2$.

We define for $v \in \mathbb{S}^1$ the set $A(v)$

$$A(v) = \{i \in \{1, \ldots, k\} : |U_i v| = \max_{j=1,\ldots,k} |U_j v|\}.$$

We conclude by Step 1, that $A(v)$ is a singleton for all but finitely many v. Let

$$G = \{i : A(v) = \{i\} \text{ on a (relatively) open subset of } \mathbb{S}^1\}$$
$$\mathcal{C} = \{U_i,\, i \in G\}.$$

We have to show that G contains at least two points. Suppose that G contains only one element, $\mathcal{C} = \{U_i\}$. Then $|U_j v| \leq |U_j v|$ for all $v \in \mathbb{S}^1$, $j \neq i$, and therefore in particular $\lambda_{\max}(U_j) \leq \lambda_{\max}(U_i)$ and $\lambda_{\min}(U_j) \leq \lambda_{\min}(U_i)$. Since $\det U_i = \det U_j$, we also have $\lambda_{\min}(U_j) \geq \lambda_{\min}(U_i)$ and this allows us to conclude $\lambda_{\min}(U_j) = \lambda_{\min}(U_i)$ as well as $\lambda_{\max}(U_j) = \lambda_{\max}(U_i)$. Moreover, the matrices $U_i^T U_i$ and $U_j^T U_j$ have the same system of eigenvectors, and thus $\mathrm{SO}(2)U_i = \mathrm{SO}(2)U_j$, a contradiction. If $M_0 = \{U_1, \ldots, U_k\} \setminus \mathcal{C} \neq \emptyset$, then

$$|U_i \boldsymbol{v}| \leq \max_{U \in \mathcal{C}} |U\boldsymbol{v}| \text{ for all } \boldsymbol{v} \in \mathbb{S}^1,\ U_i \in M_0,$$

and therefore $M_0 \subset \mathcal{A}$, independently of the choice of the set $\mathcal{C}$. It is easy to see that $(K \cup M)^{\mathrm{qc}} = K^{\mathrm{qc}}$ for all $M \subseteq K^{\mathrm{qc}}$, we may therefore assume in the sequel that $\mathcal{C} = \{U_1, \ldots U_k\}$.

Step 3: Every matrix $U_i \in \mathcal{C}$ is the end point of exactly two distinct maximal arcs. That is, for all $U_i \in \mathcal{C}$ there exist U_p, $U_q \in \mathcal{C}$, $p \neq i$, $q \neq i$, and $\boldsymbol{e}_p$, $\boldsymbol{e}_q \in \mathbb{S}^1$, $\boldsymbol{e}_p \neq \boldsymbol{e}_q$ such that

$$|U_i \boldsymbol{e}_p| = |U_p \boldsymbol{e}_p| \geq \max_{U \in \mathcal{C} \setminus \{U_i, U_p\}} |U\boldsymbol{e}_p|,\ \ |U_i \boldsymbol{e}_q| = |U_q \boldsymbol{e}_q| \geq \max_{U \in \mathcal{C} \setminus \{U_i, U_q\}} |U\boldsymbol{e}_q|.$$

Note that Proposition 2.2.4 below implies that the inequality is necessarily strict. We first show that there are at least two maximal arcs with end point U_i. For the arguments below, it is convenient to identify vectors $\boldsymbol{e} \in \mathbb{S}^1$ with angles φ in $[0, \pi)$ by $\boldsymbol{e} = \boldsymbol{v}(\varphi) = (\cos\varphi, \sin\varphi)$ or $-\boldsymbol{e} = \boldsymbol{v}(\varphi)$. We define $g : [0, \pi] \to \mathbb{R}$ by

$$g(\varphi) = |U_i \boldsymbol{v}(\varphi)| - \max_{U \in \mathcal{C} \setminus \{U_i\}} |U\boldsymbol{v}(\varphi)|.$$

Then g is continuous and periodic, and since $\ell \geq 2$ there exist at least two angles φ_p, φ_q with $0 \leq \varphi_p < \varphi_q < \pi$ and $g(\varphi_p) = g(\varphi_q) = 0$. Let $\boldsymbol{e}_p$ and $\boldsymbol{e}_q$ be the corresponding vectors. Then

$$|U_i \boldsymbol{e}_p| = \max_{U \in \mathcal{C} \setminus \{U_i\}} |U\boldsymbol{e}_p|,\ \ |U_i \boldsymbol{e}_q| = \max_{U \in \mathcal{C} \setminus \{U_i\}} |U\boldsymbol{e}_q|,$$

and there exist U_p, $U_q \in \mathcal{C} \setminus \{U_i\}$ such that

$$|U_i \boldsymbol{e}_s| = |U_s \boldsymbol{e}_s| \geq \max_{U \in \mathcal{C} \setminus \{U_i, U_s\}} |U\boldsymbol{e}_s|,\quad s = p,\ q,$$

as asserted. Assume now that there exist three maximal arcs with end point U_i, that is, there exist indices p, q, r different from i and corresponding vectors $\boldsymbol{e}_s = \boldsymbol{v}(\varphi_s)$, $s = p$, q, r, with $0 \leq \varphi_p < \varphi_q < \varphi_r < \pi$ such that

$$|U_i \boldsymbol{e}_s| = |U_s \boldsymbol{e}_s| \geq \max_{U \in \mathcal{C} \setminus \{U_i, U_s\}} |U\boldsymbol{e}_s|,\quad s = p,\ q,\ r.$$

By Proposition A.2.5, there exist $Q_s \in \mathrm{SO}(2)$ and $t_s \in \mathbb{R}$ such that

$$Q_s U_s - U_i = t_s U_i \boldsymbol{e}_s \otimes \boldsymbol{e}_s^{\perp},\quad s = p,\ q,\ r.$$

This implies for all $\boldsymbol{v} \in \mathbb{S}^1$ that

$$|U_s \boldsymbol{v}|^2 = |U_i \boldsymbol{v}|^2 + 2t_s \langle U_i \boldsymbol{v}, U_i \boldsymbol{e}_s\rangle \langle \boldsymbol{e}_s^{\perp}, \boldsymbol{v}\rangle + t_s^2 |U_i \boldsymbol{e}_s|^2 \langle \boldsymbol{e}_s^{\perp}, \boldsymbol{v}\rangle^2.$$

We define $P(s,v) = t_s^2 \, |U_i e_s|^2 \, \langle e_s^\perp, v\rangle^2 \geq 0$ and obtain

$$\begin{aligned}
|U_p e_r|^2 &= |U_i e_r|^2 + 2t_p \langle U_i e_r, U_i e_p\rangle \, \langle e_p^\perp, e_r\rangle + P(p, e_r),\\
|U_r e_p|^2 &= |U_i e_p|^2 + 2t_r \langle U_i e_p, U_i e_r\rangle \, \langle e_r^\perp, e_p\rangle + P(r, e_p),\\
|U_q e_r|^2 &= |U_i e_r|^2 + 2t_q \langle U_i e_r, U_i e_q\rangle \, \langle e_q^\perp, e_r\rangle + P(q, e_r),\\
|U_r e_q|^2 &= |U_i e_q|^2 + 2t_r \langle U_i e_q, U_i e_r\rangle \, \langle e_r^\perp, e_q\rangle + P(r, e_q),\\
|U_q e_p|^2 &= |U_i e_p|^2 + 2t_q \langle U_i e_p, U_i e_q\rangle \, \langle e_q^\perp, e_p\rangle + P(q, e_p),\\
|U_p e_q|^2 &= |U_i e_q|^2 + 2t_p \langle U_i e_q, U_i e_p\rangle \, \langle e_p^\perp, e_q\rangle + P(p, e_q).
\end{aligned}$$

By assumption, $|U_i e_s|$ always is maximal, and therefore the second term on the right hand side of the foregoing identities must be less than or equal to zero. Since

$$\begin{aligned}
&\langle e_p^\perp, e_r\rangle \geq 0, \quad \langle e_r^\perp, e_p\rangle \leq 0,\\
&\langle e_q^\perp, e_r\rangle \geq 0, \quad \langle e_r^\perp, e_q\rangle \leq 0,\\
&\langle e_q^\perp, e_p\rangle \leq 0, \quad \langle e_p^\perp, e_q\rangle \geq 0,
\end{aligned}$$

we obtain the following set of conditions:

$$t_p \langle U_i e_r, U_i e_p\rangle \leq 0, \tag{2.18a}$$
$$t_r \langle U_i e_p, U_i e_r\rangle \geq 0, \tag{2.18b}$$
$$t_q \langle U_i e_r, U_i e_q\rangle \leq 0, \tag{2.18c}$$
$$t_r \langle U_i e_q, U_i e_r\rangle \geq 0, \tag{2.18d}$$
$$t_q \langle U_i e_p, U_i e_q\rangle \geq 0, \tag{2.18e}$$
$$t_p \langle U_i e_q, U_i e_p\rangle \leq 0. \tag{2.18f}$$

Suppose now that $t_p \geq 0$ (the case $t_p \leq 0$ is analogous). Then $\langle U_i e_r, U_i e_p\rangle \leq 0$ and $t_r \leq 0$ by (2.18b). Hence $\langle U_i e_q, U_i e_r\rangle \leq 0$ by (2.18d), and we obtain from (2.18c) that $t_q \geq 0$. This implies by (2.18e) that $\langle U_i e_p, U_i e_q\rangle \geq 0$, and we conclude from (2.18f) that $t_p \leq 0$. This is only possible of $t_p = 0$ and we deduce that

$$|U_p e_r| = |U_i e_r| = |U_r e_r|.$$

This contradicts the definition of a maximal arc unless $p = r$. The analogous chain of implications shows also that $q = r$ and hence $p = q = r$. In view of Proposition A.2.5 we find $Q_s \in \mathrm{SO}(2)$ and $t_s \in \mathbb{R}$ such that

$$Q_s U_p - U_i = t_s U_i e_s \otimes e_s^\perp,$$

that is, there exist three different rank-one connections between the well $\mathrm{SO}(2)U_i$ and $\mathrm{SO}(2)U_p$. However, the equation $\det(QU_p - U_i) = 0$ has at most two solutions for $i \neq p$, and this contradiction establishes the assertion of the step.

Step 4: Definition of $\mathcal{E}$.

The results in Step 3 allow us to define a graph $\mathcal{G}$ of degree two (that is, all nodes are end points of exactly two edges in the graph) with nodes corresponding to the matrices in $\mathcal{C}$ and edges corresponding to the maximal arcs. Two nodes U_i, $U_j \in \mathcal{C}$ are connected by an edge if and only if U_i and U_j are the endpoints of a maximal arc. It is easy to see that $\mathcal{G}$ is the union of a finite number of disjoint cycles. We assert now that $\mathcal{G}$ consists of a single cycle. Otherwise we choose one cycle in the graph and denote the set of indices corresponding to the matrices in this cycle by M_1. By assumption, $M_2 = \{1, \ldots, \ell\} \setminus M_1 \neq \emptyset$ and we define

$$g(\boldsymbol{e}) = \max_{j \in M_1} |U_j \boldsymbol{e}| - \max_{j \in M_2} |U_j \boldsymbol{e}|.$$

By definition of $\mathcal{C}$, there exist $\boldsymbol{e}^{\pm}$ such that $g(\boldsymbol{e}^+) > 0$ and $g(\boldsymbol{e}^-) < 0$. The continuity of g implies the existence of a vector $\overline{\boldsymbol{e}}$ with $g(\overline{\boldsymbol{e}}) = 0$. Hence

$$\max_{j \in M_1} |U_j \overline{\boldsymbol{e}}| = \max_{j \in M_2} |U_j \overline{\boldsymbol{e}}|.$$

Let $p \in M_1$ and $q \in M_2$ be indices such that

$$|U_p \overline{\boldsymbol{e}}| = |U_q \overline{\boldsymbol{e}}| \geq \max_{U \in \mathcal{C} \setminus \{U_p, U_q\}} |U \overline{\boldsymbol{e}}|.$$

Then $C_{pq}(\overline{\boldsymbol{e}})$ is a maximal arc and consequently there exists at least three maximal arcs with end points U_p and U_q. This contradicts the assertion of Step 3. The graph $\mathcal{G}$ consists therefore of a single cycle with ℓ edges corresponding to ℓ maximal arcs given by ℓ distinct vectors $\boldsymbol{e}_1, \ldots, \boldsymbol{e}_\ell \in \mathbb{S}^1$. We define $\mathcal{E} = \{\boldsymbol{e}_1, \ldots, \boldsymbol{e}_\ell\}$.

Step 5: The set $\mathcal{A}$ is compact.

If follows from Step 3 that $\ell \geq 2$ and from Step 4 that $\mathcal{A}$ contains at least two conditions of the form $|F\boldsymbol{e}_i| \leq C_i$ with linearly independent vectors $\boldsymbol{e}_1$ and $\boldsymbol{e}_2$. Let $\widetilde{\boldsymbol{v}}_2 = \boldsymbol{e}_2 - \langle \boldsymbol{e}_1, \boldsymbol{e}_2 \rangle \boldsymbol{e}_1$ and define an orthonormal basis with $\boldsymbol{v}_1 = \boldsymbol{e}_1$ and $\boldsymbol{v}_2 = \widetilde{\boldsymbol{v}_2} / |\widetilde{\boldsymbol{v}_2}|$. Then

$$|F\boldsymbol{v}_1| \leq C_1,. \quad |F\boldsymbol{v}_2| \leq \frac{C_1 + C_2}{|\widetilde{\boldsymbol{e}_2}|} \text{ for all } F \in \mathcal{A}.$$

Thus the coordinates of the vectors formed by the rows of F are uniformly bounded in the orthonormal basis $\{\boldsymbol{v}_1, \boldsymbol{v}_2\}$ and hence $\mathcal{A}$ is bounded. Since all functions in the definition of $\mathcal{A}$ are continuous, $\mathcal{A}$ is closed and thus compact.

Step 6: The set of constrained points,

$$\mathcal{B} = \{F \in \mathcal{A} : \exists \boldsymbol{e} \in \mathbb{S}^1 \text{ such that } |F\boldsymbol{e}| = \max_{U \in \mathcal{C}} |U\boldsymbol{e}|\}$$

is the union of the maximal arcs.

As a first step, we show that $\mathcal{B}$ is in fact given by

$$\mathcal{B} = \{F \in \mathcal{A} : \exists i \in \{1, \ldots, \ell\} \text{ such that } |Fe_i| = \max_{U \in \mathcal{C}} |Ue_i|\}. \tag{2.19}$$

Assume thus that $F \in \mathcal{B} \setminus K$ and that there exists a $U_q \in \mathcal{C}$ with

$$|Fe| = |U_q e| \geq \max_{U \in \mathcal{C} \setminus \{U_q\}} |Ue| \quad \text{with } e \notin \mathcal{E}. \tag{2.20}$$

By Step 3, U_q is the end point of two maximal arcs $\Gamma_{pq}(e_p)$ and $\Gamma_{qr}(e_q)$. By construction, e_p and e_q are not parallel and therefore $\langle e_p^\perp, e_q\rangle \neq 0$. Moreover, $e \notin \mathcal{E}$ and thus $\langle e^\perp, e_p\rangle \neq 0$ and $\langle e^\perp, e_q\rangle \neq 0$. We find by Proposition A.2.5 rotations Q_F, Q_p, $Q_r \in SO(2)$ and scalars t_F, t_p, $t_r \in \mathbb{R}$ with $t_p \neq 0$ and $t_q \neq 0$ such that

$$Q_F F - U_q = t_F U_q e \otimes e^\perp, \tag{2.21a}$$
$$Q_p U_p - U_q = t_p U_q e_p \otimes e_p^\perp, \tag{2.21b}$$
$$Q_r U_r - U_q = t_r U_q e_q \otimes e_q^\perp. \tag{2.21c}$$

We now assume that $t_F \neq 0$. If we deduce a contradiction to this assumption, then $F = Q_F^T U_q \in K$. Thus the constraints with $e \notin \mathcal{E}$ are not active, and we establish that the constrained points are given by (2.19).

We first multiply (2.21a) by e_p and e_q and take the modulus. Since $F \in \mathcal{A}$ and $\Gamma_{pq}(e_p)$ and $\Gamma_{qr}(e_q)$ are maximal arcs, the inequalities

$$\begin{aligned} |Fe_p|^2 &= |U_q e_p|^2 + 2t_F \langle U_q e_p, U_q e\rangle \langle e^\perp, e_p\rangle + t_F^2 |U_q e|^2 \langle e^\perp, e_p\rangle^2, \\ |Fe_q|^2 &= |U_q e_q|^2 + 2t_F \langle U_q e_q, U_q e\rangle \langle e^\perp, e_q\rangle + t_F^2 |U_q e|^2 \langle e^\perp, e_q\rangle^2 \end{aligned}$$

imply that

$$2t_F \langle U_q e_p, U_q e\rangle \langle e^\perp, e_p\rangle < 0, \qquad 2t_F \langle U_q e_q, U_q e\rangle \langle e^\perp, e_q\rangle < 0. \tag{2.22}$$

Similarly,

$$\begin{aligned} |U_p e|^2 &= |U_q e|^2 + 2t_p \langle U_q e, U_q e_p\rangle \langle e_p^\perp, e\rangle + t_p^2 |U_q e_p|^2 \langle e_p^\perp, e\rangle^2, \\ |U_r e|^2 &= |U_q e|^2 + 2t_r \langle U_q e, U_q e_q\rangle \langle e_q^\perp, e\rangle + t_r^2 |U_q e_q|^2 \langle e_q^\perp, e\rangle^2, \end{aligned}$$

and by (2.20)

$$2t_r \langle U_q e, U_q e_q\rangle \langle e_q^\perp, e\rangle < 0, \qquad 2t_p \langle U_q e, U_q e_p\rangle \langle e_p^\perp, e\rangle < 0. \tag{2.23}$$

Finally,

$$\begin{aligned} |U_p e_q|^2 &= |U_q e_q|^2 + 2t_p \langle U_q e_q, U_q e_p\rangle \langle e_p^\perp, e_q\rangle + t_p^2 |U_q e_p|^2 \langle e_p^\perp, e_q\rangle^2, \\ |U_r e_p|^2 &= |U_q e_p|^2 + 2t_r \langle U_q e_p, U_q e_q\rangle \langle e_q^\perp, e_p\rangle + t_r^2 |U_q e_q|^2 \langle e_q^\perp, e_p\rangle^2, \end{aligned}$$

and since $\Gamma_{pq}(e_p)$ and $\Gamma_{qr}(e_q)$ are maximal arcs

$$2t_r\langle U_q e_p, U_q e_q\rangle\langle e_q^\perp, e_p\rangle < 0, \qquad 2t_p\langle U_q e_q, U_q e_p\rangle\langle e_p^\perp, e_q\rangle < 0. \tag{2.24}$$

It follows from $\langle e_p^\perp, e\rangle = -\langle e^\perp, e_p\rangle$ and (2.22), (2.23) that

$$t_F\langle U_q e_p, U_q e\rangle\langle e^\perp, e_p\rangle < 0, \qquad -t_p\langle U_q e_p, U_q e\rangle\langle e^\perp, e_p\rangle < 0.$$

We get similarly from (2.22) and (2.23)

$$t_F\langle U_q e_q, U_q e\rangle\langle e^\perp, e_q\rangle < 0, \qquad -t_r\langle U_q e_q, U_q e\rangle\langle e^\perp, e_q\rangle < 0,$$

and finally from (2.24)

$$t_r\langle U_q e_p, U_q e_q\rangle\langle e_q^\perp, e_p\rangle < 0, \qquad -t_p\langle U_q e_p, U_q e_q\rangle\langle e_q^\perp, e_p\rangle < 0.$$

The foregoing estimates yield

$$t_F t_p < 0, \quad t_F t_r < 0, \quad t_p t_r < 0,$$

and this set of inequalities has no solution. Consequently the assumption $t_F \neq 0$ cannot hold, and this establishes the formula (2.19) for the constrained points.

We may thus assume that $F \in \mathcal{B} \setminus K$ with

$$|Fe_q| = \max_{U\in\mathcal{C}} |Ue_q| \quad \text{with } e_q \in \mathcal{E}.$$

By definition of $\mathcal{E}$ there exist two matrices U_q, $U_r \in \mathcal{C}$ such that $\Gamma_{qr}(e_q)$ is a maximal arc with $|Fe_q| = |U_q e_q| = |U_r e_q|$. Moreover, there are indices p, $s \in \{1,\ldots,\ell\}$, $p \neq q$, $s \neq r$ and vectors e_p, $e_r \in \mathcal{C}$ such that $\Gamma_{pq}(e_p)$ and $\Gamma_{rs}(e_r)$ are maximal arcs. By Proposition A.2.5 we find Q_r, $Q_F \in \mathrm{SO}(2)$ and t_r, $t_F \in \mathbb{R}$, $t_r \neq 0$, $t_F \neq 0$, such that

$$Q_r U_r - U_q = t_r U_q e_q \otimes e_q^\perp, \quad Q_F F - U_q = t_F U_q e_q \otimes e_q^\perp, \tag{2.25}$$

as well as $\widetilde{Q}_q$, $\widetilde{Q}_F \in \mathrm{SO}(2)$ and $\widetilde{t}_q$, $\widetilde{t}_F \in \mathbb{R}$, $\widetilde{t}_q \neq 0$, $\widetilde{t}_F \neq 0$ with

$$\widetilde{Q}_q U_q - U_r = \widetilde{t}_q U_r e_q \otimes e_q^\perp, \quad \widetilde{Q}_F F - U_r = \widetilde{t}_F U_r e_q \otimes e_q^\perp. \tag{2.26}$$

We show below that

$$t_F \in [0, t_r] \text{ if } t_r > 0, \quad t_F \in [t_r, 0] \text{ if } t_r < 0. \tag{2.27}$$

This implies $t_F = \lambda t_r$, $\lambda = t_F/t_r \in [0,1]$, and thus

$$Q_F F = U_q + \lambda t_r U_q e_q \otimes e_q^\perp = (1-\lambda)U_q + \lambda Q_r U_r \in K^{(1)}. \tag{2.28}$$

In order to prove (2.27), we first observe that by (2.25)

$$\begin{aligned} |U_r e_p|^2 &= |U_q e_p|^2 + 2\,t_r\langle U_q e_p, U_q e_q\rangle\,\langle e_q^\perp, e_p\rangle + t_r^2\,|U_q e_q|^2\,\langle e_q^\perp, e_p\rangle^2, \\ |Fe_p|^2 &= |U_q e_p|^2 + 2\,t_F\langle U_q e_p, U_q e_q\rangle\,\langle e_q^\perp, e_p\rangle + t_F^2\,|U_q e_q|^2\,\langle e_q^\perp, e_p\rangle^2. \end{aligned}$$

Since $\Gamma_{pq}(e_p)$ is a maximal arc,

$$|U_q e_p| \geq |U_r e_p|, \quad |U_q e_p| \geq |F e_p|,$$

and we deduce

$$2\, t_r \langle U_q e_p, U_q e_q\rangle \, \langle e_q^{\perp}, e_p\rangle + t_r^2 \, |U_q e_q|^2 \, \langle e_q^{\perp}, e_p\rangle^2 \leq 0,$$
$$2\, t_F \langle U_q e_p, U_q e_q\rangle \, \langle e_q^{\perp}, e_p\rangle + t_F^2 \, |U_q e_q|^2 \, \langle e_q^{\perp}, e_p\rangle^2 \leq 0.$$

By assumption, the vectors e_p and e_q are not parallel, hence $\langle e_q^{\perp}, e_p\rangle \neq 0$. This implies

$$2\, t_r \langle U_q e_p, U_q e_q\rangle \, \langle e_q^{\perp}, e_p\rangle < 0,$$
$$2\, t_F \langle U_q e_p, U_q e_q\rangle \, \langle e_q^{\perp}, e_p\rangle < 0,$$

and we infer that $t_F \, t_r > 0$, that is, t_F and t_r have the same sign. The same arguments applied to the identity (2.26) with e_r instead of e_p show that $\widetilde{t}_F \, \widetilde{t}_q > 0$. It follows from the proof of Proposition 2.2.4 below that

$$\widetilde{t}_F = t_F - t_r, \quad \widetilde{t}_q = -t_r,$$

see formulae (2.30) and (2.31) with $\alpha_2 = t_r$, $\alpha_j = t_F$, $\widetilde{\alpha}_1 = \widetilde{t}_q$, and $\widetilde{\alpha}_j = \widetilde{t}_F$. We obtain that

$$t_F \, t_r > 0, \quad -(t_F - t_r)\, t_r > 0.$$

If $t_r > 0$, then $t_F > 0$ and $-(t_F - t_r) > 0$, hence $t_F \leq t_r$ and therefore $t_r \in [0, t_F]$. We conclude similarly if $t_r < 0$ and this finishes the proof of Step 6.

Step 7: The set $\mathcal{A}$ is contained in the lamination convex hull of K. More precisely, $\mathcal{A} \subseteq K^{(2)}$.

This is an immediate consequence of Step 6 which implies that the set of all constrained points is a subset of $K^{(1)}$. If $F \in \mathcal{A}$ is an unconstrained point, then we apply the splitting method to prove that $F \in K^{(2)}$. Choose any rank-one curve $t \mapsto F(t) = F(\mathbb{I} + t v \otimes v^{\perp})$ with $v \in \mathbb{S}^1$ and define

$$t^- = \sup\{t < 0 : \exists e \in \mathcal{E} : |F(t)e|^2 = \max_{U \in \mathcal{C}} |Ue|^2,$$
$$t^+ = \inf\{t > 0 : \exists e \in \mathcal{E} : |F(t)e|^2 = \max_{U \in \mathcal{C}} |Ue|^2.$$

Since $\mathcal{A}$ is compact, the parameters $t^{\pm}$ are finite. Let $F^{\pm} = F(t^{\pm})$. Then $F^{\pm} \in \mathcal{B} \subseteq K^{(1)}$, $F^+ - F^-$ is a matrix of rank-one, and F is contained in the rank-one segment between F^+ and F^-. Thus $F \in K^{(2)}$ and this concludes the proof of the theorem. □ □

The proof of Theorem 2.2.3 used the following fact.

Proposition 2.2.4. *Assume that* $\det U_i = \Delta > 0$ *for* $i = 1, \dots, \ell$ *with* $\ell \geq 3$, *that* $\mathrm{SO}(2)U_i \neq \mathrm{SO}(2)U_j$ *for* $i \neq j$, *and that there exists an* $\mathbf{e} \in \mathbb{S}^1$ *such that*

$$|U_1\mathbf{e}|^2 = \cdots = |U_\ell \mathbf{e}|^2.$$

Then we may relabel the matrices in such a way that there exist $\alpha_j \in \mathbb{R}$ *and* $Q_j \in \mathrm{SO}(2)$, $j = 2, \dots, \ell$, *such that* $0 < \alpha_2 < \dots < \alpha_\ell$ *and*

$$Q_j U_j - U_1 = \alpha_j U_1 \mathbf{e} \otimes \mathbf{e}^\perp, \quad j = 2, \dots, \ell.$$

In particular,

$$\{U_2, \dots, U_{\ell-1}\} \subset \big(\mathrm{SO}(2)U_1 \cup \mathrm{SO}(2)U_\ell\big)^{\mathrm{lc}}.$$

Proof. By Proposition A.2.5 there exist $\alpha_i \neq 0$ and $Q_i \in \mathrm{SO}(2)$ such that

$$Q_i U_i - U_1 = \alpha_i U_1 \mathbf{e} \otimes \mathbf{e}^\perp.$$

We first show that we may assume that $\alpha_2 < \cdots < \alpha_\ell$ and $\alpha_j \neq 0$. If $\alpha_j = \alpha_{j+1}$, $j \in \{2, \dots, \ell-1\}$, then $\mathrm{SO}(2)U_j = \mathrm{SO}(2)U_{j+1}$, and $\alpha_j = 0$ implies that $\mathrm{SO}(2)U_1 = \mathrm{SO}(2)U_j$. Both identities contradict our assumption that the wells $\mathrm{SO}(2)U_j$ are pairwise disjoint. If $\alpha_2 > 0$, then we define $\lambda_j \in (0,1)$ by

$$1 - \lambda_j = \frac{\alpha_j}{\alpha_\ell} \in (0,1), \quad j = 2, \dots, \ell - 1.$$

We obtain

$$\begin{aligned} \lambda_j U_1 + (1 - \lambda_j) Q_\ell U_\ell &= \lambda_j U_1 + (1 - \lambda_j)\big(U_1 + \alpha_\ell U_1 \mathbf{e} \otimes \mathbf{e}^\perp\big) \\ &= U_1 + \alpha_j U_1 \mathbf{e} \otimes \mathbf{e}^\perp = Q_j U_j. \end{aligned}$$

Hence

$$U_j \in \big(\mathrm{SO}(2)U_1 \cup \mathrm{SO}(2)U_\ell\big)^{\mathrm{lc}},$$

and the assertion of the proposition is immediate. If $\alpha_2 < 0$ then we use Proposition A.2.5 for U_2 instead of for U_1 and we find $\widetilde{\alpha}_i \neq 0$ and $\widetilde{Q}_i \in \mathrm{SO}(2)$ with

$$\widetilde{Q}_i U_i - U_2 = \widetilde{\alpha}_i U_2 \mathbf{e} \otimes \mathbf{e}^\perp, \quad i = 1, 3, \dots, \ell.$$

By definition,

$$Q_2 U_2 - U_1 = \alpha_2 U_1 \mathbf{e} \otimes \mathbf{e}^\perp, \quad \widetilde{Q}_1 U_1 - U_2 = \widetilde{\alpha}_1 U_2 \mathbf{e} \otimes \mathbf{e}^\perp,$$

and thus

$$\widetilde{Q}_1 Q_2 U_2 - \big[U_2 + \widetilde{\alpha}_1 U_2 \mathbf{e} \otimes \mathbf{e}^\perp\big] = \alpha_2 \widetilde{Q}_1 U_1 \mathbf{e} \otimes \mathbf{e}^\perp. \tag{2.29}$$

Multiplication of (2.29) with e from the right yields $(\widetilde{Q}_1 Q_2 U_2 - U_2)e = 0$ which is equivalent to $\widetilde{Q}_1 Q_2 U_2 - U_2 = 0$ since there are no rank-one connections in SO(2). We thus obtain from $\widetilde{Q}_1 U_1 e = U_2 e$ that $(\widetilde{\alpha}_1 + \alpha_2) U_2 e = 0$ and hence

$$\widetilde{\alpha}_1 = -\alpha_2 > 0. \tag{2.30}$$

Similarly, for $j \geq 3$,

$$Q_j U_j - U_1 = \alpha_j U_1 e \otimes e^{\perp},$$

and

$$\widetilde{Q}_j U_j - U_2 = \widetilde{\alpha}_j U_2 e \otimes e^{\perp}, \quad \widetilde{Q}_1 U_1 - U_2 = \widetilde{\alpha}_1 U_2 e \otimes e^{\perp}.$$

These equalities yield as before

$$\widetilde{Q}_j U_j - [\widetilde{Q}_1 U_1 - \widetilde{\alpha}_1 U_2 e \otimes e^{\perp}] = \widetilde{\alpha}_j U_2 e \otimes e^{\perp}$$

and

$$\widetilde{Q}_j U_j - \widetilde{Q}_1 Q_j U_j = -\alpha_j \widetilde{Q}_1 U_1 e \otimes e^{\perp} + (\widetilde{\alpha}_j - \widetilde{\alpha}_1) U_2 e \otimes e^{\perp}.$$

We conclude $-\alpha_j + \widetilde{\alpha}_j - \widetilde{\alpha}_1 = 0$ and therefore

$$\widetilde{\alpha}_j = \widetilde{\alpha}_1 + \alpha_j = -\alpha_2 + \alpha_j \geq 0. \tag{2.31}$$

This proves the assertion. □

We include two typical examples to illustrate Theorem 2.2.3. The three-well configuration is motivated by the analysis of microstructures in cubic to orthorhombic transitions, see Section 5.3. The four-well problem arises as a special case of a tetragonal to monoclinic transition in Section 5.4.

Examples. 1) A three-well problem. For $\xi > \eta > 0$ we define

$$U_1 = \begin{pmatrix} \xi+\eta & 0 \\ 0 & \xi-\eta \end{pmatrix}, \quad U_2 = \begin{pmatrix} \xi - \frac{\eta}{2} & -\frac{\sqrt{3}\eta}{2} \\ -\frac{\sqrt{3}\eta}{2} & \xi + \frac{\eta}{2} \end{pmatrix}, \quad U_3 = \begin{pmatrix} \xi - \frac{\eta}{2} & \frac{\sqrt{3}\eta}{2} \\ \frac{\sqrt{3}\eta}{2} & \xi + \frac{\eta}{2} \end{pmatrix}.$$

A short calculation shows that the matrices U_i satisfy the extremality property (2.17) with $v_1 = (1,0)$, $v_2 = \frac{1}{2}(1, -\sqrt{3})$ and $v_3 = \frac{1}{2}(1, \sqrt{3})$ (see below). The rank-one connections between the wells are given by $Q_i U_1 - U_2 = a_i \otimes n_i$, $i = 1, 2$, with

$$a_1 = \frac{\xi\eta}{2(\xi^2 - \xi\eta + \eta^2)} \begin{pmatrix} \xi - 2\eta \\ \sqrt{3}\xi \end{pmatrix}, \quad n_1 = \begin{pmatrix} 3 \\ -\sqrt{3} \end{pmatrix},$$

and

$$a_2 = \frac{\xi\eta}{2(\xi^2 + \xi\eta + \eta^2)} \begin{pmatrix} \sqrt{3}\xi \\ -(\xi + 2\eta) \end{pmatrix}, \quad n_2 = \begin{pmatrix} \sqrt{3} \\ 3 \end{pmatrix};$$

moreover, $Q_iU_1 - U_3 = \boldsymbol{a}_i \otimes \boldsymbol{n}_i$, $i = 3, 4$, with

$$\boldsymbol{a}_3 = \frac{\xi\eta}{2(\xi^2 - \xi\eta + \eta^2)} \begin{pmatrix} \xi - 2\eta \\ -\sqrt{3}\xi \end{pmatrix}, \quad \boldsymbol{n}_3 = \begin{pmatrix} 3 \\ \sqrt{3} \end{pmatrix},$$

and

$$\boldsymbol{a}_4 = \frac{\xi\eta}{2(\xi^2 + \xi\eta + \eta^2)} \begin{pmatrix} \sqrt{3}\xi \\ \xi + 2\eta \end{pmatrix}, \quad \boldsymbol{n}_4 = \begin{pmatrix} \sqrt{3} \\ -3 \end{pmatrix};$$

finally, $Q_iU_2 - U_3 = \boldsymbol{a}_i \otimes \boldsymbol{n}_i$, $i = 5, 6$, with

$$\boldsymbol{a}_5 = \frac{\xi\eta}{\xi^2 - \xi\eta + \eta^2} \begin{pmatrix} -\sqrt{3}(2\xi - \eta) \\ -3\eta \end{pmatrix}, \quad \boldsymbol{n}_5 = \begin{pmatrix} 0 \\ 1 \end{pmatrix},$$

and

$$\boldsymbol{a}_6 = \frac{\xi\eta}{\xi^2 + \xi\eta + \eta^2} \begin{pmatrix} -3\eta \\ -\sqrt{3}(2\xi + \eta) \end{pmatrix}, \quad \boldsymbol{n}_6 = \begin{pmatrix} 1 \\ 0 \end{pmatrix},$$

see Figure 2.2 for a sketch of the situation. Since

$$|U_1\boldsymbol{n}_1|^2 = |U_2\boldsymbol{n}_1|^2 = \xi^2 + \xi\eta + \eta^2 > (\xi - \eta)^2 = |U_3\boldsymbol{n}_1|^2,$$
$$|U_1\boldsymbol{n}_2|^2 = |U_2\boldsymbol{n}_2|^2 = \xi^2 - \xi\eta + \eta^2 < (\xi + \eta)^2 = |U_3\boldsymbol{n}_2|^2,$$

the extremal curve between U_1 and U_2 is defined by $\boldsymbol{n}_1$. Also

$$|U_1\boldsymbol{n}_3|^2 = |U_3\boldsymbol{n}_3|^2 = \xi^2 + \xi\eta + \eta^2 > (\xi - \eta)^2 = |U_2\boldsymbol{n}_3|^2,$$
$$|U_1\boldsymbol{n}_4|^2 = |U_3\boldsymbol{n}_4|^2 = \xi^2 - \xi\eta + \eta^2 < (\xi + \eta)^2 = |U_2\boldsymbol{n}_4|^2,$$

and

$$|U_2\boldsymbol{n}_5|^2 = |U_3\boldsymbol{n}_5|^2 = \xi^2 + \xi\eta + \eta^2 > (\xi - \eta)^2 = |U_1\boldsymbol{n}_5|^2,$$
$$|U_2\boldsymbol{n}_6|^2 = |U_3\boldsymbol{n}_6|^2 = \xi^2 - \xi\eta + \eta^2 < (\xi + \eta)^2 = |U_1\boldsymbol{n}_6|^2,$$

and therefore the extremal curves between U_1 and U_3 and U_2 and U_3 are given by $\boldsymbol{n}_3$ and $\boldsymbol{n}_5$, respectively. Thus the semiconvex hulls of this three-well problem are described by Theorem 2.2.3 with $\mathcal{E}_3 = \{\boldsymbol{n}_1, \boldsymbol{n}_3, \boldsymbol{n}_5\}$.

In view of the application to the cubic to orthorhombic transformation in Section 5.3, we consider the simple laminates ν that can be formed with the rank-one connections between the matrices in K that do not correspond to maximal arcs. The question we want to address is whether the center of mass $F = \langle \nu, id \rangle$ has more than one representation as a simple laminate. By symmetry, it suffices to consider the rank-one connection between U_1 and U_2 with $Q_1U_1 - U_2 = \boldsymbol{a}_1 \otimes \boldsymbol{n}_1$. We define

$$F_t = U_2 + (1 - t)\boldsymbol{a}_1 \otimes \boldsymbol{n}_1$$

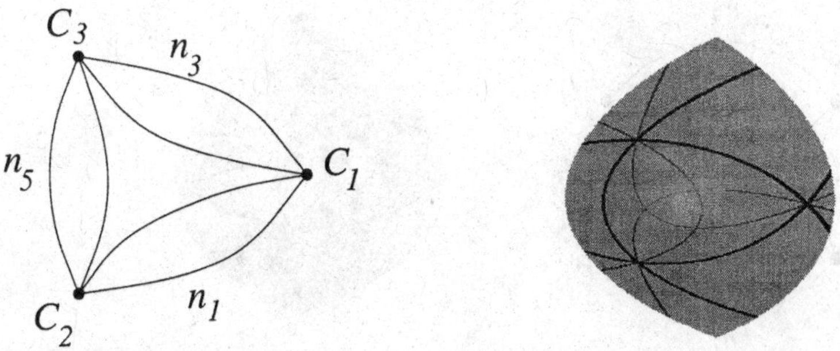

Fig. 2.2. The three-well problem in two dimensions. The left figure shows the matrices $C_i = U_i^T U_i$ and the normals $\boldsymbol{n}_i$ defining the extremal curves. The right figure (generated by a ray tracing programme) displays the convex hulls for $\xi = 2$ and $\eta = 1.5$ on the surface $\det C = \Delta^2$. In this case, the quasiconvex hull of K is given by the union of the quasiconvex hull of the two-well problems formed with pairs of wells in K.

and barycenters of laminates in the other twinning systems that do not correspond to maximal arcs,

$$G_s = U_3 + (1-s)\boldsymbol{a}_3 \otimes \boldsymbol{n}_3, \quad H_s = U_3 + (1-s)\boldsymbol{a}_5 \otimes \boldsymbol{n}_5.$$

We seek solutions t_{13}, s_{13} and t_{23}, s_{23} of

$$F_t^T F_t = G_s^T G_s \quad \text{and} \quad F_t^T F_t = H_s^T H_s,$$

respectively. A short calculation shows that

$$t_{13} = \frac{1}{3}(4 - \frac{\xi}{\eta} - \frac{\eta}{\xi}), \quad s_{13} = \frac{1}{3}(4 - \frac{\xi}{\eta} - \frac{\eta}{\xi}),$$

and

$$t_{23} = \frac{1}{3}(\frac{\xi}{\eta} + \frac{\eta}{\xi} - 1), \quad s_{23} = \frac{1}{3}(4 - \frac{\xi}{\eta} - \frac{\eta}{\xi}).$$

This shows that a simple laminate supported on $SO(2)U_1$ and $SO(2)U_2$ is uniquely determined from its center of mass $F = (1-\lambda)Q_1U_1 + \lambda U_2$ if and only if λ is not equal to one of the two values

$$\lambda_1 = \frac{1}{3}(4 - \frac{\xi}{\eta} - \frac{\eta}{\xi}), \qquad \lambda_2 = \frac{1}{3}(\frac{\xi}{\eta} + \frac{\eta}{\xi} - 1).$$

Since

$$\lambda_1 \geq \lambda_2 \quad \Leftrightarrow \quad \frac{\xi}{\eta} + \frac{\eta}{\xi} \leq \frac{5}{2},$$

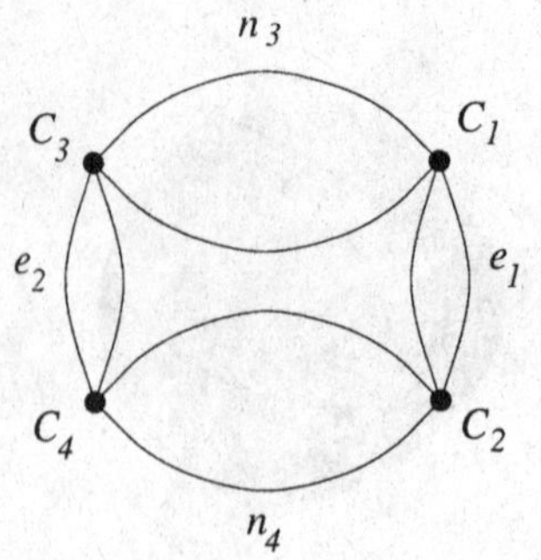

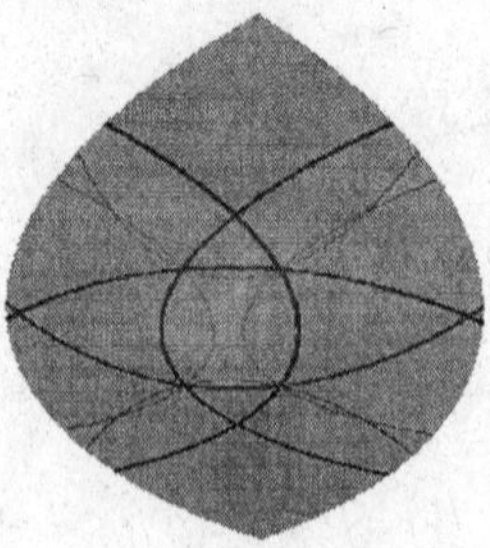

Fig. 2.3. The four-well problem. The left figure shows in a schematic sketch the matrices $C_i = U_i^T U_i$ and the normals defining the extremal curves. Here $\{\boldsymbol{e}_1, \boldsymbol{e}_2\}$ denotes the standard basis of $\mathbb{R}^2$ and $\boldsymbol{n}_3$ and $\boldsymbol{n}_4$ are parallel to $\boldsymbol{e}_1 + \boldsymbol{e}_2$ and $\boldsymbol{e}_1 - \boldsymbol{e}_2$, respectively. The right figure displays the semiconvex hulls for $\alpha = 2$, $\beta = 1.5$ and $\delta = 0.5$ on the surface $\det C = \Delta^2$.

we obtain that

$$K^{\rm qc} = \bigcup_{i \neq j} \big(\mathrm{SO}(2)U_i \cup \mathrm{SO}(2)U_j\big)^{\rm qc} \quad \Leftrightarrow \quad \frac{\xi}{\eta} + \frac{\eta}{\xi} \leq \frac{5}{2}.$$

2) A four-well problem. For $\alpha > \beta > 0$, $\delta > 0$ with $\alpha\delta - \beta^2 > 0$ we define

$$U_1 = \begin{pmatrix} \alpha & \delta \\ \delta & \beta \end{pmatrix}, \quad U_2 = \begin{pmatrix} \alpha & -\delta \\ -\delta & \beta \end{pmatrix}, \quad U_3 = \begin{pmatrix} \beta & \delta \\ \delta & \alpha \end{pmatrix}, \quad U_4 = \begin{pmatrix} \beta & -\delta \\ -\delta & \alpha \end{pmatrix}.$$

As in the case of the three-well problem, it is easy to see that the matrices U_i satisfy the maximality property (2.17). For example, taking $\widetilde{\boldsymbol{e}}$ parallel to $(1+\varepsilon, 1)$ for $\varepsilon > 0$ small enough, one obtains

$$|U_1\widetilde{\boldsymbol{e}}|^2 > \max\big\{|U_3\widetilde{\boldsymbol{e}}|^2, |U_4\widetilde{\boldsymbol{e}}|^2\big\}, \text{ and } |U_1\widetilde{\boldsymbol{e}}|^2 > |U_2\widetilde{\boldsymbol{e}}|^2 \quad \Leftrightarrow \quad \alpha^2 > \beta^2.$$

A short calculation shows that there exist rotations Q_i and vectors $\boldsymbol{a}_i \in \mathbb{R}^2$ such that

$$Q_1U_1 - U_2 = \boldsymbol{a}_1 \otimes \boldsymbol{e}_1, \quad Q_2U_1 - U_2 = \boldsymbol{a}_2 \otimes \boldsymbol{e}_2,$$
$$Q_7U_3 - U_4 = \boldsymbol{a}_7 \otimes \boldsymbol{e}_1, \quad Q_8U_3 - U_4 = \boldsymbol{a}_8 \otimes \boldsymbol{e}_2,$$

where $\{\boldsymbol{e}_1, \boldsymbol{e}_2\}$ denotes the standard basis of $\mathbb{R}^2$,

$$Q_3U_1 - U_3 = \boldsymbol{a}_3 \otimes \frac{1}{\sqrt{2}}(\boldsymbol{e}_1 + \boldsymbol{e}_2), \quad Q_4U_1 - U_3 = \boldsymbol{a}_4 \otimes \frac{1}{\sqrt{2}}(\boldsymbol{e}_1 - \boldsymbol{e}_2),$$

and

$$Q_5U_2 - U_4 = \boldsymbol{a}_5 \otimes \frac{1}{\sqrt{2}}(\boldsymbol{e}_1 - \boldsymbol{e}_2), \quad Q_6U_2 - U_4 = \boldsymbol{a}_6 \otimes \frac{1}{\sqrt{2}}(\boldsymbol{e}_1 + \boldsymbol{e}_2),$$

see Figure 2.3 for a sketch of the extremal curves.

The General Theorem. The results for k wells can easily be generalized to any compact set of matrices with fixed determinant.

Theorem 2.2.5. *Assume that* $\Delta \geq 0$, *that* $K \subset \{F \in \mathbb{M}^{2\times 2} : \det F = \Delta\}$ *is compact, and that* K *is invariant under* SO(2), *i.e.,*

$$K = \mathrm{SO}(2)K = \{F = QU : Q \in \mathrm{SO}(2),\, U \in K\}.$$

Then $K^{(2)} = K^{\mathrm{rc}} = K^{\mathrm{qc}} = K^{\mathrm{pc}}$ *and*

$$K^{\mathrm{qc}} = \{F \in \mathbb{M}^{2\times 2} : \det F = \Delta,\ |F\boldsymbol{e}|^2 \leq \max_{U\in K} |U\boldsymbol{e}|^2 \text{ for all } \boldsymbol{e} \in \mathbb{S}^1\}. \tag{2.32}$$

Proof. Let $\mathcal{A}$ denote the formula for the hulls in (2.32). Since $\mathcal{A}$ is defined by polyconvex conditions, it is clear that $K^{\mathrm{pc}} \subseteq \mathcal{A}$ and it remains to show that $\mathcal{A} \subseteq K^{(2)}$. This follows immediately form the splitting method if we prove that $F \in K^{(1)}$ whenever $F \in \mathcal{A}$ is a constrained point, i.e., F satisfies equality in one of the inequalities in the definition of $\mathcal{A}$. We consider first the case $\Delta > 0$ and mention the necessary modifications for $\Delta = 0$ at the end of the proof. Assume thus that $F \notin K$ and that for at least one $\boldsymbol{e} \in \mathbb{S}^1$

$$|F\boldsymbol{e}|^2 = \max_{U\in K} |U\boldsymbol{e}|^2 = |U_1\boldsymbol{e}|^2,\ U_1 \in K.$$

By Proposition A.2.5 there exist $\alpha_1 \in \mathbb{R}$ and $Q_1 \in \mathrm{SO}(2)$ such that

$$Q_1U_1 - F = \alpha_1 F\boldsymbol{e} \otimes \boldsymbol{e}^\perp.$$

We may assume that $\alpha_1 > 0$ (the proof for $\alpha_1 < 0$ is analogous). The idea is to show by contradiction that there exist $U_2 \in K$, $Q_2 \in \mathrm{SO}(2)$ and $\alpha_2 < 0$ such that

$$Q_2U_2 - F = \alpha_2 F\boldsymbol{e} \otimes \boldsymbol{e}^\perp.$$

This implies

$$F = \frac{-\alpha_2}{\alpha_1 - \alpha_2} Q_1U_1 + \frac{\alpha_1}{\alpha_1 - \alpha_2} Q_2U_2,$$

and thus $F \in \big(\mathrm{SO}(2)U_1 \cup \mathrm{SO}(2)U_2\big)^{(1)}$. In order to show this, assume that $\varepsilon > 0$ is small enough such that $B_\varepsilon(F) \cap K = \emptyset$. By compactness, there exists a $c_0 > 0$ with

$$\mathrm{dist}(F + sF\boldsymbol{e} \otimes \boldsymbol{e}^\perp, K) > c_0 \text{ for all } s < 0 \text{ such that } F + sF\boldsymbol{e} \otimes \boldsymbol{e}^\perp \notin B_\varepsilon(F),$$

and by continuity we may choose $t_0 > 0$ such that

$$\mathrm{dist}(F + sF\boldsymbol{e}_t \otimes \boldsymbol{e}_t^\perp, K) > \frac{c_0}{2} \tag{2.33}$$

for all $s < 0$ such that $F + sFe_t \otimes e_t^{\perp} \notin B_\varepsilon(F)$, $|t| < t_0$, where $e_t = e + te^{\perp}$. We have $\langle e_t^{\perp}, e\rangle = \langle Je_t, e\rangle = -\langle e_t, Je\rangle = -t$, where J denotes the counterclockwise rotation by $\frac{\pi}{2}$, and hence

$$\begin{aligned} |(F + sFe_t \otimes e_t^{\perp})e|^2 &= |Fe|^2 - 2st\langle Fe, Fe_t\rangle + s^2t^2|Fe_t|^2 \\ &= |Fe|^2 - 2st(|Fe|^2 + t\langle Fe, Fe^{\perp}\rangle) + s^2t^2|Fe_t|^2. \end{aligned}$$

Therefore we can find a $t_1 \in (0, t_0)$ such that

$$|(F + sFe_t \otimes e_t^{\perp})e|^2 \geq |Fe|^2 \tag{2.34}$$

for all $t \in [-t_1, 0)$, $s > 0$ with $F + sFe_t \otimes e_t^{\perp} \notin B_\varepsilon(F)$ with equality if and only if $s = 0$. We now assert that

$$|Fe_t|^2 > \max_{U \in K} |Ue_t|^2 \text{ for } t \in [-t_1, 0). \tag{2.35}$$

Assume the contrary. Suppose first that there exists a $t \in [-t_1, 0)$ and a $U_0 \in K$ such that $|Fe_t|^2 = |U_0e_t|^2$. By Proposition A.2.5 we may choose $Q_0 \in \mathrm{SO}(2)$ and $\alpha_0 \in \mathbb{R}$ with

$$Q_0U_0 - F = \alpha_0 Fe_t \otimes e_t^{\perp}. \tag{2.36}$$

Clearly $U_0 \notin B_\varepsilon(F)$ since $B_\varepsilon(F) \cap K = \emptyset$ and inequality (2.33) implies that $\alpha_0 > 0$. We obtain from (2.34) and (2.36)

$$|Q_0U_0e|^2 = |(F + \alpha_0 Fe_t \otimes e_t^{\perp})e|^2 > |Fe|^2 = \max_{U \in K} |Ue|^2,$$

a contradiction. It remains to consider the case

$$|Fe_t|^2 < \max_{U \in K} |Ue_t|^2, \quad t \in [-t_1, 0).$$

We may choose an increasing sequence $t_k < 0$, $t_k \to 0$ and $U_k = U_{t_k} \in K$ with

$$|Fe_k|^2 < |U_ke_k|^2 = \max_{U \in K} |Ue_k|^2 \tag{2.37}$$

for all k, where $e_k = e + t_ke^{\perp}$. By compactness of K we may assume that $U_k \to \overline{U}$ and therefore

$$\lim_{k\to\infty} |Fe_k|^2 \leq \lim_{k\to\infty} |U_ke_k|^2 = |\overline{U}e|^2.$$

By Proposition A.2.5 there exists $\overline{\alpha} \in \mathbb{R}$ and $\overline{Q} \in \mathrm{SO}(2)$ with

$$\overline{Q}\,\overline{U} - F = \overline{\alpha} Fe \otimes e^{\perp}. \tag{2.38}$$

By our assumptions $\overline{\alpha} > 0$, and we assert next that we may find $\alpha_k \in \mathbb{R}$, $Q_k \in \mathrm{SO}(2)$ and $v_k = e + \gamma_k e^{\perp} \in \mathbb{R}^2$ such that

$$Q_k U_k - F = \alpha_k F v_k \otimes v_k^{\perp}, \quad \text{with} \quad \alpha_k \to \overline{\alpha}, \quad Q_k \to \overline{Q}, \quad v_k \to e.$$

A short calculation shows that there exist always two rank-one connections between $SO(2)U_k$ and F,

$$Q_k^{(i)} U_k - F = a_k^{(i)} \otimes b_k^{(i)}, \quad i = 1, 2, \tag{2.39}$$

for which the rotations $Q_k^{(i)}$ do not have (for a suitable subsequence) the same limit. We may pass to the limit in (2.39) and obtain

$$\overline{Q}^{(i)} \overline{U} - F = \overline{a}^{(i)} \otimes \overline{b}^{(i)}, \quad \overline{Q}^{(1)} \neq \overline{Q}^{(2)}$$

and comparison with (2.38) establishes the assertion for a suitable subsequence since there are only two rank-one connections between $SO(2)\overline{U}$. In view of (2.34) we must have $\gamma_k > 0$. We deduce

$$\begin{aligned} |U_k e_k|^2 &= |F e_k + \alpha_k \langle v_k^{\perp}, e_k \rangle F v_k|^2 \\ &= |F e_k|^2 + 2\alpha_k \langle v_k^{\perp}, e_k \rangle \langle F e_k, F v_k \rangle + \alpha_k^2 \langle v_k^{\perp}, e_k \rangle^2 |F v_k|^2 \end{aligned}$$

and this contradicts (2.37) since $\gamma_k > 0$ and $t_k < 0$ imply

$$\langle v_k^{\perp}, e_k \rangle = \langle e^{\perp} - \gamma_k e, e + t_k e^{\perp} \rangle = t_k - \gamma_k < 0, \quad \langle F e_k, F v_k \rangle > 0$$

for k big enough. This establishes (2.35) and hence a contradiction to the assumption $F \in \mathcal{A}$. This concludes the proof of the theorem for $\Delta > 0$.

Assume now that $\Delta = 0$ and that K is thus contained in the rank-one cone $\{X : \det X = 0\}$. Suppose that $F \in \mathcal{A}$ is a constrained point, i.e., there exists an $e \in \mathbb{S}^1$ such that

$$|Fe|^2 = \max_{U \in K} |Ue|^2 = |U_1 e|^2, \quad U_1 \in K.$$

If $|Fe|^2 = 0$, then $|U_1 e|^2 = 0$ and we may assume that $F = s e^{\perp} \otimes e^{\perp}$ and $U_1 = t e^{\perp} \otimes e^{\perp}$ with $t \geq s \geq 0$. Since U_1 and $-U_1$ are contained in K, we conclude $F \in K^{(1)}$.

We now consider the case $|Fe|^2 \neq 0$. In this case the argument is identical to the one used for $\Delta > 0$ since $Fe \neq 0$ implies that $Fe_t \neq 0$ with $e_t = e + t e^{\perp}$ for t small enough. This concludes the proof of the theorem. □

An SO(2) Invariant Set with $K^{\text{rc}} \neq K^{\text{pc}}$. The next example shows the drastic differences in the complexity of hulls of sets of matrices with equal determinant and of sets without this constraint.

We assume in the following that $\alpha > \beta > 1$ or that $1 > \alpha > \beta > 0$ and that U_1 and $U_2 \in \mathbb{M}^{2\times 2}$ are given by

$$U_1 = \begin{pmatrix} \alpha & 0 \\ 0 & \beta \end{pmatrix}, \quad U_2 = \begin{pmatrix} \beta & 0 \\ 0 & \alpha \end{pmatrix}.$$

Let

$$K = \mathrm{SO}(2) \cup \mathrm{SO}(2)U_1 \cup \mathrm{SO}(2)U_2. \tag{2.40}$$

Then any $\nu \in \mathcal{M}^{\mathrm{pc}}(K)$ can be represented as

$$\nu = \lambda_0 \varrho + \lambda_1 \sigma_1 U_1 + \lambda_2 \sigma_2 U_2 \tag{2.41}$$

with $\lambda_0, \lambda_1, \lambda_2 \in [0,1]$, $\lambda_0 + \lambda_1 + \lambda_2 = 1$, and $\varrho, \sigma_1, \sigma_2 \in \mathcal{P}(\mathrm{SO}(2))$. Here $\sigma_i U_i$ denotes the measure given by $\sigma_i U_i(E) = \sigma_i(EU_i^{-1})$ where the set EU_i^{-1} is defined by $EU_i^{-1} = \{F : FU_i \in E\}$. The following theorems show that the rank-one convex hull of K is given by the union of SO(2) and the rank-one convex hull of the two compatible wells $\mathrm{SO}(2)U_1$ and $\mathrm{SO}(2)U_2$. Every element in K^{rc} is therefore given by a second order laminate. Contrary to this, the polyconvex hull is always considerably bigger than the rank-one convex hull. However, if $\nu \in \mathcal{M}^{\mathrm{pc}}(K)$ satisfies $F = \langle \nu, id \rangle \in K^{\mathrm{pc}} \setminus K^{\mathrm{rc}}$, then the mass of the support of ν on $\mathrm{SO}(2)U_1 \cup \mathrm{SO}(2)U_2$ must be bigger than a constant which depends only on α and β, see Theorem 2.2.6 below. It is exactly this property which together with Proposition 2.1.5 allows us to prove that the rank-one convex hull is given by $K^{\mathrm{rc}} = \mathrm{SO}(2) \cup (\mathrm{SO}(2)U_1 \cup \mathrm{SO}(2)U_2)^{\mathrm{rc}}$.

The following theorem summarizes our results.

Theorem 2.2.6. *Suppose that either $\alpha > \beta > 1$ or $1 > \alpha > \beta > 0$ and that K is given by (2.40). Let*

$$\lambda^* = \frac{(\alpha-\beta)^2}{(\alpha-\beta)^2 + 4(\alpha-1)(\beta-1)} \in (0,1).$$

Then there exists a polyconvex measure $\nu \in \mathcal{M}^{\mathrm{pc}}(K)$ with the representation (2.41) and $\lambda_0 \notin \{0,1\}$ if and only if $\lambda_0 \in (0, \lambda^*]$. In particular,

$$K^{\mathrm{pc}} \supsetneqq \mathrm{SO}(2) \cup (\mathrm{SO}(2)U_1 \cup \mathrm{SO}(2)U_2)^{\mathrm{pc}}.$$

Moreover, $\mathcal{M}^{\mathrm{rc}}(K)$ does not contain any measure that is supported on SO(2) *and* $\mathrm{SO}(2)U_1 \cup \mathrm{SO}(2)U_2$, *i.e.,*

$$\mathcal{M}^{\mathrm{rc}}(K) = \mathcal{M}^{\mathrm{rc}}(\mathrm{SO}(2)) \cup \mathcal{M}^{\mathrm{rc}}(\mathrm{SO}(2)U_1 \cup \mathrm{SO}(2)U_2),$$

and in particular

$$K^{\mathrm{rc}} = \mathrm{SO}(2) \cup \left(\mathrm{SO}(2)U_1 \cup \mathrm{SO}(2)U_2\right)^{\mathrm{rc}}.$$

Remark 2.2.7. For $\lambda_0 = \lambda^*$ one obtains

$$F = \lambda^* \mathbb{I} + \frac{1-\lambda^*}{2}(U_1 + U_2) = \frac{2\alpha\beta - \alpha - \beta}{\alpha + \beta - 2} \mathbb{I} \in K^{\mathrm{pc}} \setminus K^{\mathrm{rc}}.$$

It is an open problem to find a formula for K^{pc} and K^{qc}.

Proof. We begin with the assertion for polyconvex measures supported on K. The idea is to find R, S_1, $S_2 \in \mathrm{conv}(\mathrm{SO}(2))$ and λ_0, λ_1, $\lambda_2 \in (0,1)$ such that the minors relations hold, i.e.,

$$F = \lambda_0 R + \lambda_1 S_1 U_1 + \lambda_2 S_2 U_2, \quad \det F = \lambda_0 + (\lambda_1 + \lambda_2)\alpha\beta.$$

Then for arbitrary ϱ, σ_1, $\sigma_2 \in \mathcal{P}(\mathrm{SO}(2))$ with $\langle \varrho, id\rangle = R$, $\langle \sigma_1, id\rangle = S_1$, and $\langle \sigma_2, id\rangle = S_2$, the measure ν defined by $\nu = \lambda_0 \varrho + \lambda_1 \sigma_1 U_1 + \lambda_2 \sigma_2 U_2$ belongs to $\mathcal{M}^{\mathrm{pc}}(K)$. Let

$$R = \begin{pmatrix} r_1 & -r_2 \\ r_2 & r_1 \end{pmatrix}, \quad S_1 = \begin{pmatrix} s_1 & -s_2 \\ s_2 & s_1 \end{pmatrix}, \quad S_2 = \begin{pmatrix} t_1 & -t_2 \\ t_2 & t_1 \end{pmatrix}.$$

It follows from (A.2) that R, S_1, $S_2 \in \mathrm{conv}(\mathrm{SO}(2))$ if $r_1^2 + r_2^2 \leq 1$, $s_1^2 + s_2^2 \leq 1$, and $t_1^2 + t_2^2 \leq 1$. We have to solve the minors relation

$$\det F = \lambda_0 + (1 - \lambda_0)\alpha\beta,$$

where F is given by

$$F = \begin{pmatrix} \lambda_0 r_1 + \lambda_1 s_1 \alpha + \lambda_2 t_1 \beta & -\lambda_0 r_2 - \lambda_1 s_2 \beta - \lambda_2 t_2 \alpha \\ \lambda_0 r_2 + \lambda_1 s_2 \alpha + \lambda_2 t_2 \beta & \lambda_0 r_1 + \lambda_1 s_1 \beta + \lambda_2 t_1 \alpha \end{pmatrix}.$$

A short calculation shows that this is equivalent to

$$\begin{aligned} \lambda_0 + (\lambda_1 + \lambda_2)\alpha\beta &= \lambda_0^2(r_1^2 + r_2^2) + \lambda_1^2(s_1^2 + s_2^2)\alpha\beta + \lambda_2^2(t_1^2 + t_2^2)\alpha\beta \\ &\quad + \lambda_0\big(\lambda_1(r_1 s_1 + r_2 s_2) + \lambda_2(r_1 t_1 + r_2 t_2)\big)(\alpha + \beta) \\ &\quad + \lambda_1\lambda_2(s_1 t_1 + s_2 t_2)(\alpha^2 + \beta^2). \end{aligned}$$

The right hand side is quadratic in the coefficients of R, S_1 and S_2 and can therefore be made arbitrarily small. In order to show the existence of a solution of this equation we only need to derive conditions which imply that the right hand side can in fact be made greater than or equal to the left hand side. The right hand side is maximal if $R = S_1 = S_2 \in \mathrm{SO}(2)$ and we have to find λ_0, λ_1 and λ_2 such that

$$\begin{aligned} &\lambda_0 + (\lambda_1 + \lambda_2)\alpha\beta \\ &\quad \leq \lambda_0^2 + (\lambda_1^2 + \lambda_2^2)\alpha\beta + \lambda_0(\lambda_1 + \lambda_2)(\alpha + \beta) + \lambda_1\lambda_2(\alpha^2 + \beta^2). \end{aligned}$$

This is equivalent to

$$-\lambda_0(1 - \lambda_0)(\alpha - 1)(\beta - 1) + \lambda_1\lambda_2(\alpha - \beta)^2 \geq 0. \tag{2.42}$$

Let $\lambda_0 = 1 - \varepsilon$ with $\varepsilon > 0$. The left hand side is maximal for $\lambda_1 = \lambda_2 = \frac{\varepsilon}{2}$. It follows that the inequality holds for $\lambda_0 \leq 1 - \bar{\varepsilon}$ where $\bar{\varepsilon} \neq 0$ is the solution of

$$-\varepsilon(1 - \varepsilon)(\alpha - 1)(\beta - 1) + \frac{\varepsilon^2}{4}(\alpha - \beta)^2 = 0.$$

This implies

$$\bar{\varepsilon} = \frac{4(\alpha-1)(\beta-1)}{(\alpha-\beta)^2 + 4(\alpha-1)(\beta-1)}.$$

and it follows easily that $\lambda_0 \in (0, \lambda^*]$ is a necessary condition. On the other hand, if $\lambda_0 \in (0, \lambda^*]$ then inequality (2.42) holds and we may define volume fractions $\lambda_1 = \lambda_2 = (1-\lambda_0)/2 \in (0,1)$ and $r \in [0,1]$ by

$$r^2 = \frac{\lambda_0 + (\lambda_1+\lambda_2)\alpha\beta}{\lambda_0^2 + (\lambda_1^2+\lambda_2^2)\alpha\beta + \lambda_0(\lambda_1+\lambda_2)(\alpha+\beta) + \lambda_1\lambda_2(\alpha^2+\beta^2)}.$$

Consequently $R = S_1 = S_2 = r\mathbb{I} \in \operatorname{conv}(\mathrm{SO}(2))$ and if ϱ, σ_1, and σ_2 are probability measures supported on SO(2) with $\langle \varrho, id\rangle = R$, $\langle \sigma_1, id\rangle = S_1$, $\langle \sigma_2, id\rangle = S_2$, then

$$\nu = \lambda_0 \varrho \mathbb{I} + \frac{1-\lambda_0}{2}\big(\sigma_1 U_1 + \sigma_2 U_2\big) \in \mathcal{M}^{\mathrm{pc}}.$$

This proves the assertion about the polyconvex hull.

We now turn towards the proof that the rank-one convex hull is the union of SO(2) with the rank-one convex hull of the two wells. As a first step we show that the fact that (nontrivial) polyconvex measures have minimal support on $\mathrm{SO}(2)U_1 \cup \mathrm{SO}(2)U_2$ implies that

$$\operatorname{dist}(K^{\mathrm{pc}} \setminus \mathrm{SO}(2), \mathrm{SO}(2)) > 0. \tag{2.43}$$

Assume the contrary. Then there exists a sequence $F^{(n)} \in K^{\mathrm{pc}}$ with

$$F^{(n)} = \lambda_0^{(n)} R^{(n)} + \lambda_1^{(n)} S_1^{(n)} U_1 + \lambda_2^{(n)} S_2^{(n)} U_2 \notin \mathrm{SO}(2)$$

and $\operatorname{dist}(F^{(n)}, \mathrm{SO}(2)) \to 0$. It follows from the properties of the polyconvex measures proven above that

$$\lambda_1^{(n)} + \lambda_2^{(n)} \geq 1 - \lambda^*$$

and therefore we may choose a subsequence (again denoted by n) such that

$$F^{(n)} \to F^\infty = \lambda_0^\infty R^\infty + \lambda_1^\infty S_1^\infty + \lambda_2^\infty S_2^\infty \in \mathrm{SO}(2)$$

with $\lambda_0^\infty + \lambda_1^\infty + \lambda_2^\infty = 1$, and $\lambda_1^\infty + \lambda_2^\infty \geq 1 - \lambda^*$. On the other hand, the minors relations for $F^{(n)}$ imply

$$\begin{aligned} 1 = \det F^\infty &= \lim_{n\to\infty} \det F^{(n)} \\ &= \lim_{n\to\infty} \big(\lambda_0^{(n)} + (\lambda_1^{(n)} + \lambda_2^{(n)})\alpha\beta\big) \to \lambda_0^\infty + (\lambda_1^\infty + \lambda_2^\infty)\alpha\beta \neq 1, \end{aligned}$$

since $\alpha\beta \neq 1$. This is a contradiction, and we conclude that (2.43) holds.

The proof of the theorem is now a consequence of Proposition 2.1.5. In view of (2.43) we may choose compact sets C_1 and C_2 such that

$$\mathrm{SO}(2) \subset C_1,\ K^{\mathrm{rc}} \setminus \mathrm{SO}(2) \subset C_2,\ C_1 \cap C_2 = \emptyset.$$

Then (A.5) implies

$$K^{\mathrm{rc}} = (K \cap C_1)^{\mathrm{rc}} \cup (K \cap C_2)^{\mathrm{rc}} = \mathrm{SO}(2) \cup (\mathrm{SO}(2)U_1 \cup \mathrm{SO}(2)U_2)^{\mathrm{rc}}. \quad (2.44)$$

We conclude from (2.44) that $\det F \in \{1, \alpha\beta\}$ for all $F \in K^{\mathrm{rc}}$. Assume now that $\nu \in \mathcal{M}^{\mathrm{rc}}(K)$ is given by $\nu = \lambda_0 \varrho + \lambda_1 \sigma_1 U_1 + \lambda_2 \sigma_2 U_2$ with $\lambda_0 \notin \{0, 1\}$. By the minors relation for the determinant, $\det F = \lambda_0 + (1 - \lambda_0)\alpha\beta \notin \{1, \alpha\beta\}$, a contradiction. This proves the theorem. □

2.3 The Thin Film Case

Bhattacharya and James used Γ-convergence methods to derive a limiting theory for martensitic thin films. In their approach, the sets relevant for the description of the different phases are given by

$$K = \mathrm{O}(2,3)U_1 \cup \ldots \cup \mathrm{O}(2,3)U_k, \quad (2.45)$$

if the normal to the film is suitably oriented with respect to the crystalline axes. Here $\mathrm{O}(2,3)$ denotes the set of all isometries of the plane into the three-dimensional space, i.e., the set of all 3×2 matrices with $F^T F = \mathbb{I}$, and the 2×2 matrices U_i are positive definite and satisfy $\det U_i = \Delta$ for $i = 1, \ldots, k$. We define $\widehat{\pi} : \mathbb{M}^{2\times 2} \to \mathbb{M}^{3\times 2}$ by

$$\widehat{\pi}(F) = \begin{pmatrix} F_{11} & F_{12} \\ F_{21} & F_{22} \\ 0 & 0 \end{pmatrix}. \quad (2.46)$$

It is easy to see that

$$K = \mathrm{SO}(3)\widehat{\pi}(\mathrm{O}(2)U_1 \cup \ldots \cup \mathrm{O}(2)U_k),$$

and therefore it is natural to consider first sets invariant under $\mathrm{O}(2)$. Their semiconvex hulls have open interior and are therefore much bigger than the hulls for $\mathrm{SO}(2)$ invariant sets. This is due to the remarkable fact that any pair of proper and improper rotations is rank-one connected in two dimensions.

Theorem 2.3.1. *Assume that* $\Delta > 0$, *that* $K \subset \{F \in \mathbb{M}^{2\times 2} : |\det F| = \Delta\}$ *is compact, and that* K *is invariant under* $\mathrm{O}(2)$, *i.e.,*

$$K = \mathrm{O}(2)K = \{F = QU : Q \in \mathrm{O}(2),\ U \in K\}.$$

Then $K^{(3)} = K^{\mathrm{rc}} = K^{\mathrm{qc}} = K^{\mathrm{pc}}$ *and*

$$K^{\mathrm{qc}} = \{F \in \mathbb{M}^{2\times 2} : |\det F| \le \Delta,\ |Fe|^2 \le \max_{U \in K} |Ue|^2\ \forall e \in \mathbb{S}^1\}. \quad (2.47)$$

Proof. Let $\mathcal{A}$ denote the formula for the hulls in (2.47). Clearly $\mathcal{A}$ is a polyconvex set, and it suffices to prove that $F \in K^{(2)}$ for all $F \in \mathcal{A}$ with $\det F \geq 0$ and $|F\boldsymbol{e}|^2 = \max_{U\in K} |U\boldsymbol{e}|^2$ for at least one $\boldsymbol{e} \in \mathbb{S}^1$. The idea is to define a set $\widetilde{K} \subset K^{(1)}$ with $\det \widetilde{U} = \det F$ for all $\widetilde{U} \in \widetilde{K}$ such that the following assertions hold:

$$|F\boldsymbol{v}|^2 \leq \max_{\widetilde{U}\in\widetilde{K}} |\widetilde{U}\boldsymbol{v}|^2 \text{ for all } \boldsymbol{v} \in \mathbb{S}^1, \quad |F\boldsymbol{e}|^2 = \max_{\widetilde{U}\in\widetilde{K}} |\widetilde{U}\boldsymbol{e}|^2. \tag{2.48}$$

The construction of the set $\widetilde{K}$ takes advantage of the remarkable fact that $\operatorname{rank}(\mathbb{I} - R) = 1$ for all $R \in \mathrm{O}^-(2)$. This fact motivates us to choose for all $\boldsymbol{w} \in \mathbb{S}^1$ a matrix $U_{\boldsymbol{w}} \in K$ with $\det U_{\boldsymbol{w}} = \Delta$ and $|U_{\boldsymbol{w}}\boldsymbol{w}|^2 = \max_{U\in K} |U\boldsymbol{w}|^2$. We then define a matrix $\widetilde{U}_{\boldsymbol{w}}$ with $\det \widetilde{U}_{\boldsymbol{w}} = \det F$ by

$$\widetilde{U}_{\boldsymbol{w}} = (1 - \lambda_{\boldsymbol{w}})U_{\boldsymbol{w}} + \lambda_{\boldsymbol{w}} Q_{\boldsymbol{w}} U_{\boldsymbol{w}} = U_{\boldsymbol{w}} - \frac{\det U_{\boldsymbol{w}} - \det F}{(\det U_{\boldsymbol{w}})|U_{\boldsymbol{w}}^{-T}\boldsymbol{w}^\perp|^2} U_{\boldsymbol{w}}^{-T}\boldsymbol{w}^\perp \otimes \boldsymbol{w}^\perp,$$

where $\lambda_{\boldsymbol{w}} \in [0,1]$ and $Q_{\boldsymbol{w}} \in \mathrm{O}^-(2)$ are given by

$$\lambda_{\boldsymbol{w}} = \frac{\det U_{\boldsymbol{w}} - \det F}{2 \det U_{\boldsymbol{w}}}, \quad Q_{\boldsymbol{w}} = \mathbb{I} - 2\,\frac{U_{\boldsymbol{w}}^{-T}\boldsymbol{w}^\perp}{|U_{\boldsymbol{w}}^{-T}\boldsymbol{w}^\perp|} \otimes \frac{U_{\boldsymbol{w}}^{-T}\boldsymbol{w}^\perp}{|U_{\boldsymbol{w}}^{-T}\boldsymbol{w}^\perp|}.$$

Let

$$\widetilde{K} = \{Q\widetilde{U}_{\boldsymbol{w}} : \boldsymbol{w} \in \mathbb{S}^1, Q \in \mathrm{SO}(2)\} \subset K^{(1)},$$

and

$$M(\boldsymbol{v}) = \max_{U\in K} |U\boldsymbol{v}|^2 \quad \text{and} \quad \widetilde{M}(\boldsymbol{v}) = \max_{\widetilde{U}\in\widetilde{K}} |\widetilde{U}\boldsymbol{v}|^2.$$

We first assert that $M(\boldsymbol{v}) = \widetilde{M}(\boldsymbol{v})$ for all $\boldsymbol{v} \in \mathbb{S}^1$. In fact, if $\widetilde{U} = Q\widetilde{U}_{\boldsymbol{w}} \in \widetilde{K}$ with $Q \in \mathrm{O}(2)$ and $\boldsymbol{w} \in \mathbb{S}^1$, then by definition

$$|\widetilde{U}\boldsymbol{v}|^2 = |\widetilde{U}_{\boldsymbol{w}}\boldsymbol{v}|^2 = |((1 - \lambda_{\boldsymbol{w}})U_{\boldsymbol{w}} + \lambda_{\boldsymbol{w}} Q_{\boldsymbol{w}} U_{\boldsymbol{w}})\boldsymbol{v}|^2 \leq |U_{\boldsymbol{w}}\boldsymbol{v}|^2 \leq M(\boldsymbol{v})$$

and therefore $\widetilde{M}(\boldsymbol{v}) \leq M(\boldsymbol{v})$. Similarly, $\widetilde{M}(\boldsymbol{v}) \geq |\widetilde{U}_{\boldsymbol{v}}\boldsymbol{v}|^2 = |U_{\boldsymbol{v}}\boldsymbol{v}|^2 = M(\boldsymbol{v})$ and this implies the assertion. This establishes (2.48) and we deduce from the proof of Theorem 2.2.5 that $F \in \widetilde{K}^{(1)} \subset K^{(2)}$. This concludes the proof of the theorem. □

We now present the result for thin films.

Corollary 2.3.2. *Assume that K is given by (2.45). Then*

$$K^{(3)} = K^{\mathrm{rc}} = K^{\mathrm{qc}} = K^{\mathrm{pc}},$$

and

$$K^{\mathrm{qc}} = \{F \in \mathbb{M}^{3\times 2} : \det(F^T F) \leq \Delta^2, |F\boldsymbol{e}|^2 \leq \max_{U\in K} |U\boldsymbol{e}|^2 \,\forall \boldsymbol{e} \in \mathbb{S}^1\}. \tag{2.49}$$

Proof. Let $\mathcal{A}$ be the set in (2.49). For $F \in \mathbb{M}^{3\times 2}$ we define $\text{adj}_{ij}(F)$ to be the determinant of the (2×2)-matrix formed by the i-th and the j-th row of F. Since

$$\det(F^T F) = \text{adj}_{12}^2(F) + \text{adj}_{23}^2(F) + \text{adj}_{31}^2(F)$$

is a polyconvex function, we conclude that $\mathcal{A}$ is a polyconvex set and thus $K^{\text{pc}} \subseteq \mathcal{A}$. On the other hand, K is invariant under multiplication with elements in SO(3) from the left, i.e. $\text{SO}(3)K = K$, and for all $F \in \mathcal{A}$ there exists a $Q \in \text{SO}(3)$ such that $QF = \widehat{\pi}(\widehat{F})$ where $\widehat{F} = (F^T F)^{1/2}$ and $\widehat{\pi}$ is the embedding defined in (2.46). By definition, $(\det \widehat{F})^2 = \det(F^T F) \leq \Delta^2$ and $|\widehat{F}e|^2 = |Fe|^2$ for all $e \in \mathbb{S}^1$, and therefore by Theorem 2.3.1

$$\widehat{F} \in \widehat{K}^{\text{lc}}, \quad \widehat{K} = \text{O}(2)U_1 \cup \ldots \cup \text{O}(2)U_k.$$

The assertion of the corollary follows since $\widehat{\pi}(\text{O}(2)) \subset \text{O}(2,3)$. □

2.4 An Optimal Taylor Bound

In this section, we consider a situation where additional invariances are generated by a different mechanism: intersections of sets rather than unions of sets. This is typically the case in the analysis of bounds for the effective behavior of polycrystals. As an example, we study a so-called nonlinear two-variant elastic material as defined by Kohn and Niethammer [KN00] and we refer to their paper for a general discussion of polycrystals and Taylor bounds. We briefly summarize the underlying ideas. For a single crystal (in some standard orientation), the quasiconvex hull K^{qc} describes the affine boundary conditions for which the effective energy is zero. A polycrystal on the other hand consists of differently oriented grains (which are themselves single crystals) and the zero set of the effective energy for a grain rotated by $Q \in \text{SO}(2)$ with respect to the standard orientation is given by $QK^{\text{qc}}Q^T$. An inner bound for the set of affine boundary conditions for which the effective energy of the polycrystal is zero, is therefore the intersection of all these sets,

$$\mathcal{T} = \bigcap_{f \in \mathbb{S}^1} \Big(\text{SO}(2)(\alpha f \otimes f + \frac{1}{\alpha} f^\perp \otimes f^\perp) \cup \text{SO}(2)(\frac{1}{\alpha} f \otimes f + \alpha f^\perp \otimes f^\perp)\Big)^{\text{qc}}.$$

This bound is called the Taylor bound and it ensures that the affine deformation can be accommodated in each grain individually; it does not take into account compensation effects between different grains.

Theorem 2.4.1. *Assume that* $K = \text{SO}(2)U_1 \cup \text{SO}(2)U_2$ *describes a two-variant elastic material with* $U_1 = \text{diag}(\alpha, \frac{1}{\alpha})$ *and* $U_2 = \text{diag}(\frac{1}{\alpha}, \alpha)$. *Then*

$$\mathcal{T} = \big\{F \in \mathbb{M}^{2\times 2} : \sigma_i(F) \in [\frac{1}{\alpha^*}, \alpha^*], \det F = 1\big\},$$

where

$$\alpha^* = \frac{1}{\sqrt{3}}\big(B + \sqrt{B^2 - 3}\big)^{1/2}, \qquad \textit{with } B = \alpha^2 + \frac{1}{\alpha^2}.$$

Proof. We define for $\boldsymbol{f} \in \mathbb{S}^1$

$$K(\boldsymbol{f}) = \mathrm{SO}(2)\big(\alpha \boldsymbol{f} \otimes \boldsymbol{f} + \frac{1}{\alpha}\boldsymbol{f}^\perp \otimes \boldsymbol{f}^\perp\big) \cup \mathrm{SO}(2)\big(\frac{1}{\alpha}\boldsymbol{f} \otimes \boldsymbol{f} + \alpha \boldsymbol{f}^\perp \otimes \boldsymbol{f}^\perp\big).$$

With this definition, it is easy to see that for all $Q \in \mathrm{SO}(2)$

$$F \in K(\boldsymbol{f})^{\mathrm{pc}} \quad \Leftrightarrow \quad QFQ^T \in K(Q\boldsymbol{f})^{\mathrm{pc}}.$$

We will first show that $\mathcal{T}$ is invariant under SO(2), i.e., $RFQ \in \mathcal{T}$ for all $R, Q \in \mathrm{SO}(2)$ and $F \in \mathcal{T}$. Indeed, if $F \in \mathcal{T}$, then $F \in K^{\mathrm{pc}}(\boldsymbol{f})$ for all $\boldsymbol{f} \in \mathbb{S}^1$, and since the sets $K(\boldsymbol{f})^{\mathrm{pc}}$ are invariant under SO(2), we have that $QF \in K(\boldsymbol{f})^{\mathrm{pc}}$ for all $\boldsymbol{f} \in \mathbb{S}^1$ and $Q \in \mathrm{SO}(2)$. Thus $QF \in \mathcal{T}$. It thus suffices to show that $QFQ^T \in \mathcal{T}$ for all $Q \in \mathrm{SO}(2)$. Let us suppose that $F \in \mathcal{T}$, and that there exists a $Q \in \mathrm{SO}(2)$ such that $QFQ^T \notin \mathcal{T}$. Then there exists an $\boldsymbol{f} \in \mathbb{S}^1$ such that $QFQ^T \notin K(\boldsymbol{f})^{\mathrm{pc}}$, i.e.,

$$QFQ \notin \Big(\mathrm{SO}(2)\big(\alpha \boldsymbol{f} \otimes \boldsymbol{f} + \frac{1}{\alpha}\boldsymbol{f}^\perp \otimes \boldsymbol{f}^\perp\big) \cup \mathrm{SO}(2)\big(\frac{1}{\alpha}\boldsymbol{f} \otimes \boldsymbol{f} + \alpha \boldsymbol{f}^\perp \otimes \boldsymbol{f}^\perp\big)\Big)^{\mathrm{pc}}.$$

This implies that $F \notin Q^T K(\boldsymbol{f})^{\mathrm{pc}} Q = K(Q\boldsymbol{f})^{\mathrm{pc}}$, contradicting the assumption that $F \in \mathcal{T}$.

We show next that α^* is the maximal strain that can be recovered in any basis in a single crystal in standard orientation. To do so, we first consider the quasiconvex hull for the grain in its standard orientation in some detail. It follows from Theorem 2.2.3 that for $K = K(\boldsymbol{e}_1)$

$$K^{\mathrm{qc}} = \Big\{F \in \mathbb{M}^{2\times 2} : \det F = 1, \Big|F\begin{pmatrix}1\\ \pm 1\end{pmatrix}\Big|^2 \le \alpha^2 + \frac{1}{\alpha^2}\Big\}.$$

As a first step, we determine the direction of 'maximal rigidity' of the grain and the corresponding maximal principle strain, i.e., for a given basis $\{\boldsymbol{f}, \boldsymbol{f}^\perp\}$ of $\mathbb{R}^2$ we calculate the maximal strain $\lambda = \lambda(\boldsymbol{f}) \ge 1$ such that

$$\lambda \boldsymbol{f} \otimes \boldsymbol{f} + \frac{1}{\lambda}\boldsymbol{f}^\perp \otimes \boldsymbol{f}^\perp \in K^{\mathrm{qc}}.$$

We then minimize $\lambda(\boldsymbol{f})$ in $\boldsymbol{f} \in \mathbb{S}^1$, i.e., we consider the problem of finding

$$\min_{\boldsymbol{f}\in\mathbb{S}^1}\Big\{\max\Big\{\lambda : \Big|\Big(\lambda \boldsymbol{f} \otimes \boldsymbol{f} + \frac{1}{\lambda}\boldsymbol{f}^\perp \otimes \boldsymbol{f}^\perp\Big)\begin{pmatrix}1\\ \pm 1\end{pmatrix}\Big|^2 \le \alpha^2 + \frac{1}{\alpha^2}\Big\}\Big\}.$$

If we write $\boldsymbol{f} = (\cos\varphi, \sin\varphi)$, then this is equivalent to

$$\lambda^2 + \frac{1}{\lambda^2} \pm (\lambda^2 - \frac{1}{\lambda^2}) \sin\varphi \cos\varphi \leq \alpha^2 + \frac{1}{\alpha^2},$$

and therefore, for fixed φ, the maximal λ has to satisfy

$$\lambda^2 + \frac{1}{\lambda^2} + (\lambda^2 - \frac{1}{\lambda^2}) |\sin\varphi \cos\varphi| \leq \alpha^2 + \frac{1}{\alpha^2}.$$

Let $A = |\sin\varphi \cos\varphi| \in [0, \frac{1}{2}]$ and $B = \alpha^2 + \frac{1}{\alpha^2}$. The solution for fixed A and B is given by

$$\lambda_\pm^2(A) = \frac{B \pm \sqrt{B^2 + 4A^2 - 4}}{2(1+A)}$$

(the square root is always real since $B > 2$ for $\alpha > 1$). For fixed B the function $A \mapsto g(A) = \lambda_\pm^2(A)$ satisfies

$$g'(A) = \frac{4(1+A) - B^2 - B\sqrt{B^2 + 4A^2 - 4}}{2B^2 + 4A^2 - 4(1+A)^2} < \frac{4(1+A) - 4 - 4A}{2B^2 + 4A^2 - 4(1+A)^2} = 0$$

and therefore $\lambda_+^2(A)$ is minimal for $A = \frac{1}{2}$ or $\varphi \in \{\frac{\pi}{2}, \frac{3\pi}{2}\}$. We conclude that the maximal strain that can be recovered is given by α^* and that the corresponding basis is $\mathcal{F} = \{\frac{1}{\sqrt{2}}(e_1 \pm e_2)\}$.

Assume now that there exists a matrix $F = \lambda f \otimes f + \frac{1}{\lambda} f^\perp \otimes f^\perp \in \mathcal{T}$ with $\lambda > \alpha^*$. Let $Q = (f, f^\perp)$ and observe that by the invariance of $\mathcal{T}$ under SO(2) we have

$$Q^T F Q = \lambda e_1 \otimes e_1 + \frac{1}{\lambda} e_2 \otimes e_2 \in \mathcal{T} \subseteq K^{\mathrm{pc}},$$

a contradiction. For $\lambda \leq \alpha^*$ we have $\lambda f \otimes f + \frac{1}{\lambda} f^\perp \otimes f^\perp \in K^{\mathrm{pc}}$ for all $f \in \mathbb{S}^1$ and hence this matrix is also contained in all the sets $K(\tilde{f})^{\mathrm{pc}}$ for all $\tilde{f} \in \mathbb{S}^1$. This concludes the proof of the theorem. □

Remark 2.4.2. We recover the estimate Kohn and Niethammer for $\alpha = 1 + \varepsilon$ by expanding and keeping only lower order terms in ε. Indeed, $\alpha^* = 1 + \varepsilon^2$ (up to higher order terms) and in this case

$$C = F^T F = \begin{pmatrix} 1 & 2\varepsilon^2 \\ 2\varepsilon^2 & 1 \end{pmatrix}, \text{ where } F = \frac{\alpha^*}{2} \begin{pmatrix} 1 \\ 1 \end{pmatrix} \otimes \begin{pmatrix} 1 \\ 1 \end{pmatrix} + \frac{1}{2\alpha^*} \begin{pmatrix} 1 \\ -1 \end{pmatrix} \otimes \begin{pmatrix} 1 \\ -1 \end{pmatrix}.$$

2.5 Dimensional Reduction in Three Dimensions

In the following sections we consider problems in 3×3 matrices. The first situation that one encounters is the case that the structure of the three-dimensional wells is two-dimensional in the sense that the matrices have a block diagonal structure.

To be more specific, assume that the matrices U_i, $i = 1, \ldots, k$, are symmetric with a common eigenvector $\boldsymbol{v}$ and corresponding eigenvalue μ. In a suitable basis, the matrices are therefore block diagonal,

$$U_i = \begin{pmatrix} \hat{U}_i & \\ & \mu \end{pmatrix}, \quad \hat{U}_i \in \mathbb{M}^{2\times 2},$$

and the problem of characterizing the generalized convex hulls of

$$K = \mathrm{SO}(3)U_1 \cup \ldots \cup \mathrm{SO}(3)U_k$$

is equivalent to the corresponding two-dimensional problem for

$$\hat{K} = \mathrm{SO}(2)\hat{U}_1 \cup \ldots \cup \mathrm{SO}(2)\hat{U}_k.$$

The hulls for the set $\hat{K}$ have been characterized in Theorem 2.2.3 in the case of equal determinant and therefore we assume that the matrices U_i satisfy $\det U_i = \Delta$ for $i = 1, \ldots, k$ (see, however, Section 2.6 for the two well problem with different determinant). This is not too restrictive an assumption, and it is always satisfied in the analysis of martensitic phase transformations on which we focus here. Examples include the orthorhombic to monoclinic transformation with two symmetry related wells and one variant of the tetragonal to monoclinic transformations with four wells, see Section 5.5. This situation is also encountered for special cases of the cubic to orthorhombic transformation if the boundary conditions reduce the number of the wells on which the microstructure can be supported, see the analysis in Chapter 4.

Theorem 2.5.1. *Assume that $U_1, \ldots, U_k \in \mathbb{M}^{3\times 3}$ are positive definite and symmetric with $\det U_i = \Delta > 0$. Suppose that there exist $\mu > 0$ and $\boldsymbol{v} \in \mathbb{S}^2$ such that $U_i \boldsymbol{v} = \mu \boldsymbol{v}$ for $i = 1, \ldots, k$. Let $K = \mathrm{SO}(3)U_1 \cup \ldots \cup \mathrm{SO}(3)U_k$. Then $K^{(2)} = K^{\mathrm{lc}} = K^{\mathrm{rc}} = K^{\mathrm{qc}} = K^{\mathrm{pc}}$. Moreover, there exists a discrete set $\mathcal{E} = \{\boldsymbol{e}_1, \ldots, \boldsymbol{e}_\ell\} \subset \mathbb{S}^2$ such that*

$$K^{\mathrm{qc}} = \{F \in \mathbb{M}^{3\times 3} : \det F = \Delta,\ F^T F \boldsymbol{v} = \mu^2 \boldsymbol{v},$$
$$|F\boldsymbol{e}_i|^2 \le \max_{j=1,\ldots,\ell} |U_j \boldsymbol{e}_i|^2, i = 1, \ldots, \ell\}.$$

Example. The first nontrivial example of a quasiconvex hull in three dimensions was given by Ball and James. Suppose that

$$U_1 = \begin{pmatrix} \eta_2 & 0 & 0 \\ 0 & \eta_1 & 0 \\ 0 & 0 & \eta_1 \end{pmatrix},\ U_2 = \begin{pmatrix} \eta_1 & 0 & 0 \\ 0 & \eta_2 & 0 \\ 0 & 0 & \eta_1 \end{pmatrix},\ U_3 = \begin{pmatrix} \eta_1 & 0 & 0 \\ 0 & \eta_1 & 0 \\ 0 & 0 & \eta_2 \end{pmatrix}. \tag{2.50}$$

Then the quasiconvex hull of two wells, say $\widehat{K} = \mathrm{SO}(3)U_1 \cup \mathrm{SO}(3)U_2$, is given by

$$\widehat{K}^{\mathrm{qc}} = \{F : \det F = \eta_1^2 \eta_2,\ |F(\boldsymbol{e}_1 \pm \boldsymbol{e}_2)|^2 \le \eta_1^2 + \eta_2^2,\ (F^T F)\boldsymbol{e}_3 = \eta_1^2 \boldsymbol{e}_3\}.$$

2.6 The Two-well Problem in Three Dimensions

We now consider the two well problem in three dimensions, i.e., the set

$$K = \mathrm{SO}(3)U_1 \cup \mathrm{SO}(3)U_2$$

with U_1, $U_2 \in \mathbb{M}^{3\times 3}$ positive definite. Changing the dependent and independent coordinates, we may assume that $U_1 = \mathbb{I}$ and that U_2 is a diagonal matrix which we denote by $H = \mathrm{diag}(h_1, h_2, h_3)$ with h_1, h_2, $h_3 > 0$. In view of Proposition A.2.1, the two wells are rank-one connected if and only if the middle eigenvalue of H is equal to one. In this case, $\mathbb{I}$ and H have a common eigenvalue and one obtains again a reduction from the three-dimensional problem to a two-dimensional one.

Theorem 2.6.1. *Assume that*

$$K = \mathrm{SO}(3) \cup \mathrm{SO}(3)H, \quad \textit{with} \quad H = \mathrm{diag}\,(h_1, h_2, h_3) \tag{2.51}$$

with $h_1 \geq 1 \geq h_2$ *and* $h_3 = 1$. *Then*

$$K^{\mathrm{lc}} = K^{\mathrm{rc}} = K^{\mathrm{qc}} = K^{\mathrm{pc}} = \left\{ Q \begin{pmatrix} \widehat{F} & \\ & 1 \end{pmatrix} : \ Q \in \mathrm{SO}(3), \ \widehat{F} \in \widehat{K}^{\mathrm{lc}} \right\},$$

where $\widehat{K} = \mathrm{SO}(2) \cup \mathrm{SO}(2)\,\mathrm{diag}(h_1, h_2)$.

Remark 2.6.2. An explicit formula for

$$\widehat{K}^{\mathrm{lc}} = \widehat{K}^{(3)} = \widehat{K}^{\mathrm{rc}} = \widehat{K}^{\mathrm{qc}} = \widehat{K}^{\mathrm{pc}}$$

follows from Šverák's result for the two-dimensional two-well problem in Theorem 2.2.2.

Theorem 2.6.3. *Let the set K be given by (2.51) and assume that either* $h_1 \geq h_2 \geq h_3 > 1$ *or that* $h_1 \leq h_2 \leq h_3 < 1$. *Then* K^{pc} *is trivial, i.e.,* $K^{\mathrm{pc}} = K$.

We now turn to the most interesting case which shows the surprising differences between two-dimensional and three-dimensional models. Assume that the two wells $\mathrm{SO}(3)$ and $\mathrm{SO}(3)H$ are incompatible and that the assumptions in Theorem 2.6.3 are not satisfied. In this case, it turns out that the polyconvex hull is only trivial if the parameters h_i satisfy additional conditions. If these are violated, then it is possible to solve explicitly the minors relations which are now a system of nine quadratic and one cubic equation. This has to be contrasted with the fact that the rank-one convex hull is always trivial.

More precisely, we have the following results. Here we use the convention that $h_4 = h_1$ and $h_3 = h_0$.

Theorem 2.6.4. *Suppose that K is given by*

$$K = \mathrm{SO}(3) \cup \mathrm{SO}(3)H, \quad \text{with} \quad H = \mathrm{diag}\,(h_1, h_2, h_3),\ h_1 \geq h_2 \geq h_3 > 0,$$

and that K does not contain rank-1 connections, i.e., that $h_2 \neq 1$. Assume, in addition, that one the following two conditions holds:

i) there exists an i such that $(h_i - 1)(h_{i-1}h_{i+1} - 1) \geq 0$,
ii) $h_1 \geq h_2 > 1 > h_3 > \frac{1}{3}$ or $3 > h_1 > 1 > h_2 \geq h_3 > 0$.

Then

$$\mathcal{M}^{\mathrm{pc}}(K) \text{ is trivial and in particular } K^{\mathrm{pc}} = K.$$

The next theorem demonstrates that additional assumptions on the parameters h_i are needed for the assertion to be true, even if both wells have the same determinant.

Proposition 2.6.5. *Let $H = \mathrm{diag}(h^2, \frac{1}{h}, \frac{1}{h})$ with $h \geq h^*$ where h^* is the largest solution of the equation*

$$\frac{4\sqrt{h}}{1+h} - \frac{2h}{1+h^2} = 1. \tag{2.52}$$

Suppose that $K = \mathrm{SO}(3) \cup \mathrm{SO}(3)H$. Then $K^{\mathrm{pc}} \setminus K^{\mathrm{rc}} \neq \emptyset$.

Proof. We have to find $\lambda \in (0, 1)$ and R, $S \in \mathrm{conv}(\mathrm{SO}(3))$ such that $F = \lambda R + (1 - \lambda)SH$ satisfies the minors relations,

$$\begin{aligned}\mathrm{cof}\, F =& \lambda R + (1 - \lambda)S \,\mathrm{cof}\, H,\\ \det F =& \lambda + (1 - \lambda) \det H = 1.\end{aligned}$$

We fix $\lambda = \frac{1}{2}$ and seek $R = S$ in the following form:

$$H = \mathrm{diag}(h^2, \frac{1}{h}, \frac{1}{h}), \quad R = S = \mathrm{diag}(r_1, r_2, r_2).$$

Since all matrices in the construction are diagonal, the minors relations reduce to three nonlinear equations,

$$\begin{aligned}\frac{r_2^2}{4}(1 + \frac{1}{h})^2 =& \frac{r_1}{2}(1 + \frac{1}{h^2}),\\ \frac{r_1 r_2}{4}(1 + h^2)(1 + \frac{1}{h}) =& \frac{r_2}{2}(1 + h),\\ \frac{r_1 r_2^2}{8}(1 + h^2)(1 + \frac{1}{h})^2 =& 1.\end{aligned}$$

These equations can be solved explicitly for r_1 and r_2,

$$r_1 = \frac{2h}{1+h^2}, \quad r_2 = \frac{\pm 2\sqrt{h}}{1+h}.$$

It remains to check whether $R \in \text{conv}(\text{SO}(3))$. Since $r_1, r_2 \in [0,1]$ and $r_1 \le r_2$ this is in view of (A.4) true if $2r_2 - r_1 \le 1$, i.e. if $h \ge h^*$ where h^* is the solution of (2.52). □

Remark 2.6.6. We find $h \approx 8.35241$ and $h^2 \approx 69.7628$.

Remark 2.6.7. For $h = 9$, i.e. $H = \text{diag}(81, \frac{1}{9}, \frac{1}{9})$, one obtains a particularly nice matrix $F \in K^{\text{pc}} \setminus K^{\text{rc}} \neq \emptyset$. In fact,

$$F = \text{diag}(9, \frac{1}{3}, \frac{1}{3}) \in K^{\text{pc}} \setminus K^{\text{rc}}.$$

A short calculation shows that

$$R = \text{diag}(\frac{9}{41}, \frac{3}{5}, \frac{3}{5}) \in \text{conv}(\text{SO}(3)).$$

If we write $D_{(x,y,z)} = \text{diag}(x, y, z)$, then R has the representation

$$R = \frac{124}{205} D_{(1,1,1)} + \frac{8}{41} D_{(-1,-1,1)} + \frac{8}{41} D_{(-1,1-1)} + \frac{1}{205} D_{(1,-1,-1)}.$$

By construction, $F = \frac{1}{2}R + \frac{1}{2}RH$ is the center of mass of a polyconvex measure supported on eight matrices in K.

In contrary to this, the rank-one convex hull of two incompatible wells is always trivial.

Theorem 2.6.8. *Suppose that K is given by (2.51) and that K contains no rank-one connections. Then*

$$\mathcal{M}^{\text{rc}}(K) \text{ is trivial and in particular } K^{\text{rc}} = K.$$

The example of the two-well problem demonstrates the rich effects that passage from two to three dimensional situations produces. The next section presents one of the few fully three-dimensional problems for which the semiconvex hulls can be characterized.

2.7 Wells Defined by Singular Values

We now study the case of sets K given by the singular values of the matrix F, i.e.,

$$K = \{F \in \mathbb{M}^{n \times n} : \lambda_i(F) = \gamma_i, \det F = \gamma\},$$

where $\gamma_1, \ldots, \gamma_n > 0$ are given parameters and $\gamma = \gamma_1 \cdots \gamma_n$. The results in this section are motivated by the mathematical analysis of the effective behavior of nematic elastomers in Chapter 3, and they allow us to find an explicit formula for the quasiconvex hull of the free energy of the system proposed by Bladon, Terentjev, and Warner. Before we address the three-dimensional situation, we consider the two-dimensional case.

The Two-dimensional Case. We now consider sets that are invariant under multiplication by SO(2) from the left and the right. We restrict ourselves to the case of sets with constant determinant, since sets of this form are important in the analysis of nematic elastomers in Chapter 3.

Assume that K is given by

$$K = \{F \in \mathbb{M}^{2\times 2} : \lambda_1(F) = \gamma_1,\ \lambda_2(F) = \gamma_2,\ \det F = \gamma_1\gamma_2\}, \tag{2.53}$$

where $0 < \gamma_1 \leq \gamma_2$ and $0 \leq \lambda_1(F) \leq \lambda_2(F)$ are the singular values of F. Equivalently, K can be represented as

$$K = \bigcup_{e\in\mathbb{S}^1} \mathrm{SO}(2)(\gamma_1 e \otimes e + \gamma_2 e^\perp \otimes e^\perp).$$

The next theorem gives an explicit characterization of all the semiconvex hulls.

Theorem 2.7.1. *Assume that K is given by (2.53). Then*

$$\begin{aligned}\operatorname{conv}(K) = \{F : -\gamma_1 - \gamma_2 \leq \min_{R\in O^+(2)} F : R \leq \max_{R\in O^+(2)} F : R \leq \gamma_1 + \gamma_2,\\ \gamma_1 - \gamma_2 \leq \min_{R\in O^-(2)} F : R \leq \max_{R\in O^-(2)} F : R \leq -\gamma_1 + \gamma_2\}.\end{aligned}$$

Moreover, $K^{(2)} = K^{\mathrm{rc}} = K^{\mathrm{qc}} = K^{\mathrm{pc}}$ and these sets are given by

$$K^{\mathrm{pc}} = \{F : \det F = \gamma_1\gamma_2,\ \lambda_i(F) \in [\gamma_1, \gamma_2]\}.$$

In particular, $K^{\mathrm{pc}} = \operatorname{conv}(K) \cap \{\det F = \gamma_1\gamma_2\}$.

Remark 2.7.2. The case that K is defined in terms of the singular values without the determinant constraint was treated in [DcT98], see also Figure 2.4. Here $\operatorname{conv}(K) = \{F : \lambda_2 \leq \gamma_2, \lambda_1 + \lambda_2 \leq \gamma_1 + \gamma_2\}$. The two conditions in the definition of the convex hull follow immediately from general properties of singular values since $F \mapsto \lambda_1(F) + \lambda_2(F)$ is a convex function.

Proof. Let M denote the right hand side of the formula in the assertion of the theorem. In view of Proposition 2.7.8 it is clear that M is convex with $\operatorname{conv}(K) \subseteq M$ and that M is invariant under SO(2), i.e., $\mathrm{SO}(2)M\mathrm{SO}(2) = M$. It therefore suffices to show that all diagonal matrices $\Lambda = \operatorname{diag}(\mu_1, \mu_2)$ in M are contained in $\operatorname{conv}(K)$. However, if Λ satisfies all the inequalities in the definition of M, then

$$\Lambda \in \operatorname{conv}\Big\{\operatorname{diag}(\gamma_1,\gamma_2), \operatorname{diag}(-\gamma_1,-\gamma_2), \operatorname{diag}(\gamma_2,\gamma_1), \operatorname{diag}(-\gamma_2,-\gamma_1)\Big\},$$

see Figure 2.4. The remaining assertions are an immediate consequence of the characterization of the semiconvex hulls of the two-well problem in Theorem 2.2.2. In fact, $g(F) = (\det F - \gamma_1\gamma_2)^2$ is a polyconvex function and therefore $K^{\mathrm{pc}} \subset \{\det X = \gamma_1\gamma_2\}$. Moreover, if $\Lambda = \operatorname{diag}(\mu_1, \mu_2)$ is a diagonal matrix in $\operatorname{conv}(K)$, then

$$\Lambda \in \big(\mathrm{SO}(2)\operatorname{diag}(\gamma_1,\gamma_2) \cup \mathrm{SO}(2)\operatorname{diag}(\gamma_2,\gamma_1)\big)^{\mathrm{qc}},$$

and this establishes the assertion. □

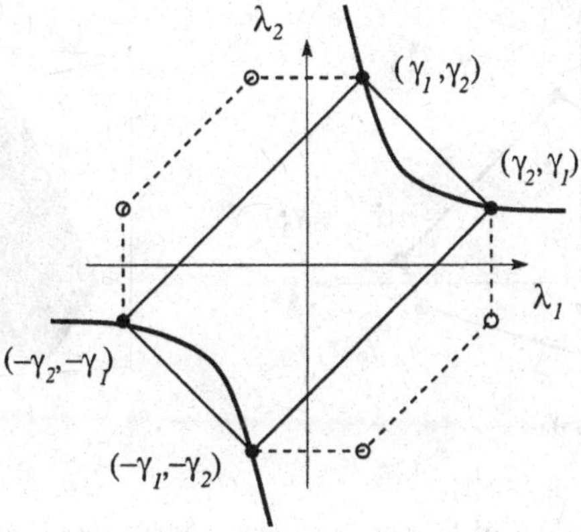

Fig. 2.4. *Semiconvex hulls of sets defined by singular values. The solid dots are the four diagonal matrices in K in (2.53), the four circles are the additional diagonal matrices in K if the condition of positive determinant is dropped. The convex hull of the set K in (2.53) is the solid rectangle, the polyconvex hull consist of the two hyperbolic arcs.*

The Three-dimensional Case. We now extend the result in Theorem 2.7.1 to the three-dimensional situation. It turns out that the convex hull of K is given by $\mathrm{SO}(3)\Sigma\mathrm{SO}(3)$ where $\Sigma = \mathrm{diag}(s_1, s_2, s_3)$ is a diagonal matrix such that the vector (s_1, s_2, s_3) is contained in the convex hull $\mathcal{P}$ of the diagonal matrices in K. As a first step we derive a formula for this set. It is important to note that $\mathcal{P}$ contains matrices with negative determinant, e.g.,

$$\frac{1}{2}\,\mathrm{diag}\left((\gamma_1 - \gamma_3, \gamma_2, -(\gamma_1 - \gamma_3)\right) \in \mathcal{P}.$$

Proposition 2.7.3. *Assume that* $0 < \gamma_1 \le \gamma_2 \le \gamma_3$ *and define the set* $\mathcal{E}$ *by*

$$\mathcal{E} = \left\{(\varepsilon_i\gamma_i, \varepsilon_j\gamma_j, \varepsilon_k\gamma_k) : \varepsilon_{i,j,k} \in \{\pm 1\},\, \varepsilon_i\varepsilon_j\varepsilon_k = 1, \{i,j,k\} = \{1,2,3\}\right\}.$$

Then $\mathrm{conv}(\mathcal{E}) = \mathcal{P}$ *where* $\mathcal{P}$ *is given by*

$$\begin{aligned}\mathcal{P} = \{\sigma \in \mathbb{R}^3 : &|\langle\sigma, \varepsilon\rangle| \le \gamma_1 + \gamma_2 + \gamma_3,\ \varepsilon_i \in \{\pm 1\},\ \varepsilon_1\varepsilon_2\varepsilon_3 = 1,\\ &|\langle\sigma, \varepsilon\rangle| \le -\gamma_1 + \gamma_2 + \gamma_3,\ \varepsilon_i \in \{\pm 1\},\ \varepsilon_1\varepsilon_2\varepsilon_3 = -1,\\ &|\sigma_i| \le \gamma_i,\ i = 1,2,3\}.\end{aligned}$$

Remark 2.7.4. If all γ_i are distinct, then $\mathcal{E}$ contains 24 points and its convex hull is the intersection of the 14 halfspaces defined by the normals $\pm e_i$ and $\frac{1}{\sqrt{3}}(\pm e_1 \pm e_2 \pm e_3)$, see Figure 2.5.

Proof. It is clear that $\mathcal{P}$ contains $\mathrm{conv}(\mathcal{E})$ since $\mathcal{P}$ is a convex set that contains all points in $\mathcal{E}$. We therefore have to show that $\mathcal{P} \subseteq \mathrm{conv}(\mathcal{E})$. Since $\mathcal{P}$ is a compact set it suffices to show that $\partial\mathcal{P} \subset \mathrm{conv}(\mathcal{E})$, where $\partial\mathcal{P}$ consist

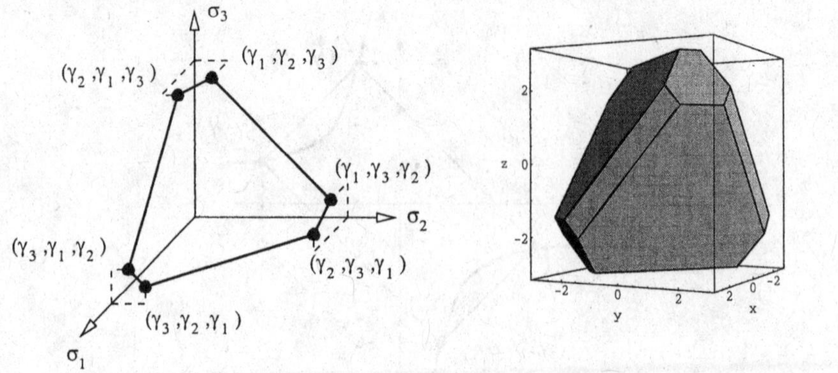

Fig. 2.5. *The convex hull of $\mathcal{E}$. The left figure shows the generation of one of the hexagons in the boundary of the convex hull, the right figure shows the convex hull of $\mathcal{E}$. The boundary of the polygon consists of six rectangles, with normals $\pm e_i$, six small hexagons, with normals $(\varepsilon_1, \varepsilon_2, \varepsilon_3)$, $\varepsilon_i \in \{\pm 1\}$ with $\varepsilon_1\varepsilon_2\varepsilon_3 = 1$, and six large hexagons, corresponding to $\varepsilon_1\varepsilon_2\varepsilon_3 = -1$, respectively.*

of all points in $\mathcal{P}$ for which at least one of the inequalities in the definition of $\mathcal{P}$ is an equality. We choose two representative cases, namely $\sigma_3 = \gamma_3$ and $\sigma_1 + \sigma_2 + \sigma_3 = \gamma_1 + \gamma_2 + \gamma_3$; all the other cases can be handled similarly.

Assume first that $\boldsymbol{p} = (\sigma_1, \sigma_2, \sigma_3) \in \mathcal{P}$ with $\sigma_3 = \gamma_3$. We assert first that

$$\boldsymbol{p} \in \mathcal{C} = \operatorname{conv}\{(\gamma_1, \gamma_2, \gamma_3), (\gamma_2, \gamma_1, \gamma_3), (-\gamma_1, -\gamma_2, \gamma_3), (-\gamma_2, -\gamma_1, \gamma_3)\}$$

(see Figure 2.5). In fact, it follows from the inequalities in the definition of $\mathcal{P}$ that

$$\begin{aligned}
\sigma_1 + \sigma_2 + \sigma_3 &\le \gamma_1 + \gamma_2 + \gamma_3 &\Leftrightarrow&& \sigma_1 + \sigma_2 &\le \gamma_1 + \gamma_2,\\
-\sigma_1 - \sigma_2 + \sigma_3 &\le \gamma_1 + \gamma_2 + \gamma_3 &\Leftrightarrow&& -\sigma_1 - \sigma_2 &\le \gamma_1 + \gamma_2,\\
-\sigma_1 + \sigma_2 + \sigma_3 &\le -\gamma_1 + \gamma_2 + \gamma_3 &\Leftrightarrow&& -\sigma_1 + \sigma_2 &\le -\gamma_1 + \gamma_2,\\
\sigma_1 - \sigma_2 + \sigma_3 &\le -\gamma_1 + \gamma_2 + \gamma_3 &\Leftrightarrow&& \sigma_1 - \sigma_2 &\le -\gamma_1 + \gamma_2.
\end{aligned}$$

These four inequalities characterize the convex hull of

$$\{(\gamma_1, \gamma_2), (\gamma_2, \gamma_1), (-\gamma_1, -\gamma_2), (-\gamma_2, -\gamma_1)\}$$

(see Figure 2.4), and this proves that $\boldsymbol{p} \in \operatorname{conv}(\mathcal{E})$.

Assume now that $\sigma_1 + \sigma_2 + \sigma_3 = \gamma_1 + \gamma_2 + \gamma_3$. In this case we show that that $\boldsymbol{p}$ is contained in the convex hull $\mathcal{C}$ of the six points

$$\{(\gamma_1, \gamma_2, \gamma_3), (\gamma_1, \gamma_3, \gamma_2), (\gamma_2, \gamma_3, \gamma_1), (\gamma_2, \gamma_1, \gamma_3), (\gamma_3, \gamma_1, \gamma_2), (\gamma_3, \gamma_2, \gamma_1)\},$$

(see Figure 2.5). The set $\mathcal{C}$ forms a hexagon in the affine subspace A given by $A = \{\sigma_1 + \sigma_2 + \sigma_3 = \gamma_1 + \gamma_2 + \gamma_3\}$, and a point in A belongs to $\mathcal{C}$ if and only if

$$\sigma_1 \le \gamma_1, \quad \sigma_2 \le \gamma_1, \quad \sigma_3 \le \gamma_1,$$

and

$$\begin{aligned} \sigma_1 + \sigma_2 - \sigma_3 &\le -\gamma_1 + \gamma_2 + \gamma_3, \\ \sigma_1 - \sigma_2 + \sigma_3 &\le -\gamma_1 + \gamma_2 + \gamma_3, \\ -\sigma_1 + \sigma_2 + \sigma_3 &\le -\gamma_1 + \gamma_2 + \gamma_3. \end{aligned}$$

These inequalities are contained in the definition of $\mathcal{P}$ and thus $\boldsymbol{p} \in \operatorname{conv}(\mathcal{E})$. All the remaining cases can be handled similarly. □

The next theorem describes the fundamental construction for the generation of the lamination convex hull of K. To simplify the notation, we assume that $\gamma_1\gamma_2\gamma_3 = 1$. We use the convention that $\lambda_0 = \lambda_3$, $\lambda_4 = \lambda_1$ and $\xi_0 = \xi_3$, $\xi_4 = \xi_1$ and we write $\Lambda(F) = \{\lambda_1(F), \lambda_2(F), \lambda_3(F)\}$ for the set of the singular values of F.

Theorem 2.7.5. *Assume that* $0 < \xi_1 \le \xi_2 \le \xi_3$ *with* $\xi_1\xi_2\xi_3 = 1$ *and that*

$$K = \{F \in \mathbb{M}^{3\times 3} : \det F = 1,\ \lambda_i(F) = \xi_i,\ i = 1,2,3\}.$$

Then the sets M_i *defined by*

$$\begin{aligned} M_i = \{F \in \mathbb{M}^{3\times 3} &: \det F = 1,\ \xi_i \in \Lambda(F), \\ &\Lambda(F) \setminus \{\xi_i\} \subset [\,\min\{\xi_{i-1}, \xi_{i+1}\}, \max\{\xi_{i-1}, \xi_{i+1}\}\,]\} \end{aligned}$$

are contained in $K^{(1)}$ *for* $i = 1, 2, 3$. *Moreover, we have*

$$K^{(1)} = M_1 \quad \text{if} \quad \xi_1 = \xi_2 \qquad \text{and} \qquad K^{(1)} = M_3 \quad \text{if} \quad \xi_2 = \xi_3.$$

Remark 2.7.6. In general the inclusion $M_1 \cup M_2 \cup M_3 \subset K^{(1)}$ is strict.

Proof. To prove the first part of the theorem, we assume that $i = 1$, and we write $M = M_1$. The argument is analogous for $i = 2$ and $i = 3$. The assertion of the proposition is now equivalent to $M \subseteq K^{(1)}$ where

$$M = \{F \in \mathbb{M}^{3\times 3} : \det F = 1,\ \lambda_1(F) = \xi_1, \lambda_2(F), \lambda_3(F) \in [\,\xi_2, \xi_3\,]\}.$$

Let $F \in M$. Since $QFR \in M$, for all $Q, R \in SO(3)$ and $F \in M$, we may suppose that F is diagonal, $F = \operatorname{diag}(\xi_1, \mu_2, \mu_3)$, with $\mu_2, \mu_3 \in [\,\xi_2, \xi_3\,]$. Note that $\mu_2\mu_3 = \xi_2\xi_3$ since $\det F = 1$. There is nothing to prove if $\mu_2 = \xi_2$ or $\mu_3 = \xi_3$, since the condition $\det F = 1$ implies in this case that $F \in K$. We now show that there exists for $\mu_2, \mu_3 \in (\xi_2, \xi_3)$ a $\delta > 0$ (which depends on μ_2 and μ_3) such that

$$F^{\pm} = \operatorname{diag}(\xi_1, \widehat{F}^{\pm}) \in K \text{ where } \widehat{F}^{\pm} = \begin{pmatrix} \mu_2 & \pm\delta \\ 0 & \mu_3 \end{pmatrix}.$$

Then

$$F^+ - F^- = 2\delta e_2 \otimes e_3 \quad \text{and} \quad F = \frac{1}{2}F^+ + \frac{1}{2}F^- \in K^{(1)}.$$

We define

$$\widehat{C}^\pm = (\widehat{F}^\pm)^T \widehat{F}^\pm = \begin{pmatrix} \mu_2^2 & \pm\delta\mu_2 \\ \pm\delta\mu_2 & \mu_3^2 + \delta^2 \end{pmatrix}.$$

The eigenvalues $t^\pm$ of $\widehat{C}^\pm$ are the solutions of

$$\det(\widehat{C}^\pm - t\mathbb{I}) = t^2 - (\mu_2^2 + \mu_3^3 + \delta^2)t + \mu_2^2\mu_3^2 = 0,$$

and the requirement that t^+ defined by

$$t^\pm = \frac{\mu_2^2 + \mu_3^2 + \delta^2}{2} \pm \sqrt{(\frac{\mu_2^2 + \mu_3^2 + \delta^2}{2})^2 - \mu_2^2\mu_3^2}$$

be equal to ξ_3^2 leads to

$$\delta = \frac{1}{\xi_3}\sqrt{\xi_3^2 - \mu_2^2}\sqrt{\xi_3^2 - \mu_3^2} > 0. \tag{2.54}$$

Since $t^+t^- = \mu_2^2\mu_3^2 = \xi_2^2\xi_3^2$, this choice of δ also yields $t^- = \xi_2^2$ and we conclude that for the value of δ given in (2.54) the matrices $F^\pm$ are contained in K and this proves the first assertion of the theorem.

It remains to prove the characterization of $K^{(1)}$ if two of the parameters in the description of K coincide. Without loss of generality we assume that $\xi_1 = \xi_2 < \xi_3$, and we have to prove that $K^{(1)} \subseteq M_1$. Suppose thus that $F \in K^{(1)} \setminus K$ and choose F_1, $F_2 \in K$ such that there exists a $\lambda \in (0,1)$ and $\boldsymbol{a}$, $\boldsymbol{n} \in \mathbb{R}^3$, $\boldsymbol{a}$, $\boldsymbol{n} \neq 0$ with

$$F = \lambda F_1 + (1-\lambda)F_2, \qquad F_1 - F_2 = \boldsymbol{a} \otimes \boldsymbol{n}.$$

Since $QFR \in K$, for all Q, $R \in \mathrm{SO}(3)$ and $F \in K$, we may choose Q and $R \in \mathrm{SO}(3)$ such that $\widetilde{F}_1 = QF_1R \in K$ is diagonal, $\widetilde{F}_1 = \mathrm{diag}(\xi_3, \xi_1, \xi_1)$. We define analogously $\widetilde{F}_2 = QF_2R \in K$, $\widetilde{F} = QFR \in K^{(1)}$, $\widetilde{\boldsymbol{a}} = Q\boldsymbol{a}$ and $\widetilde{\boldsymbol{n}} = R^T\boldsymbol{n}$. Then

$$\widetilde{F} = \lambda\widetilde{F}_1 + (1-\lambda)\widetilde{F}_2, \qquad \widetilde{F}_1 - \widetilde{F}_2 = \widetilde{\boldsymbol{a}} \otimes \widetilde{\boldsymbol{n}}.$$

The intersection of the plane spanned by the unit vectors $\boldsymbol{e}_2 = (0,1,0)$ and $\boldsymbol{e}_3 = (0,0,1)$ intersects the plane $\{\boldsymbol{w} \in \mathbb{R}^3 : \langle \boldsymbol{w}, \widetilde{\boldsymbol{n}} \rangle = 0\}$ with normal $\widetilde{\boldsymbol{n}}$ at least in a one-dimensional line through the origin parallel to some unit vector $\boldsymbol{v} \in \mathbb{S}^2$. This implies

$$\boldsymbol{0} = \langle \widetilde{\boldsymbol{n}}, \boldsymbol{v} \rangle \widetilde{\boldsymbol{a}} = (\widetilde{\boldsymbol{a}} \otimes \widetilde{\boldsymbol{n}})\boldsymbol{v} = (\widetilde{F}_1 - \widetilde{F}_2)\boldsymbol{v} = \xi_1\boldsymbol{v} - \widetilde{F}_2\boldsymbol{v},$$

and therefore $\boldsymbol{v}$ is an eigenvector of $\widetilde{F}_1$ and $\widetilde{F}_2$ with corresponding eigenvalue ξ_1. Consequently, $\widetilde{F}\boldsymbol{v} = \xi_1 \boldsymbol{v}$ and

$$\lambda_1(\widetilde{F}) = \lambda_{\min}(\widetilde{F}) = \min_{\boldsymbol{e}\in\mathbb{S}^2} |\widetilde{F}\boldsymbol{e}| \le |\widetilde{F}\boldsymbol{v}| = \xi_1.$$

To prove that ξ_1 is the smallest singular value of $\widetilde{F}$, let $(\cdot)^+$ denote the convex, nondecreasing function $t \mapsto (t)^+ = \max\{t, 0\}$. Then the functions

$$\begin{aligned} g_1(F) &= \big(\sup_{\boldsymbol{e}\in\mathbb{S}^2} |F\boldsymbol{e}| - \gamma_3\big)^+, \\ g_2(F) &= \big(\sup_{\boldsymbol{e}\in\mathbb{S}^2} |\operatorname{cof} F\boldsymbol{e}| - \frac{1}{\gamma_1}\big)^+ \end{aligned}$$

are polyconvex, and since $F \in K^{(1)} \subseteq K^{\mathrm{pc}}$ we deduce $\lambda_i(\widetilde{F}) \in [\xi_1, \xi_3]$. Therefore

$$\lambda_1(\widetilde{F}) = \xi_1 \quad \text{and} \quad \xi_1 = \min\{\xi_2, \xi_3\} \le \lambda_2(\widetilde{F}) \le \lambda_3(\widetilde{F}) \le \max\{\xi_2, \xi_3\}.$$

We obtain $\widetilde{F} \in M_1$. The matrices $\widetilde{F}$ and F have the same singular values, and hence $F \in M_1$. This concludes the proof of the theorem. □

The foregoing theorem implies immediately a formula for the semiconvex hulls in three dimensions.

Theorem 2.7.7. *Assume that* $0 < \gamma_1 \le \gamma_2 \le \gamma_3$ *with* $\gamma_1\gamma_2\gamma_3 = 1$ *and that*

$$K = \big\{F \in \mathbb{M}^{3\times 3} : \det F = 1,\ \lambda_i(F) = \gamma_i,\ i = 1, 2, 3\big\}.$$

Then

$$K^{(2)} = K^{\mathrm{lc}} = K^{\mathrm{rc}} = K^{\mathrm{qc}} = K^{\mathrm{pc}}, \tag{2.55}$$

and these sets are given by

$$K^{\mathrm{qc}} = \big\{F \in \mathbb{M}^{3\times 3} : \det F = 1,\ \lambda_i(F) \in [\gamma_1, \gamma_3],\ i = 1, 2, 3\big\}. \tag{2.56}$$

Moreover,

$$\operatorname{conv}(K) = \big\{F \in \mathbb{M}^{3\times 3} : (\sigma_1(F), \sigma_2(F), \sigma_3(F)) \in \mathcal{P}\big\},$$

where $\mathcal{P}$ has been defined in Proposition 2.7.3. In particular,

$$K^{\mathrm{pc}} \ne \operatorname{conv}(K) \cap \{F \in \mathbb{M}^{3\times 3} : \det F = 1\}.$$

Proof. Let $\mathcal{A}$ be the set given in (2.56). Since $\det F = 1$, we have

$$\lambda_{\min}(F) = \lambda_1(F) = \frac{1}{\lambda_{\max}(\operatorname{cof} F)} = \frac{1}{\lambda_3(\operatorname{cof} F)}, \tag{2.57}$$

and therefore we may rewrite the definition of $\mathcal{A}$ as

$$\mathcal{A} = \{F \in \mathbb{M}^{3\times 3} : g_1(F) \le 0,\, g_2(F) \le 0,\, g_3(F) \le 0\},$$

where g_1 and g_2 were defined in the proof of Theorem 2.7.5 and

$$g_3(F) = (\det F - 1)^2.$$

The functions g_i are polyconvex, hence $K^{\mathrm{pc}} \subseteq \mathcal{A}$. It only remains to prove that $\mathcal{A} \subseteq K^{(2)}$. We then obtain that $\mathcal{A} \subseteq K^{\mathrm{pc}} \subseteq K^{(2)} \subseteq K^{\mathrm{pc}}$ and the equation (2.55) is thus an immediate consequence. Since $QFR \in \mathcal{A}$, for all Q, $R \in \mathrm{SO}(3)$ and $F \in \mathcal{A}$ we may assume that $F \in \mathcal{A}$ is a diagonal matrix, $F = \mathrm{diag}(\mu_1, \mu_2, \mu_3)$ with $\gamma_1 \le \mu_1 \le \mu_2 \le \mu_3 \le \gamma_3$. If $\gamma_2 \le \mu_3 \le \gamma_3$, then

$$\frac{\gamma_2}{\mu_3} \le 1 \le \frac{\gamma_3}{\mu_3} \quad \text{and} \quad \gamma_2 \le \frac{\gamma_2\gamma_3}{\mu_3} \le \gamma_3.$$

By Theorem 2.7.5,

$$\begin{aligned} M_1 = \{F \in \mathbb{M}^{3\times 3} &: \det F = 1,\ \lambda_1(F) = \gamma_1, \\ &\lambda_2(F) = \min\{\frac{\gamma_2\gamma_3}{\mu_3}, \mu_3\},\ \lambda_3(F) = \max\{\frac{\gamma_2\gamma_3}{\mu_3}, \mu_3\}\} \end{aligned}$$

is contained in $K^{(1)}$. Now $\gamma_1 \le \mu_1 \le \mu_2 \le \frac{\gamma_2\gamma_3}{\mu_3}$ since

$$\mu_2 \le \frac{\gamma_2\gamma_3}{\mu_3} \quad \Leftrightarrow \quad \mu_2\mu_3 \le \gamma_2\gamma_3 \quad \Leftrightarrow \quad \gamma_1 \le \mu_1.$$

If $\gamma_2\gamma_3 \le \mu_3^2$, then $\lambda_3(F) = \mu_3$ for $F \in M_1$ while for $\gamma_2\gamma_3 > \mu_3^2$ one has $\lambda_2(F) = \mu_3$ for $F \in M_1$. Another application of Theorem 2.7.5 with $i = 3$ or $i = 2$, respectively, implies that

$$\{F \in \mathbb{M}^{3\times 3} : \lambda_i(F) = \mu_i, \} \subseteq M_1^{(1)} \subseteq K^{(2)}.$$

Suppose now that $\gamma_1 \le \mu_1 \le \mu_2 \le \mu_3 \le \gamma_2$. Then $\gamma_1 \le \frac{\gamma_1\gamma_2}{\mu_1} \le \gamma_2$, and we conclude from Theorem 2.7.5 that

$$\begin{aligned} M_3 = \{F \in \mathbb{M}^{3\times 3} &: \det F = 1,\ \lambda_3(F) = \gamma_3, \\ &\lambda_1(F) = \min\{\mu_1, \frac{\gamma_1\gamma_2}{\mu_1}\},\ \lambda_2(F) = \max\{\mu_1, \frac{\gamma_1\gamma_2}{\mu_1}\}\} \end{aligned}$$

is contained in $K^{(1)}$. In this situation, $\frac{\gamma_1\gamma_2}{\mu_1} \le \mu_2 \le \mu_3 \le \gamma_3$, since

$$\frac{\gamma_1\gamma_2}{\mu_1} \le \mu_2 \quad \Leftrightarrow \quad \gamma_1\gamma_2 \le \mu_1\mu_2 \quad \Leftrightarrow \quad \mu_3 \le \gamma_3,$$

and we can apply Theorem 2.7.5 once more (with $i = 1$ or $i = 2$) to deduce

$$\{F \in \mathbb{M}^{3\times 3} : \det F = 1,\ \lambda_i(F) = \mu_i,\ i = 1,2,3\} \subseteq M_3^{(1)} \subseteq K^{(2)}.$$

This establishes the result for the polyconvex hull.

In order to prove the formula for the convex hull, we conclude from Proposition 2.7.9 that for all $R \in \mathrm{SO}(3)$

$$\min_{U\in K} U : R = \gamma_1 - \gamma_2 - \gamma_3, \quad \max_{U\in K} U : R = \gamma_1 + \gamma_2 + \gamma_3. \tag{2.58}$$

Let $\mathcal{A} = \{F \in \mathbb{M}^{3\times 3} : (\sigma_1(F), \sigma_2(F), \sigma_3(F)) \in \mathcal{P}\}$. By polar decomposition and the definition of the (signed) singular values, $F = Q_1 \operatorname{diag}(\sigma_1, \sigma_2, \sigma_3) Q_2$ belongs to $\mathcal{A}$ for all Q_1, $Q_2 \in \mathrm{SO}(3)$ if and only if $\Lambda = \operatorname{diag}(\sigma_1, \sigma_2, \sigma_3) \in \mathcal{P}$. In view of Proposition 2.7.3, $\Lambda \in \mathcal{P}$ implies $\Lambda \in \operatorname{conv}(K)$ and consequently $\mathcal{A} \subseteq \operatorname{conv}(K)$. It remains to show that $\operatorname{conv}(K) \subseteq \mathcal{A}$ (it is not obvious that $\mathcal{A}$ is a convex set). Assume that $F \in \operatorname{conv}(K)$. Since $QKR = K$ for all Q, $R \in \mathrm{SO}(3)$, we may assume that $F = \operatorname{diag}(\sigma_1, \sigma_2, \sigma_3)$ is diagonal. We have to show that $(\sigma_1, \sigma_2, \sigma_3) \in \mathcal{P}$. The formula for $\mathcal{P}$ shows that all points $(\sigma_1, \sigma_2, \sigma_3) \in \mathcal{P}$ with $|\sigma_1| \le \sigma_2 \le \sigma_3$ are characterized by the three inequalities

$$\sigma_3 \le \gamma_3, \quad \pm\sigma_1 + \sigma_2 + \sigma_3 \le \pm\gamma_1 + \gamma_2 + \gamma_3.$$

By (2.58)

$$\min_{R\in\mathrm{SO}(3)} F : R = \sigma_1 - \sigma_2 - \sigma_3 \ge \gamma_1 - \gamma_2 - \gamma_3,$$
$$\max_{R\in\mathrm{SO}(3)} F : R = \sigma_1 + \sigma_2 + \sigma_3 \le \gamma_1 + \gamma_2 + \gamma_3,$$

and by the convexity of the norm

$$\sigma_3 = \max_{e\in\mathbb{S}^2} |Fe| \le \max_{U\in K} \max_{e\in\mathbb{S}^2} |Ue| = \gamma_3,$$

and these inequalities prove the desired estimates.

We finally show that $K^{\mathrm{pc}} \neq \operatorname{conv}(K) \cap \{F \in \mathbb{M}^{3\times 3} : \det F = \gamma_1\gamma_2\gamma_3\}$. It suffices to prove this for diagonal matrices which we identify with vectors in $\mathbb{R}^3$. Intuitively this is clear: the surface $\sigma_1\sigma_2\sigma_3 = 1$ intersects the face $\mathcal{F}$ of the convex hull of K given by $-\sigma_1 + \sigma_2 + \sigma_3 = -\gamma_1 + \gamma_2 + \gamma_3$ in $\mathbb{R}^3_+$ in a one-dimensional curve that contains the points $(\gamma_1, \gamma_2, \gamma_3)$ and $(\gamma_1, \gamma_3, \gamma_2)$ and consists of points with $\sigma_1 < \gamma_1$ between these two points. In order to make the argument precise, consider the face $\mathcal{F}$ intersected with all points that satisfy $\sigma_2 = \sigma_3$. Obviously, the two points $(\gamma_1, \frac{1}{2}(\gamma_2 + \gamma_3), \frac{1}{2}(\gamma_2 + \gamma_3))$ and $(0, \frac{1}{2}(-\gamma_1 + \gamma_2 + \gamma_3), \frac{1}{2}(-\gamma_1 + \gamma_2 + \gamma_3))$ are contained in this set. Since $\frac{1}{4}(\gamma_2+\gamma_3)^2\gamma_1 \ge \gamma_1\gamma_2\gamma_3$ and $\frac{1}{4}(-\gamma_1+\gamma_2+\gamma_3)^2 \cdot 0 = 0$ we conclude by continuity the existence of a point $(t, s, s) \in \mathcal{F}$ with $s^2 t = \gamma_1\gamma_2\gamma_3$ and $0 < t < \gamma_1$. This proves the final assertion in the theorem, since all points in the polyconvex hull of K have to satisfy the inequality $\sigma_1 \ge \gamma_1$. The proof of the corollary is now complete. □

In the characterization of the semiconvex hulls for sets depending on singular values we used the following fact. For simplicity, we first state the two-dimensional result.

Proposition 2.7.8. *Assume that $F \in \mathbb{M}^{2\times 2}$ and that $\sigma_1 \le \sigma_2$ are the signed singular values of F which satisfy $|\sigma_1| \le \sigma_2$. Then*

$$\max_{R\in O^+(2)} F:R = \sigma_1 + \sigma_2, \qquad \min_{R\in O^+(2)} F:R = -\sigma_1 - \sigma_2,$$

and

$$\max_{R\in O^-(2)} F:R = -\sigma_1 + \sigma_2, \qquad \min_{R\in O^-(2)} F:R = +\sigma_1 - \sigma_2.$$

Proof. We may assume that F is diagonal, $F = \mathrm{diag}(\sigma_1, \sigma_2)$, since for all Q_1 and $Q_2 \in SO(2)$

$$\max_{R\in O^\pm(2)} F:R = \max_{R\in O^\pm(2)} (Q_1FQ_2):R,$$
$$\min_{R\in O^\pm(2)} F:R = \min_{R\in O^\pm(2)} (Q_1FQ_2):R.$$

We first prove the formula for the maximum in the compact set O^+. Assume that $\overline{R} = e^{i\overline{\varphi}}$ (in complex notation) realizes one of the extremal values and consider for $\varepsilon > 0$ any smooth curve $\varphi : (-\varepsilon, \varepsilon) \to \mathbb{R}$ with $\varphi(0) = \overline{\varphi}$. Then

$$\frac{d}{dt}\Big|_{t=0} F : \begin{pmatrix} \cos\varphi(t) & -\sin\varphi(t) \\ \sin\varphi(t) & \cos\varphi(t) \end{pmatrix} = (-\sin\bar{\varphi})(\sigma_1 + \sigma_2) = 0.$$

If $\sigma_1 + \sigma_2 = 0$, then $F = \sigma_2\,\mathrm{diag}(1, -1)$ and $F : R = 0$ for all $R \in O^+(2)$. We may therefore assume that $\sigma_1 + \sigma_2 \ne 0$. In this case $\overline{\varphi} \in \{0, \pi\}$, and thus $\overline{R} = \pm\mathbb{I}$. This implies the assertion of the proposition. Similarly, if $\overline{R} \in O^-(2)$ realizes one of the extrema, then

$$\frac{d}{dt}\Big|_{t=0} F : \begin{pmatrix} -\cos\varphi(t) & \sin\varphi(t) \\ \sin\varphi(t) & \cos\varphi(t) \end{pmatrix} = (\sin\overline{\varphi})(\sigma_1 - \sigma_2) = 0.$$

If $\sigma_1 - \sigma_2 = 0$, then $F = \sigma_2\,\mathrm{diag}(1, 1)$ and $F : R = 0$ for all $R \in O^-(2)$. Otherwise, $\overline{\varphi} \in \{0, \pi\}$ and therefore $\overline{R} = \pm\,\mathrm{diag}(-1, 1)$ and this concludes the proof of the proposition. □

We now turn to the three-dimensional situation. To simplify the statement, we introduce some notation. Let D be any set. If $f : D \to \mathbb{R}$ and $f(x) \ge f(y)$ for all $y \in D$ then we write $x \in \mathrm{argmax}(f, D)$ or simply $x \in \mathrm{argmax}(f)$ if the domain D is clearly defined in the context. We define $\mathrm{argmin}(f)$ similarly.

Proposition 2.7.9. *Assume that $F \in \mathbb{M}^{3\times 3}$ and that $\sigma_1, \sigma_2, \sigma_3$ are the signed singular values which satisfy $|\sigma_1| \le \sigma_2 \le \sigma_3$ are the signed singular values of F. Then the following assertions hold:*

1. We have for all $F \in \mathbb{M}^{3\times 3}$

$$\max_{R\in O^+(3)} F : R = \sigma_1 + \sigma_2 + \sigma_3, \qquad \min_{R\in O^+(3)} F : R = +\sigma_1 - \sigma_2 - \sigma_3,$$

and

$$\max_{R\in O^-(3)} F : R = -\sigma_1 + \sigma_2 + \sigma_3, \qquad \min_{R\in O^-(3)} F : R = -\sigma_1 - \sigma_2 - \sigma_3.$$

2. We now assume that F *is diagonal,* $F = \mathrm{diag}(\sigma_1, \sigma_2, \sigma_3)$*, with* $\det F \geq 0$*. Assume first that* $\mathrm{rank}(F) = 1$*. Then*

$$\mathrm{argmax}(F : R, O^+(3)) = \{Q = \mathrm{diag}(\widehat{Q}, 1) : \widehat{Q} \in O^+(2)\},$$
$$\mathrm{argmin}(F : R, O^+(3)) = \{Q = \mathrm{diag}(\widehat{Q}, -1) : \widehat{Q} \in O^-(2)\}.$$

If $\mathrm{rank}(F) = 2$*, then*

$$\mathrm{argmax}(F : R, O^+(3)) = \{\mathbb{I}\},$$
$$\mathrm{argmin}(F : R, O^+(3)) = \{\mathrm{diag}(1, -1, -1)\}.$$

Finally, if $\det F > 0$ *and* $\sigma_1 = \sigma_2 = \sigma_3$ *then*

$$\mathrm{argmax}(F : R, O^+(3)) = \{\mathbb{I}\},$$
$$\mathrm{argmin}(F : R, O^+(3)) = \{Q = -\mathbb{I} + 2\boldsymbol{e} \otimes \boldsymbol{e},\ \boldsymbol{e} \in \mathbb{S}^2\},$$

while in all the other cases,

$$\mathrm{argmax}(F : R, O^+(3)) = \{\mathbb{I}\},$$
$$\mathrm{argmin}(F : R, O^+(3)) = \{\mathrm{diag}(1, -1, -1)\}.$$

3. If $\det F < 0$*, then*

$$\mathrm{argmax}(F : R, O^-(3)) = -\,\mathrm{argmin}(F : R, O^+(3)),$$

and

$$\mathrm{argmin}(F : R, O^-(3)) = -\,\mathrm{argmax}(F : R, O^+(3)).$$

Proof. As in the two-dimensional situation we may assume that F is a diagonal matrix, $F = \mathrm{diag}(\sigma_1, \sigma_2, \sigma_3)$. The assertions are obvious if $\mathrm{rank}(F) \leq 2$, i.e., $\sigma_1 \leq 0$. We may thus suppose that $\mathrm{rank}(F) = 3$. Since for every skew symmetric matrix X there exists a differentiable curve $\gamma_X : (-\varepsilon, \varepsilon) \to SO(3)$ with $\gamma(0) = \mathbb{I}$ and $\gamma'(0) = X$ (we may choose $\gamma(t) = \exp(tX)$), any $Q \in SO(3)$ realizing one of the extrema must satisfy

$$\left.\frac{d}{dt}\right|_{t=0} \gamma(t)Q : F = XQ : F = 0, \qquad \left.\frac{d}{dt}\right|_{t=0} Q\gamma(t) : F = QX : F = 0.$$

In view of the identity $\operatorname{tr}(AB) = \operatorname{tr}(BA)$ for all $A, B \in \mathbb{M}^{n\times n}$ we have (recall that F is diagonal) $XQ : F = \operatorname{tr}(FXQ) = \operatorname{tr}(QFX) = -QF : X = 0$ and $QX : F = \operatorname{tr}(FQX) = -FQ : X = 0$, and this implies that both the matrices QF and FQ must be symmetric. The resulting equations $QF = FQ^T$ and $FQ = Q^TF$ (or $Q_{ij}\sigma_j = Q_{ji}\sigma_i$ and $\sigma_i Q_{ij} = \sigma_j Q_{ji}$) can be rewritten in different ways, in particular as

$$(\sigma_i - \sigma_j)(Q_{ij} + Q_{ji}) = 0, \quad (\sigma_i + \sigma_j)(Q_{ij} - Q_{ji}) = 0, \quad (\sigma_i^2 - \sigma_j^2)Q_{ij} = 0. \tag{2.59}$$

We assume first that $\sigma_1 > 0$ and distinguish four different cases:
Case 1: $\sigma_1 = \sigma_2 = \sigma_3$. Then Q is symmetric and thus either the identity or a 180° rotation about an axis $\boldsymbol{e}$ given by $Q = -\mathbb{I} + 2\boldsymbol{e}\otimes\boldsymbol{e}$. The assertion in this case follows from $F : \mathbb{I} = \operatorname{tr}(F)$ and $F : (-\mathbb{I} + 2\boldsymbol{e}\otimes\boldsymbol{e}) = -\sigma_1$.
Case 2: $\sigma_1 < \sigma_2 = \sigma_3$. In this case, $Q_{12} = Q_{21} = Q_{13} = Q_{31} = 0$ and $Q_{23} = Q_{32}$. Thus $Q = \operatorname{diag}(\pm 1, \widehat{Q}^{\pm})$ with $Q^{\pm} \in \mathrm{O}^{\pm}(2)$. If $Q_{11} = -1$ then $F : Q = -\sigma_1$ for all $Q^- \in \mathrm{O}^-(2)$, while $Q : F = \sigma_1 + 2(\cos\varphi)\sigma_3$ for $Q_{11} = 1$. The maximum is therefore attained for $\varphi = 0$ while the minimum is realized for $Q = \operatorname{diag}(1, -1, -1)$.
Case 3: $\sigma_1 = \sigma_2 < \sigma_3$. It follows that $Q_{13} = Q_{31} = Q_{23} = Q_{32} = 0$ and that $Q_{12} = Q_{21}$. Thus $Q = \operatorname{diag}(\widehat{Q}^{\pm}, \pm 1)$ with $\widehat{Q}^{\pm} \in \mathrm{O}^{\pm}(2)$. If $Q_{33} = -1$ then $F : Q = -\sigma_3$ for all $Q^- \in \mathrm{O}^-(2)$, while $Q : F = 2(\cos\varphi)\sigma_1 + \sigma_3$ for $Q_{33} = 1$. The maximum is therefore attained for $\varphi = 0$ and the minimum for $\varphi = \pi$.
Case 4: $\sigma_1 < \sigma_2 < \sigma_3$. By (2.59) the matrix Q is diagonal and the conclusion is obvious.
Let us now assume that $\det F < 0$. Then

$$\max_{R\in\mathrm{O}^-(3)} F : R = -\min_{R\in\mathrm{O}^-(3)} F : (-R) = -\min_{R\in\mathrm{O}^+(3)} F : R.$$

This identity and the analogous one for the minimum prove the formulae in parts 1 and 3 of the proposition. □

3. Macroscopic Energy for Nematic Elastomers

One goal in the analysis of phase transformations via variational techniques is to identify the so-called effective or relaxed energy of the system,

$$W^{qc}(F) = \inf_{\substack{u \in W^{1,\infty}(\Omega;\mathbb{R}^3) \\ u(x)=Fx \text{ on } \partial\Omega}} \frac{1}{|\Omega|} \int_\Omega W(Du)\mathrm{d}x.$$

It describes the energy of the system for affine boundary conditions, if the system is allowed to form locally energetically optimal microstructures. In particular, this formulation allows infinitesimally fine structures by neglecting all higher order effects such as surface energies or the atomistic structure of the material that typically introduce finite length scales.

The energy W^{qc} governs the macroscopic behavior of the system and is - from a practical point of view - the right quantity for the computation of averaged quantities such as the macroscopic stress. Since W^{qc} is quasiconvex, the variational problem has a minimizer, and numerical schemes are expected to provide reliable and mesh independent results without oscillations on a scale comparable to the underlying triangulation. However, it is one of the unresolved challenges in the field to derive a characterization of W^{qc}. Even in the case of the cubic to tetragonal transition, which has been investigated extensively and for which Ericksen and James proposed a quartic model energy, nothing in known about its relaxation. It is surprising – or perhaps an indication of the true complexity of the matter – that the zero set of W^{qc}, the quasiconvex hull of the three martensitic wells in (5.3), has not been found yet, despite numerous attacks.

This motivates us to go in this chapter beyond the analysis of crystalline microstructures and to study a different physical system, nematic elastomers, a class of polymers that undergo a nematic to isotropic phase transformation. As a result of the isotropy of the high temperature phase, the energy in the nematic phase depends only on the singular values of the deformation gradient. We have seen in Section 2 that this invariance allows one for example a characterization of the semiconvex hulls of the zero set, and we show in this chapter how to find the macroscopic energy for the system.

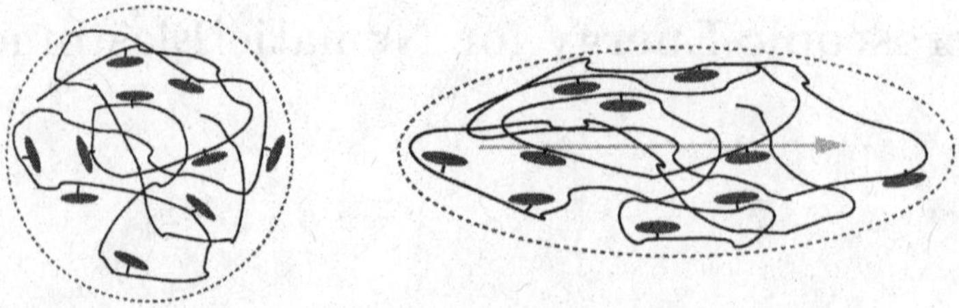

Fig. 3.1. *The isotropic-nematic phase transformation in nematic elastomers, polymers consisting of weakly cross-linked side-chain polymer liquid crystals. The nematic elements (rigid, rod-like molecules) are attached to the backbone chain. They have a random orientation in the high temperature (isotropic) phase due to thermal fluctuations. In the low temperature (nematic) phase, a local alignment of the mesogens causes a stretch of the network in direction of the director* $\boldsymbol{n}$ *(indicated by the arrow in the right figure) and a contraction in the directions perpendicular to it.*

3.1 Nematic Elastomers

Nematic elastomers are a fascinating material that combine nematic properties with the rubber elasticity of the underlying polymeric network. They are synthesized by cross-linking polymer liquid crystals close to their isotropic to nematic transformation temperature. As a result, one obtains a solid with an isotropic to nematic phase transformation at a comparable temperature. The low temperature phase is characterized by the coupling of the elastic deformation to the orientation of the nematic director, see Figure 3.1. Bladon, Terentjev and Warner proposed a formula for the free energy of nematic elastomers modeling them as cross-linked networks of anisotropic Gaussian chains within a continuum model. The state variables are the deformation gradient F and a vector field $\boldsymbol{n}$, the director, describing the orientation of the mesogens in the nematic phase. Using the undistorted high temperature phase as the reference configuration, their expression can be written as

$$W_{\mathrm{BTW}}(F,\boldsymbol{n}) = \begin{cases} \frac{\mu}{2}\Big(r^{1/3}\big[\,|F|^2 - \frac{r-1}{r}|F^T\boldsymbol{n}|^2\,\big] - 3\Big) & \text{if } \det F = 1, \\ +\infty & \text{otherwise.} \end{cases}$$

Here μ and r are two positive, temperature dependent material constants, the rubber energy scale and the backbone anisotropy parameter (i.e., the mean ratio of chain dimensions in the directions parallel and perpendicular to the director: for $r = 1$ the chain is a spherical coil, while $r > 1$ corresponds to the prolate case and $r < 1$ to the oblate one), respectively. The for elastomers typical incompressibility is incorporated into the model by assuming that the energy is infinite if the determinant of the deformation gradient is not equal to one. Above the transformation temperature, we have $r = 1$, and we recover the standard form of a neo-Hookean, isotropic free energy density,

$$W_{\text{iso}}(F) = \begin{cases} \frac{\mu}{2}\left(\lambda_1^2 + \lambda_2^2 + \lambda_3^2 - 3\right) & \text{if } \det F = 1, \\ +\infty & \text{otherwise,} \end{cases}$$

where $0 < \lambda_1 \le \lambda_2 \le \lambda_3$ are the singular values of F. Below the transformation temperature we have $r > 1$ in the prolate case which we will consider here. Since the formula for the energy does not contain derivatives of the director $\boldsymbol{n}$, we can minimize first in $\boldsymbol{n}$ and then consider the variational problem for the deformation $\boldsymbol{u}$. We define

$$W_{\text{ne}}(F) = \min_{\boldsymbol{n}\in\mathbb{S}^2} W_{\text{BTW}}(F, \boldsymbol{n})$$

and obtain from

$$\max_{\boldsymbol{n}\in\mathbb{S}^2} |F^T\boldsymbol{n}| = \lambda_3(F)$$

that

$$W_{\text{ne}}(F) = \begin{cases} \frac{\mu}{2}\left(r^{1/3}\left[\lambda_1^2 + \lambda_2^2 + \frac{1}{r}\lambda_3^2\right] - 3\right), & \text{if } \det F = 1, \\ +\infty & \text{otherwise.} \end{cases}$$

For the mathematical analysis, it is convenient to consider W_{ne} as a special case of the family of energies

$$W(F) = \begin{cases} \frac{\lambda_1^p(F)}{\gamma_1^p} + \frac{\lambda_2^p(F)}{\gamma_2^p} + \frac{\lambda_3^p(F)}{\gamma_3^p} - 3 & \text{if } \det F = 1, \\ +\infty & \text{else,} \end{cases} \tag{3.1}$$

with $0 < \gamma_1 \le \gamma_2 \le \gamma_3$, $\gamma_1\gamma_2\gamma_3 = 1$, and $p \ge 2$. The relaxation result that we prove in the subsequent sections provides us with the following formula for the relaxation of the free energy,

$$W_{\text{ne}}^{\text{qc}}(F) = \begin{cases} 0 & \text{if } F \in \text{L} \\ \frac{\mu}{2}\left(\frac{2}{r^{1/6}\lambda_1(F)} + r^{1/3}\lambda_1^2(F) - 3\right) & \text{if } F \in \text{I}_1, \\ W(F) & \text{if } F \in \text{S}, \\ +\infty & \text{else,} \end{cases}$$

where

$$\begin{aligned} \text{L} &= \left\{F \in \mathbb{M}^{3\times 3} : \det F = 1,\ \lambda_{\max}(F) \le r^{1/3}\right\}, \\ \text{I}_1 &= \left\{F \in \mathbb{M}^{3\times 3} : \det F = 1,\ \lambda_{\min}(F)\lambda_{\max}^2(F) \le r^{1/2}\right\}, \\ \text{S} &= \left\{F \in \mathbb{M}^{3\times 3} : \det F = 1,\ \lambda_{\min}(F)\lambda_{\max}^2(F) \ge r^{1/2}\right\}. \end{aligned}$$

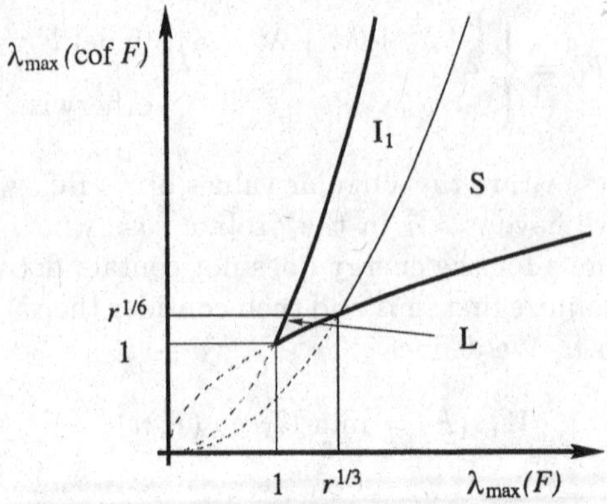

Fig. 3.2. *The macroscopic phase diagram for nematic elastomers. The phase boundary between the intermediate phase and the solid phase is given by* $\lambda_{\min}(F)\lambda_{\max}^2(F) = r^{1/2}$.

The formula for the macroscopic energy reveals three different types of mechanical response of the system to applied strains, see Figure 3.2. In the *liquid* phase L, the energy is identical to zero and the system shows no resistance to applied stretches and has a completely soft behavior. In the *intermediate* phase I_1, the expression for the energy depends only on the smallest singular value λ_1, and therefore the system behaves like a liquid along transformation paths that leave λ_1 fixed; all other deformations lead to a change in the stress. In the *solid* phase, finally, the material has the properties of a neo-Hookean rubber and all deformations change the stresses in the material. The liquid like behavior has been reported in the experimental literature (within certain limits) and first numerical experiments with the macroscopic energy $W_{\text{ne}}^{\text{qc}}$ by Conti, DeSimone, and Dolzmann show a qualitative agreement with the experimental results.

3.2 The General Relaxation Result

The goal of our analysis is to find an explicit formula for the quasiconvex envelope of the energy W given in (3.1). From a more general point of view, the ideas for the characterization of relaxed energies are similar to those developed for quasiconvex hulls of sets.

In the setting of hulls of compact sets, it was crucial to find an inner and an outer bound for the quasiconvex hull K^{qc}, i.e., sets $\mathcal{A}$ and $\mathcal{B}$ with $\mathcal{A} \subseteq K^{\text{qc}} \subseteq \mathcal{B}$, and then to prove that the inclusion $\mathcal{B} \subseteq \mathcal{A}$ holds. In this case, all the inclusions are equalities, hence $\mathcal{A} = K^{\text{qc}} = \mathcal{B}$, and this establishes a representation for K^{qc}.

In the framework of envelopes of functions, the inner and the outer bound are replaced by a lower and an upper bound for the relaxed energy, i.e., functions W_* and W^* with $W_* \leq W^{\mathrm{qc}} \leq W^*$. If it turns out that $W^* \leq W_*$, then equality holds in this chain of inequalities and one obtains a formula for W^{qc}. As in the case of semiconvex hulls, there are two canonical choices for W_* and W^*, namely the polyconvex envelope W^{pc} and the rank-one convex envelope W^{rc}, and the goal is then to prove that $W^{\mathrm{rc}} = W^{\mathrm{pc}}$. If W is a real valued function, then this identity implies that $W^{\mathrm{rc}} = W^{\mathrm{qc}} = W^{\mathrm{pc}}$ and one obtains a formula for the relaxed energy. In the situation at hand, the incompressibility constraint causes a slight difficulty since extended valued quasiconvex functions are not necessarily rank-one convex. To resolve this issue, we use a construction by Müller and Šverák to show that $W^{\mathrm{qc}} \leq W^{\mathrm{rc}}$. Since polyconvexity implies quasiconvexity for extended valued functions, we conclude $W^{\mathrm{qc}} \leq W^{\mathrm{rc}} = W^{\mathrm{pc}} \leq W^{\mathrm{qc}}$ and this demonstrates that all the envelopes are in fact equal.

Based on this approach, we prove in the subsequent sections the general relaxation result. In order to simplify the notation, we define the parameters

$$\gamma^* = \frac{\gamma_2}{\gamma_3} < 1 \quad \text{and} \quad \Gamma^* = \frac{\gamma_2}{\gamma_1} > 1, \tag{3.2}$$

and sets of matrices L, I_1, I_3, S by

$$\begin{aligned}
\mathrm{L} &= \Big\{F \in \mathbb{M}^{3\times 3} : \det F = 1,\ \lambda_{\max}(F) \leq \gamma_3,\ \lambda_{\max}(\operatorname{cof} F) \leq \frac{1}{\gamma_1}\Big\},\\
\mathrm{I}_1 &= \Big\{F \in \mathbb{M}^{3\times 3} : \det F = 1,\ \lambda_{\max}(\operatorname{cof} F) \geq \frac{1}{\gamma_1},\\
&\qquad \gamma^* \lambda_{\max}^2(F) \leq \lambda_{\max}(\operatorname{cof} F) \leq \lambda_{\max}^2(F)\Big\},\\
\mathrm{I}_3 &= \Big\{F \in \mathbb{M}^{3\times 3} : \det F = 1,\ \lambda_{\max}(F) \geq \gamma_3,\\
&\qquad \sqrt{\lambda_{\max}(F)} \leq \lambda_{\max}(\operatorname{cof} F) \leq \sqrt{\Gamma^* \lambda_{\max}(F)}\Big\},\\
\mathrm{S} &= \Big\{F \in \mathbb{M}^{3\times 3} : \det F = 1,\\
&\qquad \sqrt{\Gamma^* \lambda_{\max}(F)} \leq \lambda_{\max}(\operatorname{cof} F) \leq \gamma^* \lambda_{\max}^2(F)\Big\}.
\end{aligned}$$

We also denote by L, I_1, I_3, S the corresponding subsets in the (s,t)-plane which are formally given by replacing $\lambda_{\max}(F)$ by s and $\lambda_{\max}(\operatorname{cof} F)$ by t in the foregoing definition. It is easy to see that all matrices $F \in \mathbb{M}^{3\times 3}$ with $\det F = 1$ are contained in $\mathrm{L} \cup \mathrm{I}_1 \cup \mathrm{I}_3 \cup \mathrm{S}$. We finally define the sets E_0, E_1, and E_3 by

$$\begin{aligned}
\mathrm{E}_0 &= [0, \gamma_3] \times [0, \frac{1}{\gamma_1}] \supset L,\\
\mathrm{E}_1 &= \Big\{(s,t) : t \geq \frac{1}{\gamma_1},\ t \geq s^2\Big\},\\
\mathrm{E}_3 &= \Big\{(s,t) : s \geq \gamma_3,\ t \leq \sqrt{s}\Big\},
\end{aligned}$$

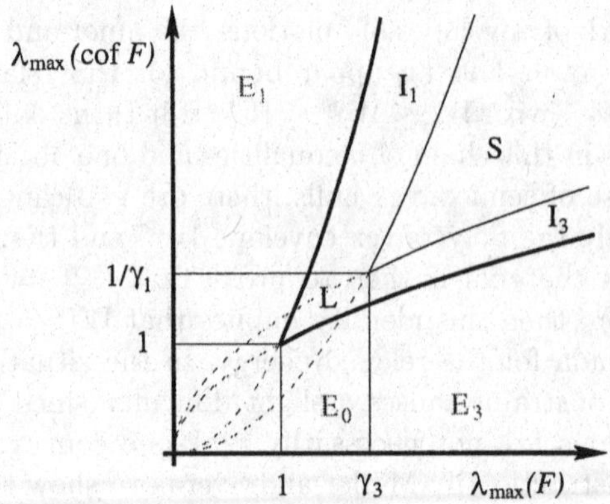

Fig. 3.3. The phase diagram for the relaxed energy.

see Figure 3.3 for a sketch of these domains in the phase plane.

We now state our relaxation result.

Theorem 3.2.1. *Suppose that* $0 < \gamma_1 \leq \gamma_2 \leq \gamma_3$ *with* $\gamma_1\gamma_2\gamma_3 = 1$ *and that* $p \geq 2$. *Let* $W : \mathbb{M}^{3\times 3} \to \mathbb{R}$ *be given by*

$$W(F) = \begin{cases} \dfrac{\lambda_1^p(F)}{\gamma_1^p} + \dfrac{\lambda_2^p(F)}{\gamma_2^p} + \dfrac{\lambda_3^p(F)}{\gamma_3^p} - 3 & \text{if } \det F = 1, \\ +\infty & \text{else.} \end{cases} \tag{3.3}$$

Then the relaxed energy W^{qc} *of the system is given by*

$$W^{\mathrm{qc}}(F) = \begin{cases} \psi(\lambda_{\max}(F), \lambda_{\max}(\mathrm{cof}\,F)) & \text{if } \det F = 1, \\ +\infty & \text{else,} \end{cases}$$

where the function $\psi : [0,\infty) \times [0,\infty) \to \mathbb{R}$ *is given by*

$$\psi(s,t) = \begin{cases} 0 & \text{if } (s,t) \in \mathrm{E}_0, \\ \left(\frac{1}{\gamma_1 t}\right)^p + 2(\gamma_1 t)^{p/2} - 3 & \text{if } (s,t) \in \mathrm{I}_1 \cup \mathrm{E}_1, \\ \left(\frac{1}{\gamma_1 t}\right)^p + \left(\frac{t}{\gamma_2 s}\right)^p + \left(\frac{s}{\gamma_3}\right)^p - 3 & \text{if } (s,t) \in \mathrm{S}, \\ \left(\frac{s}{\gamma_3}\right)^p + 2\left(\frac{\gamma_3}{s}\right)^{p/2} - 3 & \text{if } (s,t) \in \mathrm{I}_3 \cup \mathrm{E}_3. \end{cases} \tag{3.4}$$

We split the proof into several steps in which we follow the general strategy outlined above. The first step is the construction of an upper bound W^*. Then W^* is shown to be polyconvex and hence to be a lower bound and this establishes the theorem, up to the additional argument that is needed in order to show that $W^{\mathrm{qc}} \leq W^{\mathrm{rc}}$. We conclude the proof in Section 3.5.

3.3 An Upper Bound for the Relaxed Energy

The lamination method described in Chapter 2 was based on the fact that the segment $\lambda F_1 + (1-\lambda)F_2$, $\lambda \in [0,1]$, is contained in K^{rc} and hence in K^{qc} if $\operatorname{rank}(F_1 - F_2) = 1$ and if the end points F_1 and F_2 belong to K. By the same argument, $W^{\mathrm{rc}}(\lambda F_1 + (1-\lambda)F_2) \le \max\{W(F_1), W(F_2)\}$ along rank-one lines. More generally

$$W^{\mathrm{rc}}(F) \le \max_{X \in K} W(X) \quad \text{for all } F \in K^{\mathrm{rc}},$$

and the idea behind the construction of W^* is to apply this inequality to level sets of W, i.e., sets of the form $\{X : W(X) = w_0\}$, $w_0 \in \mathbb{R}$.

Proposition 3.3.1. *The rank-one convex envelope W^{rc} of W satisfies the inequality $W^{\mathrm{rc}}(F) \le W^*(F)$ where the function $W^*(F)$ is given by*

$$W^*(F) = \begin{cases} 0 & \text{if } F \in \mathrm{L}, \\ \left(\frac{\lambda_1}{\gamma_1}\right)^p + 2\left(\frac{\gamma_1}{\lambda_1}\right)^{p/2} - 3 & \text{if } F \in \mathrm{I}_1, \\ \left(\frac{\lambda_1}{\gamma_1}\right)^p + \left(\frac{\lambda_2}{\gamma_2}\right)^p + \left(\frac{\lambda_3}{\gamma_3}\right)^p - 3 & \text{if } F \in \mathrm{S}, \\ \left(\frac{\lambda_3}{\gamma_3}\right)^p + 2\left(\frac{\gamma_3}{\lambda_3}\right)^{p/2} - 3 & \text{if } F \in \mathrm{I}_3, \\ +\infty & \text{if } \det F \neq 1. \end{cases} \tag{3.5}$$

Moreover, for all $F \in \mathbb{M}^{3\times 3}$ with $\det F = 1$ there exist pairs $(\xi_i, F_i)_{i=1,\ldots,k}$, $k \le 4$, with $\xi_i \in [0,1]$, $F_i \in \mathbb{M}^{3\times 3}$, and $\det F_i = 1$ satisfying condition $\mathcal{H}_k$ such that

$$F = \sum_{i=1}^{k} \xi_i F_i, \quad \text{and} \quad W^*(F) = W(F_i), \quad i = 1, \ldots, k.$$

Remark 3.3.2. It turns out that $k = 2$ for matrices in the intermediate phases, and that $k = 4$ for matrices in the liquid phase.

Proof. It follows from the inequality between the geometric and the arithmetic mean that $W \ge 0$ and that

$$K = \{X : W(X) = 0\} = \{F : \det F = 1, \lambda_i(F) = \gamma_i \text{ for } i = 1,2,3\}.$$

By the definition of L and Theorem 2.7.7, we have $K^{\mathrm{rc}} = K^{(2)} = \mathrm{L}$ and thus

$$W^{\mathrm{rc}}(F) \le \max_{X \in K} W(X) = 0 \quad \text{for all } F \in L.$$

It therefore suffices to prove the assertion for the matrices contained in the sets I_1 and I_3. By the isotropy of W, we may assume that F is a diagonal matrix, $F = \operatorname{diag}(\mu_1, \mu_2, \mu_3)$ with $\mu_1\mu_2\mu_3 = 1$.

Assume first that $\lambda_1(F) = \mu_1 \le \gamma_1$. All diagonal matrices with this property can be parameterized by

$$t \mapsto \alpha_1(t) = \Big(\mu_1, \frac{1}{t\sqrt{\mu_1}}, \frac{t}{\sqrt{\mu_1}}\Big), \quad t \in [\mu_1^{3/2}, \mu_1^{-3/2}]$$

where the domain of t is chosen in such a way that μ_1 is the smallest eigenvalue of F. The energy along this curve is given by

$$w_1(t) = \begin{cases} \Big(\frac{\mu_1}{\gamma_1}\Big)^p + \Big(\frac{t}{\gamma_2\sqrt{\mu_1}}\Big)^p + \Big(\frac{1}{t\gamma_3\sqrt{\mu_1}}\Big)^p - 3 & \text{if } t \in [\mu_1^{3/2}, 1], \\ \Big(\frac{\mu_1}{\gamma_1}\Big)^p + \Big(\frac{1}{t\gamma_2\sqrt{\mu_1}}\Big)^p + \Big(\frac{t}{\gamma_3\sqrt{\mu_1}}\Big)^p - 3 & \text{if } t \in [1, \mu_1^{-3/2}]. \end{cases}$$

The extrema of w_1 on $(\mu_1^{3/2}, 1)$ satisfy

$$w_1'(t) = \frac{pt^{p-1}}{\gamma_2^p\mu_1^{p/2}} - \frac{p}{t^{p+1}\gamma_3^p\mu_1^{p/2}} = 0 \quad \Leftrightarrow \quad t = t_0 = \Big(\frac{\gamma_2}{\gamma_3}\Big)^{1/2}.$$

In view of $\gamma_1 \le \gamma_2 \le \gamma_3$ and $\gamma_3\gamma_2\gamma_1 = 1$,

$$\Big(\frac{\gamma_2}{\gamma_3}\Big)^{1/2} = (\gamma_2^2\gamma_1)^{1/2} \ge \gamma_1^{3/2} \ge \mu_1^{3/2}$$

and hence $t_0 \in (\mu_1^{3/2}, 1)$. Since

$$w_1''(t) = \frac{p(p-1)t^{p-2}}{\gamma_2^p\mu_1^{p/2}} + \frac{p(p+1)}{t^{p+2}\gamma_3^p\mu_1^{p/2}} > 0, \quad t \in (0,1),$$

this extremum is a minimum at

$$\alpha_1(t_0) = \Big(\mu_1, \frac{\sqrt{\gamma_3}}{\sqrt{\gamma_2\mu_1}}, \frac{\sqrt{\gamma_2}}{\sqrt{\gamma_3\mu_1}}\Big) = \Big(\mu_1, \frac{1}{\sqrt{\gamma^*\mu_1}}, \frac{\sqrt{\gamma^*}}{\mu_1}\Big)$$

with

$$w_1(t_0) = 2\Big(\frac{\gamma_1}{\mu_1}\Big)^{p/2} + \Big(\frac{\mu_1}{\gamma_1}\Big)^p - 3.$$

We have in view of the isotropy of the energy that $W(F) = w_1(t_0)$ on

$$K(t_0) = \{Q \operatorname{diag}\big(\mu_1, \frac{1}{\sqrt{\gamma^*\mu_1}}, \frac{\sqrt{\gamma^*}}{\sqrt{\mu_1}}\big)R,\ Q, R \in SO(3)\},$$

and the definition of quasiconvexity implies that $W(F) \le w_1(t_0)$ for all points in the quasiconvex hull of $K(t_0)$ which is given by Theorem 2.7.5 by

$$W(F) \le w_1(t_0) \text{ if } \lambda_1(F) = \mu_1,\ \lambda_3(F) \in [\frac{1}{\sqrt{\mu_1}}, \frac{1}{\sqrt{\gamma^* \mu_1}}].$$

Since $\mu_1(F) = \lambda_{\max}(\operatorname{cof} F)$, this establishes the formula for W^* in I_1. The remaining case, $F = \operatorname{diag}(\mu_1, \mu_2, \mu_3)$ with $\mu_3 \ge \gamma_3$ is analogous. □

3.4 The Polyconvex Envelope of the Energy

We are now going to prove that the function W^* is in fact polyconvex and therefore

$$W^{\mathrm{pc}} \le W^{\mathrm{rc}} \le W^* \le W^{\mathrm{pc}} \quad \Rightarrow \quad W^* = W^{\mathrm{pc}} = W^{\mathrm{rc}}.$$

The polyconvexity of W^* follows almost immediately from the following proposition which asserts that the function ψ is convex and nondecreasing.

Proposition 3.4.1. *The function ψ defined in (3.4) is convex and nondecreasing in its arguments.*

Proof. In order to simplify the notation we define the three functions

$$g_1 : \mathrm{E}_1 \cup \mathrm{I}_1 \to \mathbb{R}, \quad g_2 : \mathrm{S} \to \mathbb{R}, \quad g_3 : \mathrm{E}_3 \cup \mathrm{I}_3 \to \mathbb{R}$$

by

$$\begin{aligned} g_1(s,t) &= (\frac{1}{\gamma_1 t})^p + 2(\gamma_1 t)^{p/2} - 3, \\ g_2(s,t) &= (\frac{1}{\gamma_1 t})^p + (\frac{t}{\gamma_2 s})^p + (\frac{s}{\gamma_3})^p - 3, \\ g_3(s,t) &= (\frac{s}{\gamma_3})^p + 2(\frac{\gamma_3}{s})^{p/2} - 3. \end{aligned}$$

We first prove that ψ is continuous. To see this, it suffices to consider ψ on $\partial(\mathrm{E}_3 \cup \mathrm{I}_3)$ and $\partial(\mathrm{E}_1 \cup \mathrm{I}_1)$. If $s = \gamma_3$, then $g_3(\gamma_3, t) = 0$, and along the curve $t = \sqrt{\Gamma^* s}$ we have

$$g_2(s,t) = (\frac{1}{\gamma_1 \gamma_2 s})^{p/2} + (\frac{1}{\gamma_1 \gamma_2 s})^{p/2} + (\frac{s}{\gamma_3})^p - 3 = g_3(s,t).$$

Similarly, $g_1(s, \frac{1}{\gamma_1}) = 0$ and along the curve $s = \sqrt{t/\gamma^*}$ we obtain

$$g_2(s,t) = (\frac{1}{\gamma_1 t})^p + (\frac{t}{\gamma_2 \gamma_3})^{p/2} + (\frac{t}{\gamma_2 \gamma_3})^{p/2} - 3 = g_1(s,t).$$

In order to prove that ψ is nondecreasing, we calculate

$$Dg_1(s,t) = \Big(0, -\frac{p}{\gamma_1^p t^{p+1}} + p\gamma_1^{p/2} t^{p/2-1}\Big),$$
$$Dg_2(s,t) = \Big(-\frac{pt^p}{\gamma_2^p s^{p+1}} + \frac{ps^{p-1}}{\gamma_3^p}, -\frac{p}{\gamma_1^p t^{p+1}} + \frac{pt^{p-1}}{\gamma_2^p s^p}\Big),$$
$$Dg_3(s,t) = \Big(\frac{ps^{p-1}}{\gamma_3^p} + -\frac{p\gamma_3^{p/2}}{s^{p/2+1}}, 0\Big).$$

We obtain from these formulae that

$$\begin{aligned}
\partial_t g_1(s,t) \geq 0 \quad &\Leftrightarrow \quad t^{3p/2} \geq \Big(\frac{1}{\gamma_1}\Big)^{3p/2},\\
\partial_s g_2(s,t) \geq 0 \quad &\Leftrightarrow \quad s^{2p} \geq \Big(\frac{t}{\gamma^*}\Big)^{p},\\
\partial_t g_2(s,t) \geq 0 \quad &\Leftrightarrow \quad t^{2p} \geq (\Gamma^* s)^p,\\
\partial_s g_3(s,t) \geq 0 \quad &\Leftrightarrow \quad s^{3p/2} \geq \gamma_3^{3p/2}.
\end{aligned}$$

All these inequalities are satisfied in the domains of the functions g_i, and we conclude that ψ is nondecreasing in its arguments. We now show that ψ is continuously differentiable. Since

$$\partial_s g_3(\gamma_3, t) = 0, \quad \partial_t g_1(s, \frac{1}{\gamma_1}) = 0,$$

we only need to check this along the curves $t = \sqrt{\Gamma^* s}$ and $t = \gamma^* s^2$, respectively. A short calculation shows that

$$Dg_2(s, \big(\frac{\gamma_2}{\gamma_1}\big)^{1/2}\sqrt{s}) = Dg_3(s,t), \quad Dg_2(\big(\frac{\gamma_3}{\gamma_2}\big)^{1/2}\sqrt{t}, t) = Dg_1(s,t),$$

and this establishes the differentiability of the function ψ.

It remains to prove the convexity of ψ. It is clear that g_1 and g_3 are convex since the functions $s \mapsto s^q$ and $s \mapsto s^{-q}$ are convex on $\mathbb{R}_+$ for $q \geq 1$. We obtain for g_2 that

$$D^2 g_2(s,t) = \begin{pmatrix} \frac{p(p+1)t^p}{\gamma_2^p s^{p+2}} + \frac{p(p-1)s^{p-2}}{\gamma_3^p} & -\frac{p^2 t^{p-1}}{\gamma_2^p s^{p+1}} \\ -\frac{p^2 t^{p-1}}{\gamma_2^p s^{p+1}} & \frac{p(p+1)}{\gamma_1^p t^{p+2}} + \frac{p(p-1)t^{p-2}}{\gamma_2^p s^p} \end{pmatrix}$$

and thus the determinant of the matrix of the second derivatives is given by

$$\frac{p^2(p+1)^2\gamma_3^p}{s^{p+2}t^2} - \frac{p^2 t^{2p-2}}{\gamma_2^{2p} s^{2p+2}} + \frac{p^2(p^2-1)\gamma_2^p s^{p-2}}{t^{p+2}} + \frac{p^2(p-1)^2\gamma_1^p t^{p-2}}{s^2}.$$

By assumption, $\frac{1}{s^2} \leq \frac{\gamma_1^*}{t}$ and thus for $p \geq 2$ and $(p-1)^2 \geq 1$,

$$\frac{p^2(p-1)^2\gamma_1^p t^{p-2}}{s^2} - \frac{p^2 t^{2p-2}}{\gamma_2^{2p} s^{2p+2}} \geq \frac{p^2\gamma_1^p t^{p-2}}{s^2} - \frac{p^2 t^{2p-2}}{\gamma_3^p \gamma_2^p t^p s^2} = 0.$$

Since also $(D^2 g)_{11} > 0$, we conclude that g_2 is convex on its domain and this finishes the proof of the proposition. □

The next proposition is the analogue of the fact that convex and nondecreasing functions of convex functions are convex. We use this to show that $\psi(\lambda_{\max}(F), \lambda_{\max}(\operatorname{cof} F))$ is a polyconvex function since the maximal singular value of a matrix is a convex function.

Proposition 3.4.2. *Assume that* $\psi : \mathbb{R}_+^2 \to \mathbb{R}$ *is given by (3.4). Then the function* $\Psi_1 : \mathbb{M}^{3\times 3} \to \mathbb{R}$, *given by*

$$\Psi_1(F) = \psi(\lambda_{\max}(F), \lambda_{\max}(\operatorname{cof} F))$$

is polyconvex.

Proof. By definition, Ψ_1 is polyconvex if there exists a convex function

$$g : \mathbb{M}^{3\times 3} \times \mathbb{M}^{3\times 3} \times \mathbb{R} \to \mathbb{R}$$

such that $\Psi_1(F) = g(F, \operatorname{cof} F, \det F)$. We define

$$g(X, Y, \delta) = \psi(\sup_{e\in\mathbb{S}^2} |Xe|, \sup_{e\in\mathbb{S}^2} |Ye|).$$

It follows that for all matrices X_1, X_2, Y_1, $Y_2 \in \mathbb{M}^{3\times 3}$, scalars $\delta_1, \delta_2 \in \mathbb{R}$ and $\lambda \in [0,1]$

$$\begin{aligned}
&g\big(\lambda(X_1, Y_1, \delta_1) + (1-\lambda)(X_2, Y_2, \delta_2)\big)\\
&= \psi(\sup_{e\in\mathbb{S}^2} |(\lambda X_1 + (1-\lambda)Y_1)e|, \sup_{e\in\mathbb{S}^2} |(\lambda X_2 + (1-\lambda)Y_2)e|)\\
&\leq \psi(\lambda \sup_{e\in\mathbb{S}^2} |X_1 e| + (1-\lambda) \sup_{e\in\mathbb{S}^2} |Y_1 e|, \lambda \sup_{e\in\mathbb{S}^2} |X_2 e| + (1-\lambda) \sup_{e\in\mathbb{S}^2} |Y_2 e|)\\
&= \psi(\lambda(\sup_{e\in\mathbb{S}^2} |X_1 e|, \sup_{e\in\mathbb{S}^2} |X_2 e|) + (1-\lambda)(\sup_{e\in\mathbb{S}^2} |Y_1 e|, \sup_{e\in\mathbb{S}^2} |Y_2 e|))\\
&\leq \lambda\psi(\sup_{e\in\mathbb{S}^2} |X_1 e|, \sup_{e\in\mathbb{S}^2} |Y_1 e|) + (1-\lambda)\psi(\sup_{e\in\mathbb{S}^2} |X_2 e|, \sup_{e\in\mathbb{S}^2} |Y_2 e|)\\
&= \lambda g(X_1, Y_1, \delta_1) + (1-\lambda) g(X_2, Y_2, \delta_2).
\end{aligned}$$

Here we used the triangle inequality for the norm and the fact that ψ is nondecreasing in the first inequality, and the convexity of ψ for the second inequality. This establishes the polyconvexity of Ψ_1 and concludes the proof. □

We are now in a position to establish a characterization for the rank-one convex and the polyconvex envelope of W.

Theorem 3.4.3. *The rank-one convex and the polyconvex envelope of W coincide and are given by*

$$W^{\mathrm{rc}}(F) = W^{\mathrm{pc}}(F) = \begin{cases} \psi(\lambda_{\max}(F), \lambda_{\max}(\operatorname{cof} F)) & \textit{if } \det F = 1, \\ +\infty & \textit{else.} \end{cases}$$

Proof. We define

$$\Psi_2(F) = I_1(\det F) \quad \text{where} \quad I_1(t) = \begin{cases} 0 & \text{if } t = 1, \\ \infty & \text{else.} \end{cases}$$

Then $\widehat{W}(F) = \Psi_1(F) + \Psi_2(F)$ is a polyconvex function which is finite only on the set $\{F \in \mathbb{M}^{3\times 3} : \det F = 1\}$. This implies that

$$\lambda_1(F) = \lambda_{\min}(F) = \frac{1}{\lambda_{\max}(\operatorname{cof} F)} \quad \text{and} \quad \lambda_2(F) = \frac{1}{\lambda_{\min}(F)\lambda_{\max}(F)}$$

whenever the energy is finite. In view of the definition of ψ,

$$\widehat{W}(F) = \begin{cases} 0 & \text{if } F \in L, \\ (\frac{\lambda_{\min}(F)}{\gamma_1})^p + 2(\frac{\gamma_1}{\lambda_{\min} F})^{p/2} - 3 & \text{if } F \in \mathrm{I}_1, \\ W(F) & \text{if } F \in \mathrm{S}, \\ (\frac{\lambda_{\max}(F)}{\gamma_3})^p + 2(\frac{\gamma_3}{\lambda_{\max}(F)})^{p/2} - 3 & \text{if } F \in \mathrm{I}_3, \\ +\infty & \text{else,} \end{cases}$$

and a comparison with (3.5) shows that $\widehat{W} = W^* \le W^{\mathrm{rc}}$. Therefore

$$W^{\mathrm{rc}} \le W^* = \widehat{W} \le W^{\mathrm{pc}} \le W^{\mathrm{rc}},$$

and hence equality holds throughout this chain of inequalities. This proves the assertion of the theorem. □

3.5 The Quasiconvex Envelope of the Energy

The final step in the construction is to prove that the quasiconvex envelope is actually equal to the rank-one convex and the polyconvex envelope. This is accomplished using the following construction by Müller and Šverák.

Lemma 3.5.1. *Let Σ be given by*

$$\Sigma = \{F \in \mathbb{M}^{m\times n} : M(F) = t\},$$

where M is a minor of F and $t \neq 0$. Let V be an open set in Σ, let $F \in V^{\mathrm{rc}}$, and let $\varepsilon > 0$. Then there exists a piecewise linear map $\boldsymbol{u} : \Omega \subset \mathbb{R}^n \to \mathbb{R}^m$ such that $D\boldsymbol{u} \in V^{\mathrm{rc}}$ a.e. in Ω and

$$\big|\{\boldsymbol{x} : D\boldsymbol{u}(\boldsymbol{x}) \notin V\}\big| < \varepsilon|\Omega|, \qquad \boldsymbol{u}(\boldsymbol{x}) = F\boldsymbol{x} \text{ on } \partial\Omega.$$

After these preparations, we prove the relaxation result in Theorem 3.2.1.

Proof of Theorem 3.2.1. We have to construct for all $F \in \mathbb{M}^{3\times 3}$ with $\det F = 1$ and for all $\delta > 0$ a function $\varphi_{F,\delta} \in W^{1,\infty}(\Omega; \mathbb{R}^3)$ such that $\varphi_{F,\delta} = F\boldsymbol{x}$ on $\partial\Omega$ and

$$\int_\Omega W(D\varphi_{F,\delta})\mathrm{d}\boldsymbol{x} \leq |\Omega| W^{\mathrm{pc}}(F) + \mathcal{O}(\delta),$$

where $\mathcal{O}(\delta) \to 0$ as $\delta \to 0$. This implies $W^{\mathrm{qc}}(F) \leq W^{\mathrm{pc}}(F)$, and since W^{pc} is quasiconvex, we conclude $W^{\mathrm{qc}} = W^{\mathrm{pc}}$.

We give the proof for the situation that W^{pc} is obtained from W by averaging with respect to laminates within laminates. If follows from Theorem 2.7.7 that there exist pairs $(\lambda_i, F_i)_{i=1,\dots,4}$ such that

$$F = \sum_{i=1}^{4} \lambda_i F_i, \quad \text{and} \quad W^{\mathrm{pc}}(F) = W(F_i),\ i = 1, \dots, 4.$$

Moreover, $F \in \widetilde{K}^{(2)}$ where $\widetilde{K} = \{F_1, F_2, F_3, F_4\}$. We choose

$$\Sigma = \big\{F \in \mathbb{M}^{3\times 3} : \det F = 1\big\},$$

and define for $\delta > 0$

$$V_\delta = \big\{F \in \Sigma : \operatorname{dist}(F, \widetilde{K}) < \delta\big\},\ \omega_\delta = \sup\big\{W(X) : X \in V_\delta\big\} - W^{\mathrm{pc}}(F).$$

Since W is continuous on Σ we have $\omega_\delta \to 0$ as $\delta \to 0$. Lemma 3.5.1 guarantees the existence of a piecewise linear map $\varphi_{F,\delta} : \Omega \to \mathbb{R}^3$ with $D\varphi_{F,\delta}(\boldsymbol{x}) \in V_\delta^{\mathrm{rc}}$ a.e. and

$$\varphi_{F,\delta}(\boldsymbol{x}) = F\boldsymbol{x} \text{ on } \partial\Omega, \quad \text{and} \quad \big|\{\boldsymbol{x} \in \Omega : D\varphi_{F,\delta}(\boldsymbol{x}) \notin V_\delta\}\big| \leq \delta|\Omega|.$$

Therefore, if M is an upper bound for W on V_1,

$$\begin{aligned}\int_\Omega W(D\varphi_{F,\delta})\mathrm{d}\boldsymbol{x} &\leq \big|\{D\varphi_{F,\delta}(\boldsymbol{x}) \in V_\delta\}\big|\big(W^{\mathrm{pc}}(F) + \omega_\delta\big) + \delta M|\Omega| \\ &\leq |\Omega| W^{\mathrm{pc}}(F) + |\Omega|\big(\omega_\delta + \delta M\big).\end{aligned}$$

The assertion of the theorem follows as $\delta \to 0$. □

4. Uniqueness and Stability of Microstructure

The numerical analysis of finite element minimizers of nonconvex variational problems leads to the question of how one can quantitatively describe the behavior of functions with small energy. Motivated by this problem, Luskin analyzed in a sequence of papers uniqueness and stability of microstructures corresponding to gradient Young measures ν for martensitic phase transformations in two and three dimensions. The goal of this chapter is two-fold. First, we present a mathematically rigorous definition of Luskin's intuitive concept of stability. Secondly, we introduce a general framework for the analysis of stability based on an algebraic condition on the set K and the barycenter $F = \langle \nu, id \rangle$. Starting from this condition, we develop a theory that clarifies the relation between uniqueness and stability and identifies simple laminates as the only (known) class of microstructures that allows for a closed theory. In particular, this theory includes all results in the literature and extends them to an n-dimensional framework. It is based on a Young measure approach and underlines the general concepts underlying the estimates. We finally prove that our results are optimal for two-dimensional problems and we demonstrate the flexibility of our method by extending it to the setting of thin films and to higher order laminates.

The main focus of our analysis is the identification of sufficient conditions for uniqueness of microstructure. The key point of our approach is to state estimates for finite element minimizers as a corollary of the uniqueness theory. In order to motivate the algebraic condition we define below, see Definition 4.1.4, we discuss a representative example in two dimensions that actually appears in the analysis of the cubic to orthorhombic phase transformation in three dimensions in Section 5.3. For $\xi, \eta > 0$ let

$$U_1 = \begin{pmatrix} \xi - \eta & 0 \\ 0 & \xi + \eta \end{pmatrix}, \quad U_2 = \begin{pmatrix} \xi - \frac{\eta}{2} & -\frac{\sqrt{3}\eta}{2} \\ -\frac{\sqrt{3}\eta}{2} & \xi + \frac{\eta}{2} \end{pmatrix}, \quad U_3 = \begin{pmatrix} \xi - \frac{\eta}{2} & \frac{\sqrt{3}\eta}{2} \\ \frac{\sqrt{3}\eta}{2} & \xi + \frac{\eta}{2} \end{pmatrix},$$

and define $K = \mathrm{SO}(2)U_1 \cup \mathrm{SO}(2)U_2 \cup \mathrm{SO}(2)U_3$. Guided by the results in Section 2.2 and by Figure 2.2, it is clear that the only microstructures that are unique correspond to simple laminates for which the barycenter is a constrained point in K^{qc}. By Step 6 in the proof of Theorem 2.2.3 these points form the boundary of K^{qc} which consists of maximal arcs between the matrices U_i. Consequently these points satisfy for some $\boldsymbol{e} \in \mathbb{S}^1$ the inequality

$$|F\boldsymbol{e}| = |U_i\boldsymbol{e}| = |U_j\boldsymbol{e}| > |U_k\boldsymbol{e}|, \qquad \{i,j,k\} = \{1,2,3\}.$$

This is exactly the condition (C_b) in Definition 4.1.4 since in two dimensions conditions on the cofactor of a matrix are equivalent to conditions on the matrix itself.

The main results in this chapter can now be summarized as follows: We prove in Section 4.1 for bulk materials (i.e., three-dimensional models) that

$$F \in K^{qc} \text{ satisfies } (C_b) \Rightarrow \big(\nu \in \mathcal{M}^{qc}(K;F) \text{ is unique } \Leftrightarrow \nu \text{ is stable}\big).$$

In the two-dimensional situation, our results are optimal in the sense that

$$F \in K^{qc} \text{ satisfies } (C_b) \Leftrightarrow \nu \in \mathcal{M}^{qc}(K;F) \text{ is unique } \Leftrightarrow \nu \text{ is stable.}$$

We extend our theory in Section 4.4 to the analysis of thin films and we obtain again an optimal criterion for uniqueness and stability since

$$F \in K^{qc} \text{ satisfies } (C_{tf}) \Leftrightarrow \nu \in \mathcal{M}^{qc}(K;F) \text{ is unique } \Leftrightarrow \nu \text{ is stable.}$$

The most detailed analysis is given in Section 4.1 for bulk materials in two and three dimensions. We briefly sketch the underlying ideas in the framework of the three-well problem mentioned above. Suppose that

$$F = \lambda Q_2U_2 + (1-\lambda)Q_3U_3, \quad Q_2U_2 - Q_3U_3 = \boldsymbol{a} \otimes \boldsymbol{b},$$

is the center of mass of a simple laminate supported on

$$K = SO(2)U_1 \cup SO(2)U_2 \cup SO(2)U_3$$

and that the twinning system corresponds to one of the twinning systems that lead to a unique microstructure (i.e., the center of mass F is a constrained point in K^{pc}). Then

$$|F\boldsymbol{e}| = |U_2\boldsymbol{e}| = |U_3\boldsymbol{e}| > |U_1\boldsymbol{e}|$$

for some $\boldsymbol{e} \in \mathbb{S}^1$, see the example following Proposition 2.2.4. Suppose that $\nu \in \mathcal{M}^{pc}(K;F)$ and that ν is represented by

$$\nu = \lambda_1\nu_1 + \lambda_2\nu_2 + \lambda_3\nu_3, \quad \nu_i \in \mathcal{P}(SO(2)U_i),\ \lambda_i \geq 0,\ \lambda_1 + \lambda_2 + \lambda_3 = 1.$$

Then

$$|F\boldsymbol{e}| \leq \int_{\operatorname{supp}\nu} |A\boldsymbol{e}|\,d\nu(A) = \lambda_1|U_1\boldsymbol{e}| + \lambda_2|U_2\boldsymbol{e}| + \lambda_3|U_3\boldsymbol{e}| \leq |F\boldsymbol{e}| \tag{4.1}$$

and thus $\lambda_1 = 0$. Hence

$$\nu = \lambda \delta_{Q_2 U_2} + (1-\lambda)\delta_{Q_3 U_3}$$

is the unique gradient Young measure with center of mass equal to F. Now let $d(\cdot,\cdot)$ be a distance on the set of all probability measures. The foregoing observation can be stated equivalently by saying that for every $\mu \in \mathcal{M}^{\text{pc}}(K;F)$ with

$$\int_{\mathbb{M}^{2\times 2}} \operatorname{dist}(A,K)\mathrm{d}\mu(A) = 0 \quad \Rightarrow \quad d(\nu,\mu) = 0.$$

The main focus of this chapter is to generalize this result to probability measures $\mu \in \mathcal{M}^{\text{pc}}(\mathbb{M}^{2\times 2};F)$ (i.e., to probability measures supported on the space of all 2×2 matrices, not only on K) and to prove the following stability estimate: For all $\varepsilon > 0$ there exists a $\delta > 0$ such that

$$\int_{\mathbb{M}^{2\times 2}} \operatorname{dist}(A,K)\mathrm{d}\mu(A) < \delta \quad \Rightarrow \quad d(\nu,\mu) < \varepsilon. \tag{4.2}$$

The proof of this result combines two fundamental estimates:

i) If the integral in (4.2) is small, then the support of μ is close to K. Moreover, a suitable perturbation of the inequality (4.1) implies that the mass of μ close to $\mathrm{SO}(2)U_1$ is small. For the mathematical statement of this idea we define the nearest point projection $\pi : \mathbb{M}^{2\times 2} \to K$ by

$$|X - \pi(X)| = \operatorname{dist}(X,K) \quad \text{for all } X \in \mathbb{M}^{2\times 2}.$$

The goal is then to derive an estimate of the type

$$\mu(\{X \in \operatorname{supp}\mu : \pi(X) \in \mathrm{SO}(2)U_1\}) \le c \int_{\mathbb{M}^{2\times 2}} \operatorname{dist}(A,K)\mathrm{d}\mu(A).$$

In particular, if the support of μ is contained in K, then there is no mass on the well $\mathrm{SO}(2)U_1$.

ii) The second estimate provides information about the sets

$$M_\gamma = \{X \in \operatorname{supp}\mu : \pi(X) \in \mathrm{SO}(2)U_\gamma\}, \quad \text{for } \gamma = 2, 3.$$

The fact that the microstructure ν is unique implies that the points in the set $M = M_2 \cup M_3$ are close to the support of ν, i.e., to the matrices Q_2U_2 and Q_3U_3. To state this precisely, we define the so-called excess rotation $R(X) \in \mathrm{SO}(2)$ for all $X \in M$ by $\pi(X) = R(X)Q_\gamma U_\gamma$ if $X \in M_\gamma$. The crucial estimate can then be formulated by saying that R is close to the identity matrix $\mathbb{I}$ in the sense that

$$\int_M |R(A) - \mathbb{I}|^2 \mathrm{d}\nu(A) \le \int_{\mathbb{M}^{2\times 2}} \operatorname{dist}(A,K)\mathrm{d}\mu(A).$$

The precise formulations of these estimates are given in Propositions 4.1.9 and 4.1.12. In Sections 4.2-4.6 we extend these ideas to the analysis of thin film problems and second laminates in three dimensions. We conclude this chapter by a review of the development of this approach to the numerical analysis of microstructure and we relate our results to the existing literature.

4.1 Uniqueness and Stability in Bulk Materials

In this section, we present the general framework for the analysis of uniqueness and stability of microstructure supported on sets K of the from

$$K = \mathrm{SO}(n)U_1 \cup \ldots \cup \mathrm{SO}(n)U_k. \tag{4.3}$$

Throughout this section we assume that the matrices $U_i \in \mathbb{M}^{n\times n}$ are positive definite with $\det U_i = \Delta > 0$ for $i = 1, \ldots, k$. Moreover, we assume that $\mathrm{SO}(n)U_i \neq \mathrm{SO}(n)U_j$ for $i \neq j$. We present applications of our approach to several three-dimensional models of phase transformations in Chapter 5.

We begin with a precise definition of uniqueness. Since we try to identify the weakest possible conditions for uniqueness, we carry out most of the analysis for polyconvex measures which form a much larger class than gradient Young measures realizing microstructures. We define uniqueness of microstructures analogously.

Definition 4.1.1. *A polyconvex measure $\nu \in \mathcal{M}^{\mathrm{pc}}(K)$ is said to be unique if $\mathcal{M}^{\mathrm{pc}}(K; \langle \nu, id\rangle) = \{\nu\}$. Similarly, a microstructure $\nu \in \mathcal{M}^{\mathrm{qc}}(K)$ is said to be unique if $\mathcal{M}^{\mathrm{qc}}(K; \langle \nu, id\rangle) = \{\nu\}$.*

Assume now that ν is unique and let $F = \langle \nu, id\rangle$. The main motivation behind our definition of stability is to give a quantitative statement of the intuitive idea that any polyconvex measure μ with center of mass F and support close to K should be close to ν in a suitable distance d on the space of all probability measures. It turns out that a convenient way to measure the distance of the support of μ to K is the integral

$$\mathcal{E}(\mu) = \int_{\mathbb{M}^{n\times n}} \big(\mathrm{dist}(A, K) + \mathrm{dist}^{\max\{2,n-1\}}(A, K)\big) \, \mathrm{d}\mu(A).$$

Here $\mathrm{dist}(\cdot)$ denotes the Euclidean distance in $\mathbb{M}^{n\times n}$. In order to define the distance $d(\cdot,\cdot)$ on the set of all probability measures supported on $\mathbb{M}^{m\times n}$, let

$$\mathcal{M}_1 = \{\mu \in \mathcal{M}(\mathbb{M}^{m\times n}) : \|\mu\|_{\mathcal{M}} \leq 1\}$$

and let $C_0(\mathbb{M}^{m\times n})$ be the closure of $C_0^\infty(\mathbb{M}^{m\times n})$ in the supremum norm (in our applications below, we have either $m = n$ or $m = 3$ and $n = 2$). Choose a dense set $\{\Phi_i\}_{i\in\mathbb{N}}$ of Lipschitz continuous functions in $C_0(\mathbb{M}^{m\times n})$ with compact support and $\Phi_i \not\equiv 0$ for all $i \in \mathbb{N}$. We then define a metric $d(\cdot,\cdot)$ that induces the weak-$*$ topology on $\mathcal{M}_1$ by

$$d(\mu, \nu) = \sum_{i=1}^{\infty} 2^{-i} \frac{|\langle \mu, \Phi_i\rangle - \langle \nu, \Phi_i\rangle|}{\|\Phi_i\|_{1,\infty}}.$$

We are now in a position to formulate our notion of stability.

Definition 4.1.2. *A polyconvex measure $\nu \in \mathcal{M}^{\mathrm{pc}}(K)$ with underlying global deformation $F = \langle \nu, id \rangle$ is said to be stable if for all $\varepsilon > 0$ there exists a $\delta > 0$ such that for all $\mu \in \mathcal{M}^{\mathrm{pc}}(\mathbb{M}^{m\times n}; F)$ with*

$$\int_{\mathbb{M}^{m\times n}} \big(\operatorname{dist}(A,K) + \operatorname{dist}^{\max\{2,n-1\}}(A,K)\big)\,\mathrm{d}\mu(A) \le \delta$$

the estimate $d(\nu,\mu) \le \varepsilon$ holds.

The main ingredient in our analysis is the estimate in Proposition 4.1.9 below that estimates the mass of a measure $\nu \in \mathcal{M}^{\mathrm{pc}}(\mathbb{M}^{n\times n}; F)$ close to the wells $\mathrm{SO}(n)U_j$, $j = 1, \ldots, k$. Before we begin with the details, we introduce the relevant notation for measures ν satisfying condition ($\mathrm{C_b}$); here α and β are the indices in the definition of condition ($\mathrm{C_b}$), see Definition 4.1.4 below. In particular, $F = \langle \nu, id \rangle$ has a representation as the center of mass of a simple laminate supported on $\mathrm{SO}(n)U_\alpha \cup \mathrm{SO}(n)U_\beta$, i.e., $F = \lambda Q_\alpha U_\alpha + (1-\lambda) Q_\beta U_\beta$ with Q_α, $Q_\beta \in \mathrm{SO}(n)$ and $\lambda \in [0,1]$. We fix Borel measurable functions (see Proposition 4.1.3) $\pi : \mathbb{M}^{n\times n} \to K$ with

$$|\pi(X) - X| = \operatorname{dist}(X,K) \quad \text{for all } X \in \mathbb{M}^{n\times n}, \tag{4.4}$$

and $\Pi : \mathbb{M}^{n\times n} \to \{Q_\alpha U_\alpha, Q_\beta U_\beta\}$ with

$$|\Pi(X) - X| = \operatorname{dist}\big(X, \{Q_\alpha U_\alpha, Q_\beta U_\beta\}\big) \quad \text{for all } X \in \mathbb{M}^{n\times n}. \tag{4.5}$$

Moreover, we may choose the projection Π without loss of generality in such a way that $\Pi(A) \in \mathrm{SO}(n)U_\gamma$ if $\pi(A) \in \mathrm{SO}(n)U_\gamma$, for $\gamma = \alpha$, β. The existence of these functions is guaranteed by Proposition 4.1.3, and the rates in the estimates below do not depend on the specific choice. For $\nu \in \mathcal{M}^{\mathrm{pc}}(\mathbb{M}^{n\times n})$ we define the set M_ℓ of points in the support of ν that are close to $\mathrm{SO}(n)U_\ell$ in the sense that

$$M_\ell = \{A \in \operatorname{supp}\nu : \pi(A) \in \mathrm{SO}(n)U_\ell\}, \quad \ell = 1, \ldots, k,$$

see Figure 4.1. It is also convenient to introduce the sets $\widehat{M}_\gamma$ of points close to $\mathrm{SO}(n)U_\gamma$ by

$$\widehat{M}_\gamma = \{A \in \operatorname{supp}\nu : \Pi(A) \in \mathrm{SO}(n)U_\gamma\}, \quad \gamma = \alpha,\ \beta,$$

and Proposition 4.1.9 shows that the difference $\widehat{M}_\gamma \setminus M_\gamma$ has small measure. We finally introduce a local version of the sets $\widehat{M}_\gamma$ for a fixed radius $\varrho > 0$ according to

$$\widehat{M}_{\gamma,\varrho} = \{A \in \operatorname{supp}\nu : \Pi(A) \in \mathrm{SO}(n)U_\gamma,\ |A - Q_\gamma U_\gamma| < \varrho\}, \quad \gamma = \alpha,\ \beta.$$

For these sets we define the corresponding volume fractions by

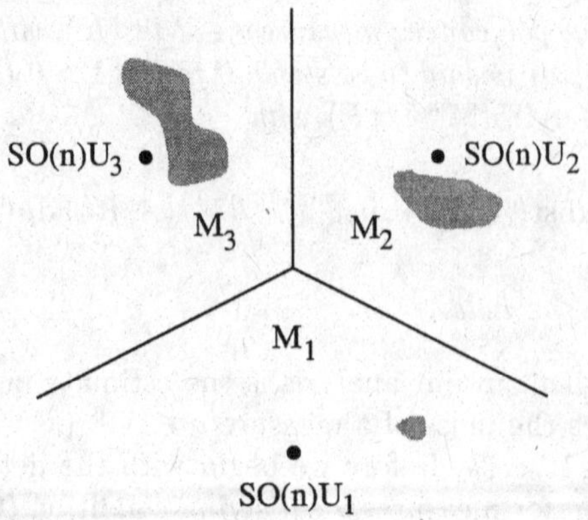

Fig. 4.1. *Sketch of the sets M_i defined by the projection π for $k = 3$ (see the example in the introduction to this chapter). The three wells are indicated by dots and the projection defines three regions in the matrix space in which the nearest point in K is contained in $\mathrm{SO}(n)U_i$. The estimates on the volume fractions τ_ℓ ensure that most of the support of a polyconvex measure (indicated by the shaded areas) is close to two of the wells.*

$$\tau_\ell = \nu(M_\ell) = \int_{M_\ell} 1 \, d\nu(A), \tag{4.6}$$

and

$$\widehat{\lambda} = \int_{\widehat{M}_\alpha} 1 \, d\nu(A), \quad \widehat{\lambda}_\varrho = \int_{\widehat{M}_{\alpha,\varrho}} 1 \, d\nu(A). \tag{4.7}$$

Finally, we set

$$M = M_\alpha \cup M_\beta, \quad N = \bigcup_{\ell \notin \{\alpha,\beta\}} M_\ell. \tag{4.8}$$

To simplify the notation, we do not indicate explicitly the dependence of ν in our formulae.

The construction of the projections π and Π used the following fact.

Proposition 4.1.3. *Assume that $K \subset \mathbb{M}^{m\times n}$ is a compact set. Then there exists a Borel measurable function $\pi : \mathbb{M}^{m\times n} \to K$ such that*

$$|\pi(F) - F| = \mathrm{dist}(F, K).$$

Proof. The following short proof is due to Kirchheim and Kristensen. Let $\mathcal{D}_k$ be the family of all dyadic cubes of side length 2^{-k} in $\mathbb{M}^{m\times n}$. We define (linear) orderings '$<_k$' on each of the families $\mathcal{D}(k)$ in such a way that

$$C_1, C_2 \in \mathcal{D}_{k+1}, \widetilde{C}_1, \widetilde{C}_2 \in \mathcal{D}_k, C_i \subset \widetilde{C}_i \text{ and } \widetilde{C}_1 <_k \widetilde{C}_2 \text{ implies } C_1 <_{k+1} C_2.$$

For a closed cube C let F_C denote its corner minimizing $|F|$. We define

$$\pi_k(F) = F_{C(k,F)},$$

where

$$C(k,F) = \min\{C \in \mathcal{D}_k : \operatorname{dist}(F,K) = \operatorname{dist}(F, K \cap C)\}.$$

Since the sets $\{F : \operatorname{dist}(F,K) = \operatorname{dist}(F, K \cap C)\}$ are closed for all dyadic cubes, it is easy to see that π_k is Borel measurable. Moreover, the definition of the linear orderings implies that $C(k+1,F) \subset C(k,F)$ and therefore the functions π_k converge uniformly to a Borel measurable function π that satisfies $|\pi(F) - F| = \operatorname{dist}(F,K)$ for all $F \in \mathbb{M}^{m\times n}$. □

Now we define first the condition (C_b) on which our estimates are based, and then we discuss the implications of the two hypotheses in (C_b).

The Condition (C_b). We begin with the definition of the condition (C_b) that is at the heart of our analysis of uniqueness and stability of microstructure underlying a global deformation F in bulk materials.

Definition 4.1.4 (Condition (C_b) for uniqueness in bulk materials). *Let $U_1, \ldots, U_k$ be positive definite $n \times n$ matrices with $\det U_i = \Delta > 0$ for $i = 1, \ldots, k$. Assume that K is given by*

$$K = \mathrm{SO}(n)U_1 \cup \ldots \cup \mathrm{SO}(n)U_k, \tag{4.9}$$

and that $F \in K^{\mathrm{pc}}$. Then F is said to satisfy condition (C_b) if F has the following properties:

i) If $n \geq 3$, then there exist two indices α and β (not necessarily distinct) and a simple laminate $\overline{\nu} \in \mathcal{M}^{\mathrm{pc}}(K;F)$ supported on $\mathrm{SO}(n)U_\alpha \cup \mathrm{SO}(n)U_\beta$.

ii) There exist two indices α, $\beta \in \{1,\ldots,k\}$ (α and β not necessarily distinct and identical to the indices in i) for $n \geq 3$) and a permutation $\{1,\ldots,k\} \setminus \{\alpha,\beta\} = \{i_1,\ldots,i_\kappa\}$ of the remaining indices with the following property: for all $j \in \{1,\ldots,\kappa\}$ there exists a vector $\boldsymbol{w}_j$ such that one of the following two sets of inequalities holds:

$$|F\boldsymbol{w}_j| = |U_\alpha \boldsymbol{w}_j| = |U_\beta \boldsymbol{w}_j| \geq \max_{\ell \in \{i_{j+1},\ldots,i_\kappa\}} |U_\ell \boldsymbol{w}_j|,$$

and

$$|F\boldsymbol{w}_j| > |U_{i_j} \boldsymbol{w}_j|,$$

or

$$|\operatorname{cof} F\boldsymbol{w}_j| = |\operatorname{cof} U_\alpha \boldsymbol{w}_j| = |\operatorname{cof} U_\beta \boldsymbol{w}_j| \geq \max_{\ell \in \{i_{j+1},\ldots,i_\kappa\}} |\operatorname{cof} U_\ell \boldsymbol{w}_j|$$

and

$$|\operatorname{cof} F\boldsymbol{w}_j| > |\operatorname{cof} U_{i_j} \boldsymbol{w}_j|.$$

Here $\kappa = k - 2$ if $\alpha \neq \beta$ and $\kappa = k - 1$ if $\alpha = \beta$.

Remark 4.1.5. Throughout the rest of the chapter we are mostly concerned with the case $\alpha \neq \beta$. The same arguments imply for $\alpha = \beta$ that the microstructure is supported on one well and thus a Dirac mass placed at a single matrix in $\mathrm{SO}(n)U_\alpha$. In this case we refer also to a single Dirac mass as a simple laminate in order to simplify the notation.

Remark 4.1.6. An apparently weaker formulation of condition ($\mathrm{C_b}$) would be to allow different indices α and β in i) and ii) for $n \geq 3$. However, the subsequent analysis shows that they are automatically identical and this justifies the formulation in Definition 4.1.4.

Remark 4.1.7. We show in Theorem 4.2.1 that ii) implies i) for $n = 2$.

In the next two propositions, we analyze the importance of the two assumptions i) and ii). The observation here is that the algebraic condition ii) on the matrix F and the set K ensures that every polyconvex measure $\nu \in \mathcal{M}^{\mathrm{pc}}(K;F)$ is supported on the two wells $\mathrm{SO}(n)U_\alpha \cup \mathrm{SO}(n)U_\beta$, see Proposition 4.1.9. The relevance of the first condition, namely that F has a representation as a barycenter of a simple laminate, lies in the fact that simple laminates supported on two wells are unique, and this is the assertion of the first proposition.

Proposition 4.1.8. *Assume that* $U_1, U_2 \in \mathbb{M}^{n\times n}$ *with* $\det U_1 = \det U_2 > 0$, *and that* $\mathrm{SO}(n)U_1 \neq \mathrm{SO}(n)U_2$. *If F is the center of mass of a simple laminate ν supported on the wells* $\mathrm{SO}(n)U_1 \cup \mathrm{SO}(n)U_2$, *then the* $\mathcal{M}^{\mathrm{pc}}(K;\langle\nu, id\rangle) = \{\nu\}$.

Proof. Assume first that $\mathrm{SO}(n)U_1$ and $\mathrm{SO}(n)U_2$ are not rank-one connected and let $F = \lambda Q_1U_1 + (1-\lambda)Q_2U_2$ be the representation of F as the barycenter of a simple laminate which exists by assumption. Since any polyconvex measure supported on two incompatible matrices has to be a single Dirac mass, we conclude $\lambda \in \{0,1\}$, and the assertion is immediate. We may thus assume that Q_1U_1 and Q_2U_2 are rank-one connected, and we have to show that Q_1, Q_2 and λ are uniquely determined.

To simplify the notation, we let $A = Q_1U_1$, $B = Q_2U_2$, and we define $\boldsymbol{a}, \boldsymbol{b} \in \mathbb{R}^n$ by

$$Q_1U_1 - Q_2U_2 = A - B = \boldsymbol{a} \otimes \boldsymbol{b}. \tag{4.10}$$

Changing coordinates, we may associate with ν in a unique way a polyconvex measure $\widetilde{\nu}$ such that

$$\operatorname{supp}\widetilde{\nu} \subset \mathrm{SO}(n) \cup \mathrm{SO}(n)AB^{-1}, \quad \widetilde{F} = FB^{-1} = \lambda AB^{-1} + (1-\lambda)\mathbb{I} = \langle\widetilde{\nu}, id\rangle.$$

Step 1: The matrix AB^{-1} has one as an eigenvalue with algebraic multiplicity at least equal to $n-2$, i.e., there exist $n-2$ orthonormal eigenvectors $\boldsymbol{v}_i$, $i = 1,\dots,n-2$, with

$$AB^{-1}\boldsymbol{v}_i = \boldsymbol{v}_i, \quad i = 1,\dots,n-2.$$

Moreover, the vector $\boldsymbol{a}$ in (4.10) is orthogonal to $\boldsymbol{v}_1, \ldots, \boldsymbol{v}_{n-2}$ and satisfies $|\widetilde{F}\boldsymbol{a}|^2 = |\boldsymbol{a}|^2 = |AB^{-1}\boldsymbol{a}|^2$.

It follows from $\det A = \det(B + \boldsymbol{a} \otimes \boldsymbol{b}) = (\det B)(1 + \langle B^{-1}\boldsymbol{a}, \boldsymbol{b}\rangle)$ that $\langle B^{-1}\boldsymbol{a}, \boldsymbol{b}\rangle = 0$ and thus

$$\begin{aligned} 0 &= \langle B^{-1}\boldsymbol{a}, \boldsymbol{b}\rangle \boldsymbol{a} = \langle \boldsymbol{a}, B^{-T}\boldsymbol{b}\rangle \boldsymbol{a} \\ &= (\boldsymbol{a} \otimes \boldsymbol{b})B^{-1}\boldsymbol{a} = (A - B)B^{-1}\boldsymbol{a} = AB^{-1}\boldsymbol{a} - \boldsymbol{a}, \end{aligned}$$

and hence

$$\widetilde{F}\boldsymbol{a} = \lambda AB^{-1}\boldsymbol{a} + (1 - \lambda)\boldsymbol{a} = \boldsymbol{a}.$$

We conclude $|\widetilde{F}\boldsymbol{a}| = |\boldsymbol{a}| = |AB^{-1}\boldsymbol{a}|$ and it only remains to construct the eigenvectors $\boldsymbol{v}_1, \ldots, \boldsymbol{v}_{n-2}$ orthogonal to $\boldsymbol{a}$. By the polar decomposition theorem, there exist R, $Q \in \mathrm{SO}(n)$ such that

$$RQAB^{-1}R^T = \widetilde{\Lambda} = \mathrm{diag}(\lambda_1, \ldots, \lambda_n)$$

with $0 < \lambda_1 \le \lambda_2 \le \cdots \le \lambda_n$. Let

$$\widetilde{Q} = RQR^T, \quad \widetilde{\boldsymbol{a}} = RQ\boldsymbol{a}, \quad \widetilde{\boldsymbol{b}} = RB^{-T}\boldsymbol{b}.$$

Then $A - B = \boldsymbol{a} \otimes \boldsymbol{b}$ is equivalent to $\widetilde{\Lambda} - \widetilde{Q} = \widetilde{\boldsymbol{a}} \otimes \widetilde{\boldsymbol{b}}$ or $\widetilde{Q} = \widetilde{\Lambda} - \widetilde{\boldsymbol{a}} \otimes \widetilde{\boldsymbol{b}}$ and

$$\widetilde{Q}^T\widetilde{Q} = \mathbb{I} = \widetilde{\Lambda}^T\widetilde{\Lambda} - \widetilde{\Lambda}^T\widetilde{\boldsymbol{a}} \otimes \widetilde{\boldsymbol{b}} - \widetilde{\boldsymbol{b}} \otimes \widetilde{\Lambda}^T\widetilde{\boldsymbol{a}} + |\widetilde{\boldsymbol{a}}|^2\widetilde{\boldsymbol{b}} \otimes \widetilde{\boldsymbol{b}}.$$

Suppose first that $\widetilde{\boldsymbol{b}}$ is parallel to $\widetilde{\Lambda}^T\widetilde{\boldsymbol{a}}$. Then $\widetilde{\Lambda}^T\widetilde{\Lambda} - \mathbb{I}$ is a matrix of rank one and $\widetilde{\Lambda}^T\widetilde{\Lambda}$ has at least $n-1$ eigenvalues equal to one. Since $\det \widetilde{\Lambda} = 1$ it follows that $\widetilde{\Lambda} = \mathbb{I}$ and this contradicts the assumption that the wells $\mathrm{SO}(n)A$ and $\mathrm{SO}(n)B$ are distinct. We may thus assume that $\widetilde{\boldsymbol{b}}$ and $\widetilde{\Lambda}^T\widetilde{\boldsymbol{a}}$ are not parallel. In this situation, let

$$\widetilde{L} = \mathrm{span}\{\widetilde{\boldsymbol{b}}, \widetilde{\Lambda}^T\widetilde{\boldsymbol{a}}\}, \quad \widetilde{H} = \widetilde{L}^\perp.$$

Then $\widetilde{\Lambda}^T\widetilde{\Lambda}\widetilde{\boldsymbol{v}} = \widetilde{\boldsymbol{v}}$ if and only if $\widetilde{\boldsymbol{v}} \in \widetilde{H}$, and since $\widetilde{\Lambda}$ is diagonal with positive entries in the diagonal, we obtain

$$\widetilde{\Lambda}^T\widetilde{\Lambda}\widetilde{\boldsymbol{v}} = \widetilde{\boldsymbol{v}} \quad \Leftrightarrow \quad \widetilde{\Lambda}\widetilde{\boldsymbol{v}} = \widetilde{\boldsymbol{v}} \quad \Leftrightarrow \quad \widetilde{\boldsymbol{v}} \in \widetilde{H}.$$

Therefore $\widetilde{\Lambda}$ has at least $n-2$ eigenvalues equal to one, and since there exists a rank-one connection between $\widetilde{\Lambda}$ and $\mathrm{SO}(n)$ we conclude that

$$\widetilde{\Lambda} = \mathrm{diag}(\lambda_1, 1, \ldots, 1, \lambda_n), \quad \lambda_1 \le 1 \le \lambda_n.$$

Moreover,

$$\widetilde{L} = \mathrm{span}\,\{\widetilde{\boldsymbol{b}}, \widetilde{\Lambda}^T\widetilde{\boldsymbol{a}}\} \perp \widetilde{H} = \mathrm{span}\,\{\boldsymbol{e}_2, \ldots, \boldsymbol{e}_{n-1}\},$$

where $\widetilde{H}$ is (a subspace of) the eigenspace corresponding to the eigenvalue one for $\widetilde{\Lambda}$. This implies

$$(RQAB^{-1}R^T)\widetilde{v} = \widetilde{v} \text{ or } (QAB^{-1})(R^T\widetilde{v}) = R^T\widetilde{v} \text{ for all } \widetilde{v} \in \widetilde{H},$$

and thus

$$QAB^{-1}v = v \quad \forall v \in H = R^T\widetilde{H} \perp L = R^T\widetilde{L} = \operatorname{span}\{B^{-T}Aa, B^{-T}b\}.$$

Hence $B^{-T}b$ is perpendicular to H, and since $QAB^{-1} - Q = Qa \otimes B^{-T}b$, we deduce from the foregoing formula that

$$v = QAB^{-1}v = Qv \quad \forall v \in H.$$

Therefore Q is the identity on H and hence maps L onto L. It remains to show that a is perpendicular to H. To see this, recall that $\widetilde{\Lambda}\widetilde{v} = \widetilde{v}$ for all $\widetilde{v} \in \widetilde{H}$, and thus

$$\langle \widetilde{a}, \widetilde{v}\rangle = \langle \widetilde{a}, \widetilde{\Lambda}\widetilde{v}\rangle = \langle \widetilde{\Lambda}^T\widetilde{a}, \widetilde{v}\rangle = 0 \quad \forall \widetilde{v} \in \widetilde{H},$$

i.e., $\widetilde{a}$ is perpendicular to $\widetilde{H}$ and hence in $\widetilde{L}$. Therefore $R^T\widetilde{a} = Qa \in L$ and the fact that Q maps L onto L implies that $Qa \in L$ if and only if $a \in L$. The choice of an orthonormal basis $v_1, \ldots, v_{n-2}$ of H concludes the proof of Step 1.

Step 2: Let $\widetilde{K} = \mathrm{SO}(n)\widetilde{U}_1 \cup \mathrm{SO}(n)\widetilde{U}_2$ and $\widetilde{F} \in \widetilde{K}^{\mathrm{pc}}$ be the center of mass of a simple laminate $\widetilde{\nu}$ supported on $\widetilde{K}$. Assume that $\widetilde{U}_1$ and $\widetilde{U}_2$ are positive definite with $\det\widetilde{U}_1 = \det\widetilde{U}_2$. Suppose that $\widetilde{U}_1$ and $\widetilde{U}_2$ have $n-2$ orthonormal common eigenvectors v_i, $i = 1, \ldots, n-2$, with corresponding eigenvalues μ_i. Assume, furthermore, that there exists an $e \in \mathbb{S}^{n-1}$ orthogonal to $v_1, \ldots, v_{n-2}$ with

$$|\widetilde{F}e|^2 = |\widetilde{U}_1e|^2 = |\widetilde{U}_2e|^2. \tag{4.11}$$

Then $\mathcal{M}^{\mathrm{pc}}(\widetilde{K}; \widetilde{F}) = \{\widetilde{\nu}\}$.

We first show that $\{\widetilde{F}v_1, \ldots, \widetilde{F}v_{n-2}, \widetilde{F}e\}$ is an orthogonal system. It follows from $\widetilde{U}_iv_i = \mu_iv_i$ and the minors relations that

$$|\widetilde{F}v_i|^2 \le \mu_i^2, \quad |\operatorname{cof}\widetilde{F}v_i|^2 \le \frac{(\det\widetilde{U}_1)^2}{\mu_i^2}, \qquad i = 1, \ldots, n-2,$$

and consequently

$$0 \le \Big|\frac{1}{\mu_i}\widetilde{F}v_i - \frac{\mu_i}{\det\widetilde{U}_1}\operatorname{cof}\widetilde{F}v_i\Big|^2 = \frac{1}{\mu_i^2}|\widetilde{F}v_i|^2 - 2 + \frac{\mu_i^2}{(\det\widetilde{U}_1)^2}|\operatorname{cof}\widetilde{F}v_i|^2 \le 0.$$

This implies

$$\widetilde{F}\boldsymbol{v}_i = \frac{\mu_i^2}{\det \widetilde{U}_1} \operatorname{cof} \widetilde{F}\boldsymbol{v}_i,$$

thus $\widetilde{F}^T\widetilde{F}\boldsymbol{v}_i = \mu_i^2\boldsymbol{v}_i$ and hence

$$\langle \widetilde{F}\boldsymbol{v}_i, \widetilde{F}\boldsymbol{v}_j\rangle = \mu_i^2\delta_{ij}, \quad \langle \widetilde{F}\boldsymbol{v}_i, \widetilde{F}\boldsymbol{e}\rangle = 0, \quad i, j = 1, \dots, n-2.$$

Therefore

$$|\widetilde{F}\boldsymbol{v}_i|^2 = \mu_i^2 = |A\boldsymbol{v}_i|^2$$

for $i = 1, \dots, n-2$ and $A \in \operatorname{supp}\widetilde{\nu}$, and thus

$$\begin{aligned} 0 &\leq \int_{\operatorname{supp}\nu} |\widetilde{F}\boldsymbol{v}_i - A\boldsymbol{v}_i|^2 \mathrm{d}\nu(A) \\ &= |\widetilde{F}\boldsymbol{v}_i|^2 - 2\int_{\operatorname{supp}\nu} \langle \widetilde{F}\boldsymbol{v}_i, A\boldsymbol{v}_i\rangle \mathrm{d}\nu(A) + \int_{\operatorname{supp}\nu} |A\boldsymbol{v}_i|^2 \mathrm{d}\nu(A) \qquad (4.12) \\ &= -|\widetilde{F}\boldsymbol{v}_i|^2 + \int_{\operatorname{supp}\nu} |A\boldsymbol{v}_i|^2 \mathrm{d}\nu(A) = 0. \end{aligned}$$

In addition we obtain from (4.11) that

$$|\widetilde{U}_1\boldsymbol{e}|^2 = |\widetilde{F}\boldsymbol{e}|^2 \leq \int_{\operatorname{supp}\nu} |A\boldsymbol{e}|^2 \mathrm{d}\nu(A) \leq |\widetilde{U}_1\boldsymbol{e}|^2,$$

and hence equality holds throughout this inequality. We get

$$0 \leq \int_{\operatorname{supp}\nu} |\widetilde{F}\boldsymbol{e} - A\boldsymbol{e}|^2 \mathrm{d}\nu(A) = 0. \qquad (4.13)$$

Equations (4.12) and (4.13) show that for a.e. $A \in \operatorname{supp}\nu$ the identities

$$A\boldsymbol{v}_i = \widetilde{F}\boldsymbol{v}_i, \quad A\boldsymbol{e} = \widetilde{F}\boldsymbol{e}$$

hold. Assume first that $A = Q\widetilde{U}_1$ with $Q \in \mathrm{SO}(n)$. Then Q satisfies

$$Q\widetilde{U}_1\boldsymbol{v}_i = \widetilde{F}\boldsymbol{v}_i, \quad Q\widetilde{U}_1\boldsymbol{e} = \widetilde{F}\boldsymbol{e},$$

i.e., the rotation Q maps the $n-1$ orthogonal vectors $\widetilde{U}_1\boldsymbol{v}_i, \widetilde{U}_1\boldsymbol{e}$ onto the $n-1$ orthogonal vectors $\widetilde{F}\boldsymbol{v}_i, \widetilde{F}\boldsymbol{e}$. This uniquely determines Q and a similar argument holds for $A = Q\widetilde{U}_2$. Consequently the support of ν is uniquely determined and ν is given by $\nu = \lambda\delta_{Q_1\widetilde{U}_1} + (1-\lambda)\delta_{Q_2\widetilde{U}_2}$ with $\lambda \in [0,1]$ and $Q_1, Q_2 \in \mathrm{SO}(n)$. To prove uniqueness of the volume fraction λ, assume that there exist $\mu \in [0,1]$, $\mu \neq \lambda$, such that $\nu = \mu\delta_{Q_1\widetilde{U}_1} + (1-\mu)\delta_{Q_2\widetilde{U}_2}$. Then $(\lambda-\mu)(Q_1\widetilde{U}_1 - Q_2\widetilde{U}_2) = 0$. Hence $\mathrm{SO}(n)\widetilde{U}_1 = \mathrm{SO}(n)\widetilde{U}_2$ and this contradicts the assumption that $\mathrm{SO}(n)\widetilde{U}_1 \neq \mathrm{SO}(n)\widetilde{U}_2$. The proof of the assertion of the second step is thus complete.

The assertion of the proposition is now an immediate consequence of Steps 1 and 2. □

We now prove that hypothesis ii) in condition (C_b) implies that the support of any $\nu \in \mathcal{M}^{pc}(K;F)$ is contained in the two wells $SO(n)U_\alpha \cup SO(n)U_\beta$. This result is a special case of the proposition below since $\mathcal{E}(\nu)=0$ for measures ν with support in K. The more general statement which we provide here is an important ingredient in the applications to stability and error estimates for finite element minimizers in Section 4.5.

Proposition 4.1.9. *Assume that K is given by (4.3), that F satisfies condition ii) in (C_b) and that $\nu \in \mathcal{M}^{pc}(\mathbb{M}^{n\times n};F)$. Then there exists a constant c that depends only on F and n, but not on ν such that*

$$\tau_\ell \le c\mathcal{E}(\nu) \quad \textit{for all } \ell \in \{1,\dots,k\}\setminus\{\alpha,\beta\}. \tag{4.14}$$

In particular, if $\nu \in \mathcal{M}^{pc}(K;F)$ then ν is supported on $SO(n)U_\alpha \cup SO(n)U_\beta$.

Remark 4.1.10. The proof shows that the estimate for the volume fractions can be improved to

$$\tau_\ell \le c\int_{\operatorname{supp}\nu}\big(\operatorname{dist}(A,K)+\operatorname{dist}^2(A,K)\big)\mathrm{d}\nu(A),\ \ell\in\{1,\dots,k\}\setminus\{\alpha,\beta\}$$

if the condition on the cofactor in condition (C_b) is not used in the estimates. The additional term involving $\operatorname{dist}^{n-1}(A,K)$ reflects the fact that the cofactor is a polynomial of degree $n-1$ in the entries of A. The same is true for the estimates below that are based on these estimates for the volume fractions.

Proof. By definition of τ_ℓ, we have for all $\boldsymbol{w} \in \mathbb{S}^{n-1}$

$$\begin{aligned}\sum_{\ell=1}^{k}\tau_\ell\big(|F\boldsymbol{w}|^2-|U_\ell\boldsymbol{w}|^2\big) &= \int_{\operatorname{supp}\nu}\big(|F\boldsymbol{w}|^2-|\pi(A)\boldsymbol{w}|^2\big)\mathrm{d}\nu(A)\\ &= \int_{\operatorname{supp}\nu}\big(2\langle F\boldsymbol{w},A\boldsymbol{w}-\pi(A)\boldsymbol{w}\rangle-|\pi(A)\boldsymbol{w}-F\boldsymbol{w}|^2\big)\mathrm{d}\nu(A),\end{aligned}$$

and thus

$$\begin{aligned}\sum_{\ell=1}^{k}\tau_\ell\big(|F\boldsymbol{w}|^2-|U_\ell\boldsymbol{w}|^2\big)&+\int_{\operatorname{supp}\nu}|\pi(A)\boldsymbol{w}-F\boldsymbol{w}|^2\mathrm{d}\nu(A)\\ &\le 2|F\boldsymbol{w}|\int_{\operatorname{supp}\nu}\operatorname{dist}(A,K)\mathrm{d}\nu(A)\le c\mathcal{E}(\nu).\end{aligned} \tag{4.15}$$

We obtain similarly from the minors relation for the cofactor that

$$\begin{aligned}&\sum_{\ell=1}^{k}\tau_\ell\big(|\operatorname{cof}F\boldsymbol{w}|^2-|\operatorname{cof}U_\ell\boldsymbol{w}|^2\big)\\ &=\int_{\operatorname{supp}\nu}\big(2\langle\operatorname{cof}F\boldsymbol{w},\operatorname{cof}A\boldsymbol{w}-\operatorname{cof}(\pi(A))\boldsymbol{w}\rangle-|\operatorname{cof}(\pi(A))\boldsymbol{w}-\operatorname{cof}F\boldsymbol{w}|^2\big)\mathrm{d}\nu(A).\end{aligned}$$

In view of the expansion

$$\operatorname{cof}(A+B)=\sum_{i=0}^{n-1}L_i(A,B)=\operatorname{cof}B+\sum_{i=1}^{n-1}L_i(A,B)$$

with functions $L_i(A,B)$ that are homogeneous of degree i in A and of degree $n-1-i$ in B we deduce that

$$\begin{aligned}\big|\operatorname{cof}(\pi(A))-\operatorname{cof}A\big| &= \big|\operatorname{cof}(\pi(A))-\operatorname{cof}\big(A-\pi(A)+\pi(A)\big)\big|\\ &\le c\sum_{i=1}^{n-1}\big|A-\pi(A)\big|^i\,\big|\pi(A)\big|^{n-1-i}.\end{aligned}$$

Since $a^i\le a+a^{n-1}$ for $i=2,\dots,n-2$ and $a\ge 0$, we conclude

$$\sum_{\ell=1}^{k}\tau_\ell\big(|\operatorname{cof}Fw|^2-|\operatorname{cof}U_\ell w|^2\big)+\int_{\operatorname{supp}\nu}|\operatorname{cof}\pi(A)w-\operatorname{cof}Fw|^2\mathrm{d}\nu(A)\le c\mathcal{E}(\nu). \tag{4.16}$$

The proof follows now by induction. We may assume that $\alpha=k-1$, $\beta=k$ and $i_j=j$ for $j=1,\dots,k-2$. Suppose that for $j=1$ the assumption on F in condition ii) in $(\mathrm{C_b})$ is satisfied. Thus there exists a vector w_1 such that

$$|Fw_1|^2-|U_1w_1|^2>0 \text{ and } |Fw_1|^2-|U_jw_1|^2\ge 0,\quad j=2,\dots,k.$$

It follows from (4.15) that $\tau_1\le c\mathcal{E}(\nu)$ with a constant c that depends only on K and F. We conclude similarly with inequality (4.16) if the assumption on the cofactor holds. Assume now that the assertion has been proved for $\ell=1,\dots,j-1<k-2$, and that the assumption on F in condition ii) in $(\mathrm{C_b})$ is satisfied for the index j. Thus there exists a vector w_j such that

$$|Fw_j|>|U_jw_j| \text{ and } |Fw_j|>|U_\ell w_j|,\quad \ell=j+1,\dots,k.$$

It follows from (4.15) and the estimates for τ_ℓ, $\ell=1,\dots,j-1$ that

$$\sum_{\ell=j}^{k}\tau_\ell\big(|Fw|^2-|U_\ell w|^2\big)\le c\mathcal{E}(\nu),$$

and hence $\tau_j\le c\mathcal{E}(\nu)$. The conclusion is analogous if the condition on the cofactor holds, and this proves the assertion of the proposition.

The proof of the proposition is now complete. □

Uniqueness of Microstructure. After these preparations, we can now state and prove our uniqueness theorem for polyconvex measures in bulk materials.

Theorem 4.1.11. *Let K be given by (4.3). Suppose that $\nu \in \mathcal{M}^{\mathrm{pc}}(K)$ and that $F = \langle \nu, id \rangle$ satisfies condition* ($\mathrm{C_b}$). *Then ν is unique, and*

$$\mathcal{M}^{\mathrm{pc}}(K; \langle \nu, id \rangle) = \{\nu\} = \{\lambda \delta_{Q_\alpha U_\alpha} + (1-\lambda)\delta_{Q_\beta U_\beta}\}$$

with $\lambda \in [0,1]$ and $Q_\alpha U_\alpha - Q_\beta U_\beta = \boldsymbol{a} \otimes \boldsymbol{b}$.

Proof. By Proposition 4.1.9,

$$\tau_\ell \leq \int_{\mathbb{M}^{n\times n}} \big(\operatorname{dist}(A,K) + \operatorname{dist}^{\max\{2,n-1\}}(A,K)\big)\,\mathrm{d}\nu(A) = 0$$

and hence ν is supported on the two wells $\mathrm{SO}(n)U_\alpha$ and $\mathrm{SO}(n)U_\beta$. The assertion of the theorem is an immediate consequence of Proposition 4.1.8. □

Stability of Microstructure. We now turn towards the analysis of stability of microstructure. Throughout this section we assume that $n \geq 3$, i.e., we concentrate on the situation with the additional hypothesis in ($\mathrm{C_b}$) that $F \in K^{\mathrm{pc}}$ has a representation as the barycenter of a simple laminate $\overline{\nu}$ given by

$$\overline{\nu} = \overline{\lambda}\delta_{Q_\alpha U_\alpha} + (1-\overline{\lambda})\delta_{Q_\beta U_\beta}, \qquad Q_\alpha U_\alpha - Q_\beta U_\beta = \boldsymbol{a} \otimes \boldsymbol{b} \tag{4.17}$$

with Q_α, $Q_\beta \in \mathrm{SO}(n)$ and $\overline{\lambda} \in [0,1]$. Furthermore we tacitly assume that $\alpha \neq \beta$ and $\lambda \notin \{0,1\}$. If $\alpha = \beta$, then the polyconvex measure is supported on $\mathrm{SO}(n)U_\alpha$ and hence a Dirac mass placed at a single matrix. The adaption of the proof in this case is obvious.

Suppose now that $F = \langle \overline{\nu}, id \rangle$ with $\overline{\nu}$ as in (4.17) satisfying condition ($\mathrm{C_b}$) and that $\nu \in \mathcal{M}^{\mathrm{pc}}(\mathbb{M}^{n\times n}; F)$. The stability analysis for $\overline{\nu}$ requires two ingredients. Proposition 4.1.9 provides estimates for the volume fractions τ_ℓ and shows that τ_ℓ is bounded by $\mathcal{E}(\nu)$ for $\ell \notin \{\alpha, \beta\}$. In addition to this, it is important to control the support of ν on the two wells $\mathrm{SO}(n)U_\alpha$ and $\mathrm{SO}(n)U_\beta$ and to have information about the part of the support that is not close to the two points $Q_\alpha U_\alpha$ and $Q_\beta U_\beta$. This is done in the next proposition which provides bounds on the excess rotation $R(A) \in \mathrm{SO}(n)$ defined by

$$\pi(A) = R(A)\Pi(A) \quad \text{on } M = M_\alpha \cup M_\beta.$$

Proposition 4.1.12. *Assume that K is given by (4.3) and that F satisfies condition* ($\mathrm{C_b}$). *Let $\nu \in \mathcal{M}^{\mathrm{pc}}(\mathbb{M}^{n\times n}; F)$. Then*

$$\int_{\operatorname{supp}\nu} |R(A) - \mathbb{I}|^2\,\mathrm{d}\nu(A) \leq c\,\mathcal{E}(\nu). \tag{4.18}$$

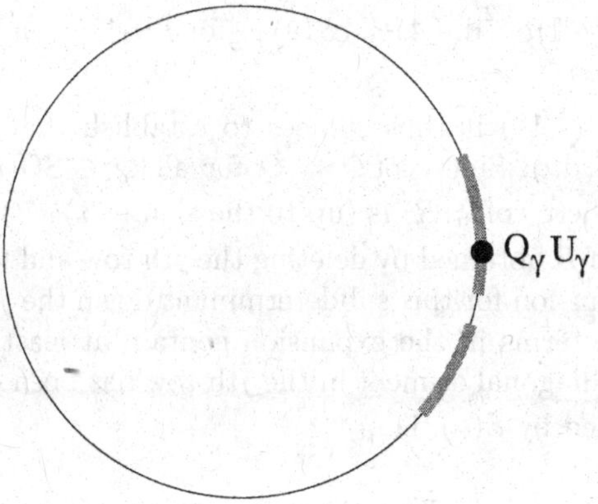

Fig. 4.2. *Sketch of the estimate for the excess rotation. The set* $SO(n)U_\gamma$ *is represented by a circle. The projection* π *maps the points in* M_γ *onto* $SO(n)U_\gamma$. *Proposition 4.1.12 ensures that the projected points are close to* $Q_\gamma U_\gamma$, *the points in the support of the unique microstructure realizing the center of mass.*

Proof. Let $\overline{\nu}$ be the simple laminate with center of mass equal to F in (4.17). Choose an orthonormal basis $\{\boldsymbol{w}_1, \ldots, \boldsymbol{w}_{n-1}, \boldsymbol{b}\}$ of $\mathbb{R}^n$ where $\boldsymbol{b}$ is the vector in the representation (4.17) for $\overline{\nu}$, and let

$$W = \operatorname{span}\{F\boldsymbol{w}_1, \ldots, F\boldsymbol{w}_{n-1}\}.$$

Fix a rotation $Q \in SO(n)$ such that

$$QW = \{\boldsymbol{x} \in \mathbb{R}^n : x_n = 0\},$$

and let $\widetilde{R}(A) = QR(A)Q^T$. It suffices to estimate $|\widetilde{R}(A) - \mathbb{I}|$ since

$$\int_{\operatorname{supp}\nu} |\widetilde{R}(A) - \mathbb{I}|^2 \,\mathrm{d}\nu(A) = \int_{\operatorname{supp}\nu} |R(A) - \mathbb{I}|^2 \,\mathrm{d}\nu(A).$$

In view of the estimate (4.14), it suffices to prove this estimate on M. Since $\boldsymbol{w}_i$ is orthogonal to $\boldsymbol{b}$, we have $F\boldsymbol{w}_i = Q_\alpha U_\alpha \boldsymbol{w}_i = Q_\beta U_\beta \boldsymbol{w}_i$ and therefore $R(A)F\boldsymbol{w}_i = \pi(A)\boldsymbol{w}_i$ on M. Then (4.15) implies

$$\begin{aligned}
\int_M \left|(\widetilde{R}(A) - \mathbb{I})QF\boldsymbol{w}_i\right|^2 \mathrm{d}\nu(A) &= \int_M \left|(QR(A)Q^T - QQ^T)QF\boldsymbol{w}_i\right|^2 \mathrm{d}\nu(A) \\
&= \int_M \left|(R(A) - \mathbb{I})F\boldsymbol{w}_i\right|^2 \mathrm{d}\nu(A) \\
&\leq \int_M \left|(\pi(A) - F)\boldsymbol{w}_i\right|^2 \mathrm{d}\nu(A) \\
&\leq c\mathcal{E}(\nu),
\end{aligned}$$

and by the definition of W,

$$\int_M \left|(\widetilde{R}(A) - \mathbb{I})e_i\right|^2 \mathrm{d}\nu(A) \leq c\mathcal{E}(\nu) \quad \text{for } i = 1, \ldots, n-1. \tag{4.19}$$

In order to prove (4.18) it thus suffices to establish that $(\widetilde{R}(A) - \mathbb{I})e_n$ is bounded in $L^2(M, \mathrm{d}\nu)$. Since $\operatorname{cof} Q = Q$ for all $Q \in \mathrm{SO}(n)$, we have that $\widetilde{R}_{jn} = \operatorname{cof}_{jn}(\widetilde{R})$ where $\operatorname{cof}_{jn}(\widetilde{R})$ is (up to the sign $(-1)^{j+n}$) the determinant of the submatrix of $\widetilde{R}$ obtained by deleting the jth row and the nth column of $\widetilde{R}$. A cofactor expansion for this subdeterminant down the jth column shows that for $j < n$ all terms in the expansion contain at least one off-diagonal element of $\widetilde{R}$ (the diagonal element in the jth row has been crossed out), and thus $\widetilde{R}_{jn}$ is bounded by $\mathcal{E}(\nu)$, i.e.,

$$\int_M \widetilde{R}_{jn}^2(A)\mathrm{d}\nu(A) \leq c\mathcal{E}(\nu).$$

For $j = n$, $\operatorname{cof}_{nn} \widetilde{R}$ is given as a sum of s terms p_k, each of which is a product of $n-1$ factors. Let $p_1 = \widetilde{R}_{11} \cdots \widetilde{R}_{(n-1)(n-1)}$ be the term corresponding to the product of the first $n-1$ diagonal entries in $\widetilde{R}$. We may estimate

$$|\widetilde{R}_{nn} - 1|^2 = \Big|\sum_{k=1}^{s} p_k - 1\Big|^2 \leq c|p_1 - 1|^2 + c\sum_{k=2}^{s} |p_k|^2. \tag{4.20}$$

Then

$$|p_1 - 1|^2 \leq c\left|(\widetilde{R}_{11} - 1)\cdots(\widetilde{R}_{(n-1)(n-1)} - 1)\right|^2 + \ldots + c|R_{(n-1)(n-1)} - 1|^2,$$

where the dots stand for products with less than $n-1$ terms. In view of (4.19) we deduce

$$\int_M |p_1 - 1|^2 \mathrm{d}\nu(A) \leq c\mathcal{E}(\nu)$$

and the assertion follows now from (4.20) since the products p_k with $k \geq 2$ contain at least one off-diagonal entry in $\widetilde{R}$ which is again bounded by (4.19) □

With the estimates for the volume fractions and the excess rotation at hand, we can prove the following list of estimates for microstructures with small energy.

Theorem 4.1.13. *Suppose that K is given by (4.3) and that F satisfies condition $(\mathrm{C_b})$. Let $\overline{\nu} \in \mathcal{M}^{\mathrm{pc}}(K; F)$ given by (4.17) be the unique polyconvex measure with center of mass F. Then the following estimates hold for all $\nu \in \mathcal{M}^{\mathrm{pc}}(\mathbb{M}^{n\times n}; F)$:*

i) Estimate orthogonal to the layering normal: for $w \in \mathbb{S}^{n-1}$ with $\langle w, b\rangle = 0$,

$$\int_{\operatorname{supp}\nu} |Aw - Fw|^2 \mathrm{d}\nu(A) \leq c\mathcal{E}(\nu).$$

ii) Estimate for the distance to the support of $\overline{\nu}$*:*

$$\int_{\operatorname{supp}\nu} |A - \Pi(A)|^2 \mathrm{d}\nu(A) \le c\mathcal{E}(\nu).$$

iii) Let $\widehat{\lambda}$ *and* $\widehat{\lambda}_\varrho$ *be defined as in (4.7). Then*

$$|\widehat{\lambda} - \overline{\lambda}| \le c\mathcal{E}^{1/2}(\nu), \qquad |\widehat{\lambda}_\varrho - \overline{\lambda}| \le \frac{c}{\varrho}\mathcal{E}^{1/2}(\nu).$$

iv) For all $f \in W^{1,\infty}(\mathbb{M}^{n\times n};\mathbb{R})$,

$$|\langle \nu, f\rangle - \langle \overline{\nu}, f\rangle| \le c\operatorname{Lip}(f)\mathcal{E}^{1/2}(\nu).$$

Proof. To prove i), note that by assumption $|U_\alpha \boldsymbol{w}| = |U_\beta \boldsymbol{w}| = |F\boldsymbol{w}|$ for all $\boldsymbol{w} \in \mathbb{S}^{n-1}$ with $\langle \boldsymbol{w}, \boldsymbol{b}\rangle = 0$. Thus by (4.15) and the estimates for the volume fractions

$$\int_{\operatorname{supp}\nu} |(\pi(A) - F)\boldsymbol{w}|^2 \mathrm{d}\nu(A) \le c\mathcal{E}(\nu) + \sum_{\ell=1}^{k} \tau_\ell \big||F\boldsymbol{w}|^2 - |U_k\boldsymbol{w}|^2\big| \le c\mathcal{E}(\nu),$$

and the assertion follows by the triangle inequality,

$$\begin{aligned}\int_{\operatorname{supp}\nu} &|(A - F)\boldsymbol{w}|^2 \mathrm{d}\nu(A)\\ &\le 2\int_{\operatorname{supp}\nu} \big(|(A - \pi(A))\boldsymbol{w}|^2 + |(\pi(A) - F)\boldsymbol{w}|^2\big)\mathrm{d}\nu(A) \le c\mathcal{E}(\nu).\end{aligned}$$

Here we use the fact that by definition $|A - \pi(A)|^2 = \operatorname{dist}^2(A, K)$.

We observe for the proof of ii) that by the triangle inequality

$$\int_{\operatorname{supp}\nu} |A - \Pi(A)|^2 \mathrm{d}\nu(A) \le 2\int_{\operatorname{supp}\nu} \big(|A - \pi(A)|^2 + |\pi(A) - \Pi(A)|^2\big)\mathrm{d}\nu(A).$$

By the definition of M and N in (4.8) and the estimates for the volume fractions, $\nu(N) \le c\mathcal{E}(\nu)$. Thus by (4.18)

$$\begin{aligned}&\int_{\operatorname{supp}\nu} |\pi(A) - \Pi(A)|^2 \mathrm{d}\nu(A)\\ &= \int_M |(R(A) - \mathbb{I})\Pi(A)|^2 \mathrm{d}\nu(A) + \int_N |\pi(A) - \Pi(A)|^2 \mathrm{d}\nu(A) \le c\mathcal{E}(\nu),\end{aligned}$$

and ii) follows easily.

The proof of iii) is an immediate consequence of ii). In fact,

$$\begin{aligned}&|\widehat{\lambda} - \overline{\lambda}|\,|Q_\alpha U_\alpha - Q_\beta U_\beta| = |\widehat{\lambda} Q_\alpha U_\alpha + (1 - \widehat{\lambda})Q_\beta U_\beta - F|\\ &= \Big|\int_{\operatorname{supp}\nu} (\Pi(A) - A)\mathrm{d}\nu(A)\Big| \le \Big(\int_{\operatorname{supp}\nu} |\Pi(A) - A|^2 \mathrm{d}\nu(A)\Big)^{1/2}.\end{aligned}$$

To prove the second assertion in iii), we observe that

$$
\begin{aligned}
&|\widehat{\lambda}_\varrho - \overline{\lambda}|\,|Q_\alpha U_\alpha - Q_\beta U_\beta| \\
&\qquad \le |\widehat{\lambda} - \overline{\lambda}|\,|Q_\alpha U_\alpha - Q_\beta U_\beta| + |\widehat{\lambda}_\varrho - \widehat{\lambda}|\,|Q_\alpha U_\alpha - Q_\beta U_\beta|.
\end{aligned}
$$

By definition,

$$\widehat{\lambda} - \widehat{\lambda}_\varrho = \nu(\{A \in \operatorname{supp}\nu : \Pi(A) \in \mathrm{SO}(n)U_\alpha, |A - Q_\alpha U_\alpha| \ge \varrho\}),$$

and therefore

$$\widehat{\lambda} - \widehat{\lambda}_\varrho \le \frac{1}{\varrho}\int_{\widehat{M}_\alpha} |A - Q_\alpha U_\alpha| \mathrm{d}\nu(A) \le \frac{c}{\varrho}\,\mathcal{E}^{1/2}(\nu).$$

To prove iv), we estimate

$$
\begin{aligned}
\big|\langle \nu, f\rangle - \langle \overline{\nu}, f\rangle\big| &\le \big|\langle \nu, f\rangle - \langle \nu, f \circ \Pi\rangle\big| + \big|\langle \nu, f\circ\Pi\rangle - \langle \overline{\nu}, f\rangle\big| \\
&\le \mathrm{Lip}(f)\langle \nu, |id - \Pi|^2\rangle^{1/2} + \big|\langle \nu, f\circ\Pi\rangle - \langle \overline{\nu}, f\rangle\big|.
\end{aligned}
$$

The first term is bounded by ii), and the second can be estimated by

$$
\begin{aligned}
&\big|\langle \nu, f\circ\Pi\rangle - \langle \overline{\nu}, f\rangle\big| \\
&\le \big|\widehat{\lambda} f(Q_\alpha U_\alpha) + (1-\widehat{\lambda}) f(Q_\beta U_\beta) - \overline{\lambda} f(Q_\alpha U_\alpha) - (1-\overline{\lambda}) f(Q_\beta U_\beta)\big| \\
&\le |\overline{\lambda} - \widehat{\lambda}|\,\mathrm{Lip}(f)\,|Q_\alpha U_\alpha - Q_\beta U_\beta|.
\end{aligned}
$$

The assertion follows now from the two foregoing inequalities and iii). □

The following stability theorem is an immediate consequence of the foregoing estimates.

Theorem 4.1.14. *Suppose that $\overline{\nu} \in \mathcal{M}^{\mathrm{pc}}(K)$ and that $F = \langle \overline{\nu}, id\rangle$ satisfies condition* $(\mathrm{C_b})$. *Then $\overline{\nu}$ is stable.*

Proof. We have to prove that for all $\varepsilon > 0$ there exists a $\delta > 0$ such that for all $\nu \in \mathcal{M}^{\mathrm{pc}}(\mathbb{M}^{n\times n}; F)$ with $\mathcal{E}(\nu) \le \delta$ the estimate $d(\nu, \overline{\nu}) \le \varepsilon$ holds. By the definition of $d(\cdot,\cdot)$ and Theorem 4.1.13,

$$
\begin{aligned}
d(\nu, \overline{\nu}) &= \sum_{i=1}^{\infty} 2^{-i}\,\frac{|\langle \nu, \Phi_i\rangle - \langle \overline{\nu}, \Phi_i\rangle|}{\|\Phi_i\|_{1,\infty}} \\
&\le c\sum_{i=1}^{\infty} 2^{-1}\,\frac{\|\Phi_i\|_{1,\infty}\,\mathcal{E}^{1/2}(\nu)}{\|\Phi_i\|_{1,\infty}} \le c\mathcal{E}^{1/2}(\nu).
\end{aligned}
$$

The assertion follows therefore for $\delta \le (\varepsilon/c)^2$. □

Equivalence of uniqueness and stability of microstructure is now an immediate consequence of the foregoing analysis.

Corollary 4.1.15. *Assume that K is given by (4.3), and that ν is a microstructure supported on K, i.e., $\nu \in \mathcal{M}^{qc}(K;F)$. Then*

$$\nu \text{ is stable} \quad \Rightarrow \quad \nu \text{ is unique.}$$

Remark 4.1.16. If F satisfies condition (C_b), then we obtain that uniqueness and stability are equivalent.

Proof. Assume the contrary. Then there exists a gradient Young measures $\nu \in \mathcal{M}^{qc}(K;F)$ that is stable, but not unique. We may thus choose $\mu \in \mathcal{M}^{qc}(K;F)$ with $\nu \neq \mu$. Since μ has compact support, we find by Zhang's Theorem on the truncation of sequences generating gradient Young measures a sequence of Lipschitz functions $\boldsymbol{u}_i$ with uniformly bounded Lipschitz constants such that the sequence $D\boldsymbol{u}_i$ generates the Young measure μ. Let $\mu_i = \operatorname{Av} \delta_{D\boldsymbol{u}_i(\cdot)}$. By construction, $\mu_i \overset{*}{\rightharpoonup} \mu$ as $i \to \infty$. It is easy to see that for all $p \geq 1$

$$\int_\Omega \operatorname{dist}^p(D\boldsymbol{u}_i, K) \mathrm{d}\boldsymbol{x} \to 0 \quad \text{for } i \to \infty.$$

Hence $\mathcal{E}(\mu_i)$ tends to zero as $i \to \infty$. By assumption $\mu \neq \nu$, and thus there exists an $\varepsilon > 0$ such that

$$d\big(\operatorname{Av} \delta_{D\boldsymbol{u}_i(\cdot)}, \nu\big) \geq \varepsilon$$

for i sufficiently large. This contradicts the definition of stability and concludes the proof. □

4.2 Equivalence of Uniqueness and Stability in 2D

The analysis of uniqueness and stability of microstructure in Section 4.1 relied on the result in Proposition 4.1.8 that simple laminates supported on two wells are uniquely determined from their center of mass. In general, it is an open problem to decide whether uniqueness of a microstructure ν implies that ν is a simple laminate. However, this is true in two dimensions and in this case we obtain an optimal characterization of stability. A further simplification is that the condition (C_b) reduces to conditions on F since $|\operatorname{cof} F\boldsymbol{e}| = |F\boldsymbol{e}^\perp|$. We assume in this section that

$$K = \mathrm{SO}(2)U_1 \cup \ldots \cup \mathrm{SO}(2)U_k \tag{4.21}$$

where the matrices U_i are positive definite with $\det U_i = \Delta$ for $i = 1, \ldots, k$.

Theorem 4.2.1. *Suppose that $\nu \in \mathcal{M}^{pc}(K)$. Then ν is unique if and only if $F = \langle \nu, id \rangle$ satisfies condition* (C_b).

Proof. In view of Theorem 4.1.11, we only need to prove that uniqueness of a polyconvex measure ν implies that $F = \langle \nu, id \rangle$ satisfies condition (C_b). Assume thus that ν is unique. Then F cannot be an unconstrained point in K since the splitting method generates a family of distinct polyconvex measures with barycenter F by decomposing F into simple laminates supported on the set of constrained points along directions in the rank-one cone.

Suppose now that ν is a constrained point. Let $\mathcal{C}$ be the set of corners as defined in (2.17), and set $\mathcal{U} = \{U_1 \ldots, U_k\}$. By Step 6 in the proof of Theorem 2.2.3 we may assume that F is contained in a maximal arc $\Gamma_{pq}(e)$ with U_p, $U_q \in \mathcal{C}$ and $e \in \mathbb{S}^1$. It suffices to show that uniqueness of ν implies

$$|U_p e| = |U_q e| = |Fe| > \max_{U \in \mathcal{U} \setminus \{U_p, U_q\}} |Ue|.$$

If strict inequality does not hold, then there exists an $U_r \in \mathcal{U} \setminus \{U_p, U_q\}$ such that

$$|U_p e| = |U_q e| = |U_r e| = |Fe|,$$

and Proposition 2.2.4 implies that we may assume without loss of generality that U_r is contained in the polyconvex hull of $\mathrm{SO}(2)U_p \cup \mathrm{SO}(2)U_q$. We then find by the construction in (2.28) in the proof of Theorem 2.2.3 rotations Q_q, Q_r, $Q_F \in \mathrm{SO}(2)$ and scalars t_r, $t_F \in [0,1]$ with $t_r \neq 0$, $t_F \neq 0$, such that

$$Q_F F = (1 - \lambda_F)U_p + \lambda_F Q_q U_q, \quad Q_r U_r = (1 - \lambda_r)U_p + \lambda_r Q_q U_q.$$

If $\lambda_r = \lambda_F$, then $F = Q_F^T Q_r U_r$ and we are done. If $\lambda_r \neq \lambda_F$, we may assume that $\lambda_F < \lambda_r$ and in this case

$$Q_F F = \frac{\lambda_r - \lambda_F}{\lambda_r} U_p + \frac{\lambda_F}{\lambda_r} Q_r U_r,$$

and we conclude again that ν is not unique. This establishes the assertion of the theorem. □

We summarize the results in the following corollary.

Corollary 4.2.2. *Let K be given by (4.21) and $\nu \in \mathcal{M}^{qc}(K)$. Then*

$$\nu \text{ stable} \quad \Leftrightarrow \quad \nu \text{ unique} \quad \Leftrightarrow \quad F = \langle \nu, id \rangle \text{ satisfies } (C_b).$$

Proof. This is an immediate consequence of Theorems 4.1.11 and 4.2.1. □

4.3 The Case of O(2) Invariant Sets

We have seen in Section 2.3 that the analysis of thin film problems is closely related to the analysis of sets invariant under O(2). Therefore we begin with the extension of the analysis of uniqueness and stability to sets with a multi-well structure invariant under O(2). We assume that

$$K = \mathrm{O}(2)U_1 \cup \ldots \cup \mathrm{O}(2)U_k \tag{4.22}$$

where the matrices U_i are positive definite and satisfy $\det U_i = \Delta > 0$ for $i = 1, \ldots, k$. Furthermore we suppose that $\mathrm{O}(2)U_i \neq \mathrm{O}(2)U_j$ for $i \neq j$. As a first step we formulate the analogue of condition $(\mathrm{C_b})$ for bulk materials for the case of O(2) invariant sets.

Definition 4.3.1. *Assume that K is given by (4.22) and that $F \in \mathbb{M}^{2\times 2}$. Then F is said to satisfy condition $(\widetilde{\mathrm{C}}_\mathrm{b})$ if one of the following criteria is satisfied:*

i) $|\det F| = \Delta$ and $(\mathrm{C_b})$ holds for $\widetilde{F}$ with $K = \mathrm{SO}(2)U_1 \cup \ldots \cup \mathrm{SO}(2)U_k$ where $\widetilde{F} = F$ if $\det F = \Delta$ and $\widetilde{F} = \mathrm{diag}(-1,1)F$ if $\det F = -\Delta$;

ii) there exists an $\alpha \in \{1, \ldots, k\}$ and an $\boldsymbol{e} \in \mathbb{S}^1$ such that

$$|F\boldsymbol{e}| = |U_\alpha \boldsymbol{e}| > \max_{j\neq\alpha} |U_j \boldsymbol{e}|;$$

iii) may relabel the matrices such that there exists an $\ell \in \{1, \ldots, k\}$, $\ell \geq 2$, and an $\boldsymbol{e} \in \mathbb{S}^1$ such that

$$|F\boldsymbol{e}| = |U_1 \boldsymbol{e}| = \ldots = |U_\ell \boldsymbol{e}| > \max_{j\geq \ell+1} |U_j \boldsymbol{e}|.$$

Additionally, for some $\alpha \in \{1, \ldots, \ell\}$ there exists a vector $\boldsymbol{v} \in \mathbb{S}^1$ with

$$|U_\alpha \boldsymbol{v}| > \max_{j\in\{1,\ldots,k\},\, j\neq\alpha} |U_j \boldsymbol{v}|, \quad \textit{and} \quad \langle U_\alpha \boldsymbol{e}, U_\alpha \boldsymbol{e}^\perp \rangle = \langle F\boldsymbol{e}, F\boldsymbol{e}^\perp \rangle.$$

Remark 4.3.2. The three cases are not exclusive, for example $F = U_i$ satisfies the assumptions in i), ii), and iii).

As in the case of bulk materials, our analysis reveals that uniqueness and stability of polyconvex measures is closely related to simple laminates. We show that $\overline{\nu} \in \mathcal{M}^{\mathrm{pc}}(K; F)$ is a simple laminate if $\overline{\nu}$ satisfies condition $(\widetilde{\mathrm{C}}_\mathrm{b})$. We represent this simple laminate by

$$\overline{\nu} = \overline{\lambda}\delta_{Q_\alpha U_\alpha} + (1-\overline{\lambda})\delta_{Q_\beta U_\beta}, \quad Q_\alpha U_\alpha - Q_\beta U_\beta = \boldsymbol{a} \otimes \boldsymbol{b}$$

with $\overline{\lambda} \in [0,1]$, Q_α, $Q_\beta \in \mathrm{O}(2)$, and α and β not necessarily distinct since the support of $\overline{\nu}$ can be contained in $\mathrm{O}(2)U_\alpha$. Throughout this section, we assume that $\overline{\lambda} \notin \{0,1\}$, i.e., that $\overline{\nu}$ is not a Dirac mass placed at a single matrix. The modifications in the case of a Dirac mass placed at one matrix in K are obvious since all the three cases are included in i).

By Proposition 4.1.3 we may choose as before Borel measurable projections $\pi : \mathbb{M}^{2\times 2} \to K$ with

$$|\pi(X) - X| = \operatorname{dist}(X, K) \quad \text{for all } X \in \mathbb{M}^{2\times 2},$$

and $\Pi : \mathbb{M}^{2\times 2} \to \{Q_\alpha U_\alpha, Q_\beta U_\beta\}$ with

$$|\Pi(X) - X| = \operatorname{dist}\big(X, \{Q_\alpha U_\alpha, Q_\beta U_\beta\}\big) \quad \text{for all } X \in \mathbb{M}^{2\times 2}.$$

Moreover, we may assume that $\Pi(A) \in SO(2)U_\gamma$ if $\pi(A) \in SO(2)Q_\gamma U_\gamma$, for $\gamma = \alpha, \beta$. As before, the rates in the estimates below do not depend on our specific choice. For $\nu \in \mathcal{M}^{pc}(\mathbb{M}^{2\times 2})$ we define the sets (see also Section 4.1)

$$\begin{aligned} M_\ell^\pm &= \{A \in \operatorname{supp}\nu : \pi(A) \in O(2)U_\ell, \det \pi(A) = \pm\Delta\}, && \ell = 1, \ldots, k, \\ \widehat{M}_\gamma &= \{A \in \operatorname{supp}\nu : \Pi(A) \in SO(2)Q_\gamma U_\gamma\}, && \gamma = \alpha, \beta, \\ \widehat{M}_{\gamma,\varrho} &= \{A \in \operatorname{supp}\nu : \Pi(A) \in SO(2)Q_\gamma U_\gamma, |A - Q_\gamma U_\gamma| < \varrho\}, && \gamma = \alpha, \beta, \end{aligned}$$

and the volume fractions

$$\tau_\ell^\pm = \nu(M_\ell^\pm), \qquad \tau_\ell = \tau_\ell^+ + \tau_\ell^-,$$

and

$$\widehat{\lambda} = \int_{\widehat{M}_\alpha} 1 \,d\nu(A), \quad \widehat{\lambda}_\varrho = \int_{\widehat{M}_{\alpha,\varrho}} 1 \,d\nu(A).$$

To simplify the notation, we do not indicate explicitly the dependence on ν in our formulae. Finally, we set

$$\begin{aligned} M_\gamma = \{A \in \operatorname{supp}\nu : \pi(A) \in SO(2)Q_\gamma U_\gamma\}, &\quad \gamma = \alpha, \beta, \\ \sigma_\gamma = \nu(M_\gamma), &\quad \gamma = \alpha, \beta, \end{aligned}$$

and

$$M = M_\alpha \cup M_\beta, \quad N = \operatorname{supp}\nu \setminus M.$$

The energy $\mathcal{E}$ is now given by

$$\mathcal{E}(\nu) = \int_{\operatorname{supp}\nu} \big(\operatorname{dist}(A, K) + \operatorname{dist}^2(A, K)\big) d\nu(A),$$

and we let

$$\eta = \begin{cases} 0 & \text{if condition i) or ii) holds in } (\widetilde{C}_b), \\ 1 & \text{if condition iii) holds in } (\widetilde{C}_b). \end{cases}$$

Proposition 4.3.3. *Suppose that* $\nu \in \mathcal{M}^{pc}(\mathbb{M}^{2\times 2}; F)$ *and that* F *satisfies condition* $(\widetilde{C}_b)$. *Then*

$$\tau \leq c\big(\mathcal{E}(\nu) + \eta\mathcal{E}^{1/2}(\nu)\big) \quad \textit{for} \quad \tau \in \{\tau_1^\pm, \ldots, \tau_k^\pm\} \setminus \{\sigma_\alpha, \sigma_\beta\}.$$

Moreover, $\{\sigma_\alpha, \sigma_\beta\} = \{\tau_\alpha^\pm, \tau_\beta^\pm\}$ *if assumption i) holds in condition* $(\widetilde{\mathrm{C}}_\mathrm{b})$ *with* $\det F = \pm\Delta$ *and* $\{\sigma_\alpha, \sigma_\beta\} = \{\tau_\alpha^\pm\}$ *if assumptions ii) and iii) hold, respectively. Finally,*

$$\int_{\operatorname{supp}\nu} |R(A) - \mathbb{I}|^2 \mathrm{d}\nu(A) \leq c\big(\mathcal{E}(\nu) + \eta\mathcal{E}^{1/2}(\nu)\big).$$

Proof. The estimates are based on three fundamental inequalities. The first inequality was already derived in (4.15) and yields for all $w \in \mathbb{S}^1$ that

$$\sum_{j=1}^k \tau_j\big(|Fw|^2 - |U_j w|^2\big) + \int_{\operatorname{supp}\nu} |\pi(A)w - Fw|^2 \mathrm{d}\nu(A) \leq c\mathcal{E}(\nu). \tag{4.23}$$

The second estimate relies on the fact that the determinant is a null-Lagrangian and that therefore

$$\sum_{j=1}^k \tau_j^\pm(\pm \det U_j) - \det F = \int_{\operatorname{supp}\nu} \big(\det \pi(A) - \det(A)\big)\mathrm{d}\nu(A).$$

We deduce from the expansion

$$\begin{aligned} \big|\det(A) - \det \pi(A)\big| &\leq \big|\det\big(A - \pi(A)\big)\big| + \big|\operatorname{cof}(\pi(A)) : (A - \pi(A))\big| \\ &\leq c\big(|A - \pi(A)|^2 + |A - \pi(A)|\big), \end{aligned}$$

that

$$\Big|\sum_{j=1}^k \tau_j^\pm(\pm \det U_j) - \det F\Big| \leq c\mathcal{E}(\nu). \tag{4.24}$$

The third estimate is in the same spirit as the two foregoing ones, but not based on a null-Lagrangian. Assume that iii) holds in condition $(\widetilde{\mathrm{C}}_\mathrm{b})$. Then

$$\begin{aligned} &\Big|\sum_{j=1}^k \tau_j\langle U_j e, U_j e^\perp\rangle - \langle Fe, Fe^\perp\rangle\Big| \\ =&\Big|\int_{\operatorname{supp}\nu} \Big(|\langle \pi(A)e, \pi(A)e^\perp\rangle - \langle Fe, Ae^\perp\rangle\Big)\mathrm{d}\nu(A)\Big| \\ \leq&\int_{\operatorname{supp}\nu} \big|\langle(\pi(A) - F)e, \pi(A)e^\perp\rangle\big| + \big|\langle Fe, (A - \pi(A))e^\perp\rangle\big|\mathrm{d}\nu(A) \\ \leq& c\Big(\int_{\operatorname{supp}\nu} |(\pi(A) - F)e|^2 \mathrm{d}\nu(A)\Big)^{1/2} + c\int_{\operatorname{supp}\nu} |A - \pi(A)|\mathrm{d}\nu(A). \end{aligned}$$

We obtain from the definition of $\mathcal{E}(\nu)$ and (4.23) that

$$\Big|\sum_{j=1}^{k} \tau_j \big(\langle U_j \boldsymbol{e}, U_j \boldsymbol{e}^\perp\rangle - \langle F\boldsymbol{e}, F\boldsymbol{e}^\perp\rangle\big)\Big| \le c\big(\mathcal{E}(\nu) + \mathcal{E}^{1/2}(\nu)\big).$$

We now distinguish three cases corresponding to the three conditions in $(\widetilde{\mathrm{C}}_\mathrm{b})$.

Case 1: Condition i) holds in $(\widetilde{\mathrm{C}}_\mathrm{b})$. We obtain from (4.23) that $\tau_j \le \mathcal{E}(\nu)$ for $j \notin \{\alpha, \beta\}$ and from (4.24) and the arguments in Section 4.1, see in particular the proof of Theorem 4.1.9, that $\tau_j^\mp \le c\mathcal{E}(\nu)$ if $\det F = \pm\Delta$. This proves the assertion.

Case 2: Condition ii) holds in $(\widetilde{\mathrm{C}}_\mathrm{b})$. In this case we conclude from (4.23) that $\tau_j \le \mathcal{E}(\nu)$ for $j \ne \alpha$.

Case 3: Condition iii) holds in $(\widetilde{\mathrm{C}}_\mathrm{b})$. We first derive some algebraic information about the matrices $U_1, \ldots, U_\ell$ (here ℓ has been defined in the statement of condition $(\widetilde{\mathrm{C}}_\mathrm{b})$; we relabel the matrices again, if necessary). Proposition 2.2.4 ensures the existence of rotations $Q_j \in \mathrm{SO}(2)$ and scalars $\alpha_j \in \mathbb{R}$, $j = 2, \ldots, \ell$ such that

$$Q_j U_j - U_1 = \alpha_j U_1 \boldsymbol{e} \otimes \boldsymbol{e}^\perp, \quad 0 < \alpha_2 < \ldots < \alpha_\ell.$$

In particular, $U_2, \ldots, U_{\ell-1} \in (\mathrm{SO}(2)U_1 \cup \mathrm{SO}(2)U_\ell)^{\mathrm{qc}}$ and

$$\langle U_j \boldsymbol{e}, U_j \boldsymbol{e}^\perp\rangle = \langle Q_j U_j \boldsymbol{e}, Q_j U_j \boldsymbol{e}^\perp\rangle = \langle U_1 \boldsymbol{e}, U_1 \boldsymbol{e}^\perp\rangle + \alpha_j |U_1 \boldsymbol{e}|^2.$$

We now turn to the estimates for the volume fractions. By (4.23)

$$\tau_j \le \mathcal{E}(\nu) \quad \text{for } j = \ell + 1, \ldots, k.$$

The assumption $|U_\alpha \boldsymbol{v}| > \max_{j \ne \alpha} |U_j \boldsymbol{v}|$ implies that $\alpha = 1$ or $\alpha = \ell$. We may assume that $\alpha = 1$. Then $\langle U_j \boldsymbol{e}, U_j \boldsymbol{e}^\perp\rangle > \langle F\boldsymbol{e}, F\boldsymbol{e}^\perp\rangle$ for $j = 2, \ldots, \ell$ and thus

$$\begin{aligned}
&\sum_{j=2}^{\ell} \tau_j \big|\langle U_j \boldsymbol{e}, U_j \boldsymbol{e}^\perp\rangle - \langle F\boldsymbol{e}, F\boldsymbol{e}^\perp\rangle\big| \\
&= \Big|\sum_{j=2}^{\ell} \tau_j \big(\langle U_j \boldsymbol{e}, U_j \boldsymbol{e}^\perp\rangle - \langle F\boldsymbol{e}, F\boldsymbol{e}^\perp\rangle\big)\Big| \\
&\le \Big|\sum_{j=1}^{k} \tau_j \big(\langle U_j \boldsymbol{e}, U_j \boldsymbol{e}^\perp\rangle - \langle F\boldsymbol{e}, F\boldsymbol{e}^\perp\rangle\big)\Big| + c\mathcal{E}(\nu) \\
&\le c\big(\mathcal{E}(\nu) + \mathcal{E}^{1/2}(\nu)\big).
\end{aligned}$$

It remains to prove the estimate for the excess rotations R. This is analogous to the proof of Proposition 4.1.12. By assumption,

$$F\boldsymbol{b}^\perp = Q_\alpha U_\alpha \boldsymbol{b}^\perp = Q_\beta U_\beta \boldsymbol{b}^\perp = \Pi(A)\boldsymbol{b}^\perp \quad \text{for } A \in M$$

and thus by (4.23)

$$\int_M |(R(A)-\mathbb{I})Fb^\perp|^2 \mathrm{d}\nu(A) = \int_M |(\pi(A)-F)b^\perp|^2 \mathrm{d}\nu(A) \le c\mathcal{E}(\nu).$$

Since $|(R(A)-\mathbb{I})(Fb^\perp)^\perp| = |(R(A)-\mathbb{I})Fb^\perp|$ and $\nu(N) \le c(\mathcal{E}(\nu)+\mathcal{E}^{1/2}(\nu))$, the assertion follows easily. □

We now state the main result that ensures that condition $(\widetilde{\mathrm{C}}_\mathrm{b})$ is necessary and sufficient for uniqueness.

Theorem 4.3.4. *Let K be given by (4.22). Then $\nu \in \mathcal{M}^{\mathrm{pc}}(K)$ is unique if and only if $F = \langle \nu, id\rangle$ satisfies condition $(\widetilde{\mathrm{C}}_\mathrm{b})$.*

Proof. We assume first that F satisfies condition $(\widetilde{\mathrm{C}}_\mathrm{b})$.

Case 1: Condition i) holds in $(\widetilde{\mathrm{C}}_\mathrm{b})$. We may assume that $\det F = \Delta$, and thus we are in the setting of the bulk materials in Section 4.1; the uniqueness follows as in Theorem 4.1.11.

Case 2: Condition ii) holds in $(\widetilde{\mathrm{C}}_\mathrm{b})$. It follows from $\mathcal{E}(\nu) = 0$ and Proposition 4.3.3 that $\operatorname{supp}\nu \in \mathrm{O}(2)U_\alpha$. Moreover,

$$|Fe| = \left|\int_{\mathrm{O}(2)U_\alpha} Ae\mathrm{d}\nu(A)\right| \le \int_{\mathrm{O}(2)U_\alpha} |Ae|\mathrm{d}\nu(A) \le |U_\alpha e|,$$

and therefore in view of the assumption $|Fe| = |U_\alpha e|$

$$0 \le \int_{\operatorname{supp}\nu} |Fe - Ae|^2 \mathrm{d}\nu(A) \le 0.$$

We deduce that $Ae = Fe$ for almost all $A \in \operatorname{supp}\nu$. There are exactly two matrices $Q^\pm \in \mathrm{O}(2)$ with $Q^\pm U_\alpha e = Fe$ and consequently

$$\nu = \lambda\delta_{Q^+U_\alpha} + (1-\lambda)\delta_{Q^-U_\alpha}.$$

The minors relation $\det F = 2\lambda - 1$ uniquely determines λ and hence ν is unique.

Case 3: Condition iii) holds in $(\widetilde{\mathrm{C}}_\mathrm{b})$. This is analogous to Case 2 since by Proposition 4.3.3 the measure ν is supported on $\mathrm{O}(2)U_\alpha$.

We now turn to the converse implication that uniqueness of ν implies the validity of condition $(\widetilde{\mathrm{C}}_\mathrm{b})$ for $F = \langle\nu, id\rangle$. By the splitting method, ν cannot be an unconstrained point. We may thus assume that F is a constrained point. For $|\det F| = \Delta$ the argument is identical to one in the proof of Theorem 4.2.1 for bulk materials. Suppose now that $|\det F| < \Delta$, and that there exists an $\ell \in \{1,\ldots,k\}$, $\ell \ge 2$, and an $e \in \mathbb{S}^1$ such that

$$|Fe| = |U_1 e| = \ldots = |U_\ell e| > \max_{j\ge\ell+1} |U_j e|.$$

Note that we only need to consider the case $\ell \ge 2$ since $\ell = 1$ corresponds to case ii) in condition $(\widetilde{\mathrm{C}}_\mathrm{b})$. We argue by contradiction. Thus assume that there does not exist an $i \in \{1,\ldots,\ell\}$ and a $v \in \mathbb{S}^1$ such that

$$|U_i v| > \max_{j \neq i} |U_j v|, \quad \text{and} \quad \langle U_i e, U_i e^\perp \rangle = \langle F e, F e^\perp \rangle. \tag{4.25}$$

As in the proof of Proposition 2.2.4 we find $Q_j \in \mathrm{SO}(2)U_j$ and $\alpha_j \in \mathbb{R}$ for $j = 2, \ldots, \ell$ with $0 < \alpha_2 < \ldots < \alpha_\ell$ and

$$Q_j U_j - U_1 = \alpha_j U_1 e \otimes e^\perp, \ \langle U_j e, U_j e^\perp \rangle = \langle U_1 e, U_1 e^\perp \rangle + \alpha_j |U_1 e|^2. \tag{4.26}$$

Finally there exists $Q_F \in \mathrm{SO}(2)$ and $\alpha_F \in \mathbb{R}$ such that

$$Q_F F - U_1 = \alpha_F U_1 e \otimes e^\perp, \qquad \langle F e, F e^\perp \rangle = \langle U_1 e, U_1 e^\perp \rangle + \alpha_F |U_1 e|^2.$$

This implies $U_2, \ldots, U_{\ell-1} \in (\mathrm{SO}(2)U_1 \cup \mathrm{SO}(2)U_\ell)^{\mathrm{qc}}$, and $\alpha_F \in (0, \alpha_\ell)$. Indeed, if $\alpha_F = 0$ or $\alpha_F = \alpha_\ell$, then

$$\langle F e, F e^\perp \rangle = \langle U_1 e, U_1 e^\perp \rangle \text{ or } \langle F e, F e^\perp \rangle = \langle U_\ell e, U_\ell e^\perp \rangle,$$

respectively. Since U_1 and U_ℓ are corners of the set K in the sense of Theorem 2.2.3, we conclude that U_1 or U_ℓ satisfy (4.25), contradicting our assumption that none of the matrices satisfies this condition. The goal is now to construct a one-parameter family of polyconvex measures supported on $\mathrm{SO}(2)U_1 \cup \mathrm{SO}(2)U_\ell$ with center of mass F. There are unique rotations $Q_1^\pm, Q_\ell^\pm \in \mathrm{O}(2)$ such that $Q_1^\pm U_1 e = F e$ and $Q_\ell^\pm U_1 e = F e$. Moreover, for all $\lambda_1^\pm, \lambda_\ell^\pm \in [0,1]$ we find vectors a, b, $c \in \mathbb{R}^2$ with

$$Q_1^+ U_1 - Q_1^- U_1 = a \otimes e^\perp, \qquad Q_\ell^+ U_\ell - Q_\ell^- U_\ell = b \otimes e^\perp,$$

and

$$V_1 - V_\ell = c \otimes e^\perp,$$

where

$$V_j = \left(\frac{\lambda_j^+}{\lambda_j^+ + \lambda_j^-} Q_j^+ U_j + \frac{\lambda_j^-}{\lambda_j^+ + \lambda_j^-} Q_j^- U_j \right), \qquad j = 1, \ell.$$

Consequently,

$$\nu = \lambda_1^+ \delta_{Q_1^+ U_1} + \lambda_1^- \delta_{Q_1^- U_1} + \lambda_\ell^+ \delta_{Q_\ell^+ U_\ell} + \lambda_\ell^- \delta_{Q_\ell^- U_\ell}$$

is a polyconvex measure. We now assert that

$$\begin{aligned} \mathcal{M}^{\mathrm{pc}}(F) = \{ \mu = {} & \lambda_1^+ \delta_{Q_1^+ U_1} + \lambda_1^- \delta_{Q_1^- U_1} + \lambda_\ell^+ \delta_{Q_\ell^+ U_\ell} + \lambda_\ell^- \delta_{Q_\ell^- U_\ell} : \\ & \lambda_1^\pm, \lambda_\ell^\pm \in [0,1], \lambda_1^+ + \lambda_1^- + \lambda_\ell^+ + \lambda_\ell^- = 1, \\ & (\lambda_1^+ + \lambda_\ell^+)\delta - (\lambda_1^- + \lambda_\ell^-)\delta = \det F, \\ & (\lambda_1^+ + \lambda_1^-)\langle U_1 e, U_1 e^\perp \rangle + (\lambda_\ell^+ + \lambda_\ell^-)\langle U_\ell e, U_\ell e^\perp \rangle = \langle F e, F e^\perp \rangle \} \end{aligned}$$

is a one-parameter family of measures. By construction, all measures are second laminates and thus polyconvex measures. We only need to show that $\langle \nu, id \rangle = F$. Let $X = \langle \nu, id \rangle$. Then

$$Xe = Fe, \quad \langle Xe^\perp, Fe \rangle = \langle Fe^\perp, Fe \rangle, \quad \det X = \det F,$$

and consequently

$$\langle Xe^\perp, (Fe)^\perp \rangle = \langle Xe^\perp, (Xe)^\perp \rangle = \det X = \det F = \langle Fe^\perp, (Fe)^\perp \rangle.$$

We obtain that $Xe^\perp = Fe^\perp$ and hence $X = F$. The assertion follows now if we can solve the linear system

$$\begin{pmatrix} \delta & \delta & -\delta & -\delta \\ \mu_1 & \mu_\ell & \mu_1 & \mu_\ell \\ 1 & 1 & 1 & 1 \end{pmatrix} \begin{pmatrix} \lambda_1^+ \\ \lambda_\ell^+ \\ \lambda_1^- \\ \lambda_\ell^- \end{pmatrix} = \begin{pmatrix} \det F \\ \mu_F \\ 1 \end{pmatrix}$$

where we used the notation $\mu_x = \langle U_x e, U_x e^\perp \rangle$ for $x \in \{1, \ell, F\}$. The augmented matrix of this system is row equivalent to

$$\begin{pmatrix} 1 & 0 & 0 & -1 & 1 - \gamma_1 - \gamma_2 \\ 0 & 1 & 0 & 1 & \gamma_1 \\ 0 & 0 & 1 & 1 & \gamma_2 \end{pmatrix}$$

with

$$\gamma_1 = \frac{\mu_F - \mu_1}{\mu_\ell - \mu_1} \in (0,1), \quad \gamma_2 = \frac{1}{2}\,(1 - \frac{\det F}{\Delta}) \in (0,1),$$

and the general solution is given by

$$s \mapsto \begin{pmatrix} 1 - \gamma_1 - \gamma_2 \\ \gamma_1 \\ \gamma_2 \\ 0 \end{pmatrix} + s \begin{pmatrix} 1 \\ -1 \\ -1 \\ 1 \end{pmatrix}.$$

The solutions $\lambda_1^\pm$, $\lambda_\ell^\pm$ satisfy the constraints $\lambda_1^\pm, \lambda_\ell^\pm \in [0,1]$ for

$$s \in [\,\max\{0, \gamma_1 + \gamma_2 - 1\}, \min\{\gamma_1, \gamma_2\}\,]$$

and this concludes the proof of the theorem. □

4.4 Applications to Thin Films

In this section, we extend the results presented so far for bulk materials to thin films. If the normal to the thin film is suitably oriented with respect to the crystallographic directions in the material, then the set K is given by

$$K = \mathrm{O}(2,3)U_1 \cup \ldots \cup \mathrm{O}(2,3)U_k \tag{4.27}$$

where $U_1, \ldots, U_\ell \in \mathbb{M}^{2\times 2}$ satisfy $\det U_i = \Delta > 0$ for $i = 1, \ldots, k$. We recall from Section 2.3 that all $F \in K^{\mathrm{qc}}$ can be written as

$$F = Q\widehat{\pi}(\widehat{F}) \quad \text{with } Q \in \mathrm{SO}(3),\ \widehat{F} \in \big(\mathrm{SO}(2)U_1 \cup \ldots \cup \mathrm{SO}(2)U_k\big)^{\mathrm{qc}}$$

and $\widehat{\pi} : \mathbb{M}^{2\times 2} \to \mathbb{M}^{3\times 2}$ defined by

$$\widehat{F} = \begin{pmatrix} F_{11} & F_{12} \\ F_{21} & F_{22} \end{pmatrix} \mapsto \widehat{\pi}(\widehat{F}) = \begin{pmatrix} F_{11} & F_{12} \\ F_{21} & F_{22} \\ 0 & 0 \end{pmatrix}.$$

The results in this section show that as is the case of the two-dimensional theories in Sections 4.2 and 4.3 uniqueness and stability are equivalent for thin films. The main difference is the at first sight surprising result that the microstructures underlying globally affine deformations $F \in K^{\mathrm{qc}}$ are not unique unless F is area preserving. This is a consequence of the extraordinarily rich folding patterns for thin films in three dimensions and can be nicely illustrated by the following example.

Let $U_1 = \mathrm{diag}(\beta, \alpha)$ and $F = \widehat{\pi}(\mathrm{diag}(\beta, t\alpha))$ with $t \in (-1, 1)$. We define $\lambda \in (0,1)$ by $2\lambda - 1 = t$ and $\varphi \in (0, \pi)$ by $\cos\varphi = t$. Then

$$F = \lambda \begin{pmatrix} \beta & 0 \\ 0 & \alpha \\ 0 & 0 \end{pmatrix} + (1-\lambda) \begin{pmatrix} \beta & 0 \\ 0 & -\alpha \\ 0 & 0 \end{pmatrix} = \frac{1}{2} \begin{pmatrix} \beta & 0 \\ 0 & \alpha\cos\varphi \\ 0 & \alpha\sin\varphi \end{pmatrix} + \frac{1}{2} \begin{pmatrix} \beta & 0 \\ 0 & \alpha\cos\varphi \\ 0 & -\alpha\sin\varphi \end{pmatrix},$$

and we find two laminates supported on K with center of mass F since

$$\begin{pmatrix} \beta & 0 \\ 0 & \alpha\cos\varphi \\ 0 & \pm\alpha\sin\varphi \end{pmatrix} = \begin{pmatrix} 1 & 0 \\ 0 & \cos\varphi \\ 0 & \pm\sin\varphi \end{pmatrix} U_1 \in \mathrm{O}(2,3)U_1.$$

The first construction is a two-dimensional one, but the second construction is the limit of genuinely three-dimensional folding patterns. This demonstrates that the behavior of thin films is qualitatively different from the two-dimensional setting in which

$$\begin{pmatrix} \beta & 0 \\ 0 & t\alpha \end{pmatrix} \in \left(\mathrm{O}(2,3)U_1\right)^{\mathrm{qc}}$$

determines a unique microstructure.

We begin our analysis with the definition of condition ($\mathrm{C_{tf}}$) for thin films which replaces condition ($\mathrm{C_b}$) for bulk materials.

Definition 4.4.1. *Assume that K is given by (4.27). Then $F \in K^{\mathrm{pc}}(K)$ is said to satisfy condition* ($\mathrm{C_{tf}}$) *if* $\det(F^TF) = \Delta^2$ *and if there exists an* $\mathbf{e} \in \mathbb{S}^1$ *and* $\alpha, \beta \in \{1,\ldots,k\}$ *such that*

$$|F\mathbf{e}| = |U_\alpha \mathbf{e}| = |U_\beta \mathbf{e}| > \max_{j\in\{1,\ldots,k\}\setminus\{\alpha,\beta\}} |U_j\mathbf{e}|.$$

We define Borel measurable projections π and Π as well as the volume fractions with the obvious modifications analogously to (4.4)-(4.8). More precisely, if $\pi(A) \in \mathrm{O}(2,3)U_\gamma$, then we define the excess rotation in SO(3) in the following way: let $S = \pi(A)Q_\gamma^{-1}$, then $R(A)$ is the matrix with the columns

$$R(A) = (S\widehat{\mathbf{e}}_1, S\widehat{\mathbf{e}}_2, S\widehat{\mathbf{e}}_1 \times S\widehat{\mathbf{e}}_2).$$

In this section, we define the energy $\mathcal{E}$ by

$$\mathcal{E}(\nu) = \int_{\operatorname{supp}\nu} \left(\operatorname{dist}(A,K) + \operatorname{dist}^2(A,K)\right)\mathrm{d}\nu(A).$$

Proposition 4.4.2. *Suppose that* $\nu \in \mathcal{M}^{\mathrm{pc}}(\mathbb{M}^{3\times 2})$ *and that* $F = \langle \nu, id\rangle$ *satisfies condition* ($\mathrm{C_{tf}}$)*. Then*

$$\tau_\ell \le c\mathcal{E}(\nu) \quad \textit{for } \ell \in \{1,\ldots,k\}\setminus\{\alpha,\beta\},$$

and

$$\int_{\operatorname{supp}\nu} |R(A) - \mathbb{I}|^2 \mathrm{d}\nu(A) \le c\left(\mathcal{E}(\nu) + \mathcal{E}^{1/2}(\nu)\right).$$

Moreover, the microstructure underlying F is unique, i.e.,

$$\mathcal{M}^{\mathrm{pc}}(K;F) = \{\lambda\delta_{Q\widehat{\pi}(Q_\alpha U_\alpha)} + (1-\lambda)\delta_{Q\widehat{\pi}(Q_\beta U_\beta)}$$

with $Q \in \mathrm{SO}(3)$, Q_α, $Q_\beta \in \mathrm{SO}(2)$ *such that* $\operatorname{rank}(Q_\alpha U_\alpha - Q_\beta U_\beta) = 1$, *and* $\lambda \in [0,1]$.

Proof. We may assume that $F = \widehat{\pi}(\widehat{F})$ with $\widehat{F} \in \mathbb{M}^{2\times 2}$ and $\det \widehat{F} = \Delta$. As in the proof of Proposition 4.1.9,

$$\sum_{\ell=1}^{k} \tau_\ell\left(|F\mathbf{w}|^2 - |U_\ell \mathbf{w}|^2\right) + \int_{\operatorname{supp}\nu} |(\pi(A) - F)\mathbf{w}|^2 \mathrm{d}\nu(A) \le c\mathcal{E}(\nu),$$

and the bounds on the volume fractions follow from (C_{tf}). If $\overline{\nu} \in \mathcal{M}^{pc}(K;F)$, then $\mathcal{E}(\overline{\nu}) = 0$ and $\overline{\nu}$ is supported on $O(2,3)U_\alpha \cup O(2,3)U_\beta$. We first show that

$$\operatorname{supp}\overline{\nu} \subseteq \widehat{\pi}(SO(2)U_\alpha \cup SO(2)U_\beta). \tag{4.28}$$

Let $\operatorname{adj}_{ij}(A)$ denote the determinant of the 2×2 matrix formed by the ith and the jth row in $A \in \mathbb{M}^{3\times 2}$. Then

$$\Delta = \operatorname{adj}_{12}(F) \leq \int_{\operatorname{supp}\overline{\nu}} |\operatorname{adj}_{12}(A)| \mathrm{d}\overline{\nu}(A) \leq \Delta,$$

and thus $\operatorname{adj}_{12}(A) = \Delta$ for almost all $A \in \operatorname{supp}\overline{\nu}$. A short calculation shows that

$$\det(F^TF) = \operatorname{adj}_{12}^2(F) + \operatorname{adj}_{23}^2(F) + \operatorname{adj}_{13}^2(F) \quad \text{for all } F \in \mathbb{M}^{3\times 2},$$

and hence $F \mapsto \det(F^TF)$ is a polyconvex function. We obtain

$$\Delta^2 = \det(F^TF) \leq \int_{\operatorname{supp}\overline{\nu}} \left(\operatorname{adj}_{12}^2(A) + \operatorname{adj}_{23}^2(A) + \operatorname{adj}_{13}^2(A)\right) \mathrm{d}\overline{\nu}(A) \leq \Delta^2,$$

and consequently $\operatorname{adj}_{13}(A) = \operatorname{adj}_{23}(A) = 0$ for almost all $A \in \operatorname{supp}\overline{\nu}$. We deduce that for almost all $A \in \operatorname{supp}\overline{\nu}$ the third row of A has to be parallel to the first and the second row and therefore the third row must be equal to the zero vector. This establishes (4.28). The analysis has therefore been reduced to the case of two-dimensional bulk materials and we conclude by Theorem 4.1.1 that the polyconvex measures are unique,

$$\overline{\nu} = \overline{\lambda}\delta_{\widehat{\pi}(Q_\alpha U_\alpha)} + (1-\overline{\lambda})\delta_{\widehat{\pi}(Q_\beta U_\beta)}, \qquad Q_\alpha U_\alpha - Q_\beta U_\beta = \boldsymbol{a}\otimes\boldsymbol{b} \tag{4.29}$$

with $\boldsymbol{a}, \boldsymbol{b} \in \mathbb{R}^2$, $\overline{\lambda} \in [0,1]$ and $Q_\gamma \in SO(2)$ for $\gamma = \alpha, \beta$.

We now turn to the proof of the estimate for the excess rotation $R(A)$. For simplicity, we frequently write R instead of $R(A)$. We write $\boldsymbol{e}_i$ for the standard basis in $\mathbb{R}^3$ and $\widehat{\boldsymbol{e}}_i$ for the standard basis in $\mathbb{R}^2$. Let $\widehat{Q} \in SO(2)$ be the rotation with $\widehat{Q}\widehat{F}\boldsymbol{b}^\perp = |\widehat{F}\boldsymbol{b}^\perp|\widehat{\boldsymbol{e}}_1$, and define $Q = \operatorname{diag}(\widehat{Q},1) \in SO(3)$. As in the proof of Proposition 4.1.12, it suffices to estimate $|\widetilde{R} - \mathbb{I}|$ where $\widetilde{R} = QRQ^T$. By definition, $\widehat{F}\boldsymbol{b}^\perp = Q_\alpha U_\alpha \boldsymbol{b}^\perp = Q_\beta U_\beta \boldsymbol{b}^\perp$ and thus we deduce that $R(A)F\boldsymbol{b}^\perp = \pi(A)\boldsymbol{b}^\perp$ on M and that

$$\begin{aligned} |\widehat{F}\boldsymbol{b}^\perp|^2 &\int_M |(\widetilde{R}(A) - \mathbb{I})\boldsymbol{e}_1|^2 \mathrm{d}\nu(A) \\ &= \int_M |(QR(A)Q^T - \mathbb{I})QF\boldsymbol{b}^\perp|^2 \mathrm{d}\nu(A) \\ &= \int_M |(R(A) - \mathbb{I})F\boldsymbol{b}^\perp|^2 \mathrm{d}\nu(A) \\ &= \int_M |(\pi(A) - F)\boldsymbol{b}^\perp|^2 \mathrm{d}\nu(A) \leq \mathcal{E}(\nu). \end{aligned} \tag{4.30}$$

Since $\widetilde{R}e_1 \in \mathbb{S}^2$, this estimate can be improved to an L^1 estimate for $\widetilde{R}_{11} - 1$. In fact,

$$1 - \widetilde{R}_{11} = \frac{1}{2}\big[(\widetilde{R}_{11} - 1)^2 + 1 - \widetilde{R}_{11}^2\big] = \frac{1}{2}\big[(\widetilde{R}_{11} - 1)^2 + \widetilde{R}_{21}^2 + \widetilde{R}_{31}^2\big],$$

and thus by (4.30)

$$\int_M |1 - \widetilde{R}_{11}| \mathrm{d}\nu(A) \leq \tfrac{1}{2} \int_M |(\widetilde{R} - \mathbb{I})e_1|^2 \mathrm{d}\nu(A) \leq c\mathcal{E}^{1/2}(\nu). \qquad (4.31)$$

The crucial estimate that requires some care is the estimate for $(\widetilde{R} - \mathbb{I})e_2$. We suppress the dependence on A in the following calculations. Since Q is a block diagonal matrix,

$$\begin{aligned}\widetilde{R}_{11}\widetilde{R}_{22} - \widetilde{R}_{21}\widetilde{R}_{12} &= \operatorname{cof}_{33}(\widetilde{R}) \\ &= \det\left(\widehat{Q}\begin{pmatrix} R_{11} & R_{12} \\ R_{21} & R_{22}\end{pmatrix}\widehat{Q}^T\right) \\ &= (\det \widehat{Q})^2\big(R_{11}R_{22} - R_{21}R_{12}\big),\end{aligned}$$

and

$$\begin{aligned}\operatorname{adj}_{12}(\pi(A)) &= \det\left(\begin{pmatrix} R_{11} & R_{12} \\ R_{21} & R_{22}\end{pmatrix}\widehat{Q}_\gamma U_\gamma\right) \\ &= \big(R_{11}R_{22} - R_{12}R_{21}\big)\Delta \quad \text{on } M.\end{aligned}$$

Hence we infer from $\det \widehat{Q} = 1$ that

$$\widetilde{R}_{11}\widetilde{R}_{22} - \widetilde{R}_{21}\widetilde{R}_{12} = \frac{1}{\Delta}\operatorname{adj}_{12}(\pi(A)) \quad \text{on } M,$$

and hence

$$\operatorname{adj}_{12} F = \Delta \geq \Delta R_{33} = \Delta\big(R_{11}R_{22} - R_{12}R_{21}\big) = \operatorname{adj}_{12}\pi(A).$$

We obtain from the fact that $\operatorname{adj}_{12}(\cdot)$ is a null-Lagrangian that

$$\begin{aligned}&\int_M |\operatorname{adj}_{12} F - \operatorname{adj}_{12}\pi(A)| \mathrm{d}\nu(A) \\ &= \int_{\operatorname{supp}\nu} \big(\operatorname{adj}_{12} F - \operatorname{adj}_{12}\pi(A)\big)\mathrm{d}\nu(A) - \int_N \big(\operatorname{adj}_{12} F - \operatorname{adj}_{12}\pi(A)\big)\mathrm{d}\nu(A) \\ &= \int_{\operatorname{supp}\nu} \big(\operatorname{adj}_{12} A - \operatorname{adj}_{12}\pi(A)\big)\mathrm{d}\nu(A) - \int_N |\operatorname{adj}_{12} F - \operatorname{adj}_{12}\pi(A)|\mathrm{d}\nu(A) \\ &\leq \int_{\operatorname{supp}\nu} |\operatorname{adj}_{12} A - \operatorname{adj}_{12}\pi(A)|\mathrm{d}\nu(A) + c\sum_{\ell\notin\{\alpha,\beta\}} \tau_\ell.\end{aligned}$$

We have by (C.3) for all $A, B \in \mathbb{M}^{2\times 2}$ that

$$\begin{aligned}\det A - \det B &= \det\big((A-B)+B\big) - \det B\\ &= \det(A-B) + \operatorname{cof}(A-B) : B,\end{aligned}$$

and we therefore conclude that

$$\begin{aligned}&\int_M |\operatorname{adj}_{12} F - \operatorname{adj}_{12}\pi(A)| \mathrm{d}\nu(A)\\ &\quad\le \int_{\operatorname{supp}\nu} \big(|A-\pi(A)|^2 + |\pi(A)|\,|A-\pi(A)|\big)\mathrm{d}\nu(A) + c\mathcal{E}(\nu) \qquad (4.32)\\ &\quad\le c\mathcal{E}(\nu).\end{aligned}$$

In view of

$$1 - (\widetilde{R}_{11}\widetilde{R}_{22} - \widetilde{R}_{12}\widetilde{R}_{21}) = (1-\widetilde{R}_{22}) + \widetilde{R}_{22}(1-\widetilde{R}_{11}) + \widetilde{R}_{12}\widetilde{R}_{21},$$

we infer

$$\begin{aligned}1 - \widetilde{R}_{22} &= -\widetilde{R}_{22}(1-\widetilde{R}_{11}) - \widetilde{R}_{12}\widetilde{R}_{21} + 1 - (\widetilde{R}_{11}\widetilde{R}_{22} - \widetilde{R}_{12}\widetilde{R}_{21})\\ &= -\widetilde{R}_{22}(1-\widetilde{R}_{11}) - \widetilde{R}_{12}\widetilde{R}_{21} + \frac{1}{\Delta}\big(\operatorname{adj}_{12}(F) - \operatorname{adj}_{12}\pi(A)\big).\end{aligned}$$

We conclude in view of (4.32) and (4.31) that

$$\begin{aligned}\int_M |\widetilde{R}_{22} - 1| \mathrm{d}\nu(A) &\le \frac{1}{\Delta}\int_M |\operatorname{adj}_{12} F - \operatorname{adj}_{12}\pi(A)|\mathrm{d}\nu(A)\\ &\quad + \int_M \Big(|\widetilde{R}_{11} - 1| + \frac{1}{4}\widetilde{R}_{12}^2 + \widetilde{R}_{21}^2\Big)\mathrm{d}\nu(A) \qquad (4.33)\\ &\le c\big(\mathcal{E}(\nu) + \mathcal{E}^{1/2}(\nu)\big) + \int_M \big(\tfrac{1}{4}\widetilde{R}_{12}^2 + \widetilde{R}_{21}^2\big)\mathrm{d}\nu(A).\end{aligned}$$

On the other hand, $\widetilde{R}e_2 \in \mathbb{S}^2$, and therefore

$$\begin{aligned}\int_M (\widetilde{R}_{12}^2 + \widetilde{R}_{32}^2)\mathrm{d}\nu(A) &= \int_M (1-\widetilde{R}_{22}^2)\mathrm{d}\nu(A) \le 2\int_M |1-\widetilde{R}_{22}|\mathrm{d}\nu(A)\\ &\le c\big(\mathcal{E}(\nu) + \mathcal{E}^{1/2}(\nu)\big) + \int_M \big(\tfrac{1}{2}\widetilde{R}_{12}^2 + 2\widetilde{R}_{21}^2\big)\mathrm{d}\nu(A).\end{aligned}$$

The term involving $\widetilde{R}_{12}^2$ on the right hand side can now be absorbed on the left hand side and we infer in view of (4.30)

$$\int_M \big(\tfrac{1}{2}\widetilde{R}_{12}^2 + \widetilde{R}_{32}^2\big)\mathrm{d}\nu(A) \le c\big(\mathcal{E}(\nu) + \mathcal{E}^{1/2}(\nu)\big).$$

This in turn implies, again in view of (4.33) that

$$\int_M |\widetilde{R}_{22} - 1|^2 \mathrm{d}\nu(A) \leq 2 \int_M |\widetilde{R}_{22} - 1| \mathrm{d}\nu(A) \leq c(\mathcal{E}(\nu) + \mathcal{E}^{1/2}(\nu)),$$

and

$$\int_M |(\widetilde{R} - \mathbb{I})e_2|^2 \mathrm{d}\nu(A) \leq c(\mathcal{E}(u) + \mathcal{E}^{1/2}(u)).$$

Finally,

$$\begin{aligned}
\int_M |(\widetilde{R} - \mathbb{I})e_3|^2 \mathrm{d}\nu(A) &= \int_M |\widetilde{R}e_1 \times \widetilde{R}e_2 - e_1 \times e_2|^2 \mathrm{d}\nu(A) \\
&= \int_M |(\widetilde{R}e_1 - e_1) \times \widetilde{R}e_2 + e_1 \times (\widetilde{R}e_2 - e_2)|^2 \mathrm{d}\nu(A) \\
&\leq c \int_M (|\widetilde{R}e_1 - e_1|^2 + |\widetilde{R}e_2 - e_2|^2) \mathrm{d}\nu(A) \\
&\leq c(\mathcal{E}(\nu) + \mathcal{E}^{1/2}(\nu)).
\end{aligned}$$

This concludes the proof of the estimate for the excess rotation. □

Theorem 4.4.3. *Let K be given by (4.27) and $\nu \in \mathcal{M}^{\mathrm{pc}}(K)$. Then*

$$F = \langle \nu, id \rangle \text{ satisfies } (\mathrm{C_{tf}}) \quad \Leftrightarrow \quad \nu \text{ is unique.}$$

Proof. We already proved in Proposition 4.4.2 that ν is unique if F satisfies condition ($\mathrm{C_{tf}}$). Assume thus that ν is unique and let $F = \langle \nu, id \rangle$. We may suppose that $F = \widetilde{\pi}(\widehat{F})$ with $\widehat{F} \in \mathbb{M}^{2\times 2}$. It follows from the discussion at the beginning of this section that $\det(F^T F) = \Delta^2$ and we may assume that $\det \widehat{F} = \Delta$. The proof of Proposition 4.4.1 shows that

$$\operatorname{supp} \nu \subseteq \widetilde{\pi}(\widehat{K}) \quad \text{with} \quad \widehat{K} = \mathrm{SO}(2)U_\alpha \cup \mathrm{SO}(2)U_\beta,$$

and we are therefore in the two-dimensional situation. In this case, condition ($\mathrm{C_{tf}}$) is equivalent to ($\mathrm{C_b}$) for $\widehat{F}$ and $\widehat{K}$, and the assertion follows from Theorem 4.2.1. □

Based on the estimates for the volume fractions and the excess rotation in Proposition 4.4.2, it is easy to deduce the following stability result.

Corollary 4.4.4. *Let K be given by (4.27) and $\nu \in \mathcal{M}^{\mathrm{qc}}(K)$. Then*

$$F = \langle \nu, id \rangle \text{ satisfies } (\mathrm{C_{tf}}) \quad \Leftrightarrow \quad \nu \text{ is unique} \quad \Leftrightarrow \quad \nu \text{ is stable.}$$

4.5 Applications to Finite Element Minimizers

In this section, we discuss applications of the uniqueness and stability results to finite element methods for nonconvex variational problems. Here we focus on bulk materials since the adaption of the techniques to the case of thin films is straight forward.

Suppose that Ω is a polygonal domain and that we want to minimize

$$\mathcal{J}(u) = \frac{1}{|\Omega|}\int_\Omega W(Du)\mathrm{d}x \tag{4.34}$$

in the class of admissible functions

$$\mathcal{A}_F = \{u \in W^{1,\max\{2,n-1\}}(\Omega;\mathbb{R}^n) : u(x) = Fx \text{ on } \partial\Omega\}.$$

We assume furthermore that the zero set K of W has a multi-well structure (4.3) and that W satisfies the coercivity condition

$$W(X) \geq \kappa\big(\operatorname{dist}(X,K) + \operatorname{dist}^{\max\{2,n-1\}}(X,K)\big), \quad \kappa > 0. \tag{4.35}$$

The simplest finite element method for the numerical solution of the minimization problem is obtained by choosing a triangulation $\mathcal{T}_h$ of Ω and by minimizing $\mathcal{J}$ in $\mathcal{A}_F \cap \mathcal{S}_h$ where $\mathcal{S}_h$ is the space of all continuous functions that are piecewise affine on the elements in $\mathcal{T}_h$. The coercivity assumption (4.35) implies the existence of a finite element minimizer $u_h \in \mathcal{S}_h$ and the goal of the numerical analysis is to describe the qualitative behavior of u_h. In order to obtain rigorous estimates, one needs uniqueness of the minimizer and this leads naturally to the situation in Section 4.1. We thus assume that F satisfies condition ($\mathrm{C_b}$). Consequently, the infimum of $\mathcal{J}$ is not attained in $\mathcal{A}_F$, but the underlying microstructure $\overline{\nu}$ is unique by Theorem 4.1.11. The question is therefore whether the oscillations in the finite element minimizer Du_h have the statistics recorded in $\overline{\nu}$. In order to compare Du_h and $\overline{\nu}$ we follow Collins, Kinderlehrer, and Luskin and associate with Du_h the gradient Young measure

$$\nu_h = \{\delta_{Du_h(x)}\}_{x\in\Omega}$$

and pass to its average $\operatorname{Av}\nu_h$ which is defined via duality by

$$\begin{aligned}\langle \operatorname{Av}\nu_h, \varphi\rangle &= \langle \operatorname{Av}\delta_{Du_h(\cdot)}, \varphi\rangle = \frac{1}{|\Omega|}\int_\Omega \langle \delta_{Du_h(x)}, \varphi\rangle \mathrm{d}x \\ &= \frac{1}{|\Omega|}\int_\Omega \varphi(Du_h(x))\mathrm{d}x \text{ for all } \varphi \in C_0(\mathbb{M}^{n\times n}).\end{aligned}$$

It turns out that the stability results in Section 4.1 together with some standard interpolation results lead to explicit error estimates for u_h, see Corollary 4.5.2 below. We begin with a more general statement about functions u with small energy that does not require that u be contained in a finite element space. We define for $\gamma = \alpha, \beta$ the sets

$$\begin{aligned}\Omega_\gamma &= \{x \in \Omega : \Pi(Du(x)) \in \mathrm{SO}(3)U_\gamma\}, \\ \Omega_{\gamma,\varrho} &= \{x \in \Omega : \Pi(Du(x)) \in \mathrm{SO}(3)U_\gamma, |Du(x) - Q_\gamma U_\gamma| < \varrho\}.\end{aligned}$$

Theorem 4.5.1. *Let $\delta > 0$ and suppose that F satisfies condition* ($\mathrm{C_b}$). *Assume that $\boldsymbol{u} \in \mathcal{A}_F$ with $\mathcal{J}(\boldsymbol{u}) \leq \delta$. Then there exists a constant c that depends only on F, n, and κ such that the following assertions hold:*
i) (estimates for directional derivatives tangential to the layering direction): for all $\boldsymbol{w} \in \mathbb{S}^{n-1}$ with $\langle \boldsymbol{w}, \boldsymbol{b} \rangle$,

$$\frac{1}{|\Omega|} \int_\Omega \left|(D\boldsymbol{u}(\boldsymbol{x}) - F)\boldsymbol{w}\right|^2 \mathrm{d}\boldsymbol{x} \leq c\delta,$$

ii) (estimates for the deformation)

$$\frac{1}{|\Omega|} \int_\Omega |\boldsymbol{u}(\boldsymbol{x}) - F\boldsymbol{x}| \mathrm{d}\boldsymbol{x} \leq c\delta,$$

iii) (total distance form the wells)

$$\frac{1}{|\Omega|} \int_\Omega |D\boldsymbol{u} - \Pi(A)|^2 \mathrm{d}\boldsymbol{x} \leq c\delta,$$

iv) (estimates for volume fractions)

$$\left|\frac{|\Omega_{\alpha,\varrho}|}{|\Omega|} - \overline{\lambda}\right| \leq \frac{c}{\varrho}\,\delta^{1/2}, \quad \left|\frac{|\Omega_\alpha|}{|\Omega|} - \overline{\lambda}\right| \leq c\delta^{1/2},$$

v) (weak convergence, averages): the following estimate holds for all $\omega \subset \Omega$ with Lipschitz boundary,

$$\left|\int_\omega (D\boldsymbol{u} - F)\mathrm{d}\boldsymbol{x}\right| \leq c(\omega, \Omega)\delta^{1/2}.$$

vi) (weak convergence, functions): for all $f \in W^{1,\infty}(\mathbb{M}^{n\times n})$

$$\left|\frac{1}{|\Omega|} \int_\Omega \big(f(D\boldsymbol{u}) - [\lambda f(Q_\alpha U_\alpha) + (1-\lambda) f(Q_\beta U_\beta)]\big)\mathrm{d}\boldsymbol{x}\right| \leq c\|f\|_{1,\infty}\delta^{1/2}.$$

Proof. Let $\nu = \operatorname{Av} \delta_{D\boldsymbol{u}(\boldsymbol{x})}$ be the homogeneous gradient Young measure associated with ν. Then

$$\begin{aligned}
\mathcal{E}(\nu) &= \int_{\operatorname{supp}\nu} \big(\operatorname{dist}(A,K) + \operatorname{dist}^{\max\{2,n-1\}}(A,K)\big)\mathrm{d}\nu(A) \\
&= \frac{1}{|\Omega|} \int_\Omega \big(\operatorname{dist}(D\boldsymbol{u},K) + \operatorname{dist}^{\max\{2,n-1\}}(D\boldsymbol{u},K)\big)\mathrm{d}\boldsymbol{x} \qquad (4.36)\\
&\leq \frac{1}{\kappa}\frac{1}{|\Omega|} \int_\Omega W(D\boldsymbol{u})\mathrm{d}\boldsymbol{x} \leq \frac{\delta}{\kappa}.
\end{aligned}$$

Proof of i) By statement i) in Theorem 4.1.13,

$$\frac{1}{|\Omega|} \int_\Omega |(D\boldsymbol{u} - F)\boldsymbol{w}|^2 \mathrm{d}\boldsymbol{x} = \int_{\operatorname{supp}\nu} |A\boldsymbol{w} - F\boldsymbol{w}|^2 \mathrm{d}\nu(A) \leq c\mathcal{E}(\nu) \leq c\delta.$$

Proof of ii) Since Ω is bounded in every direction, we may use Poincaré's inequality

$$\int_\Omega |u|^2 \mathrm{d}x \le c \int_\Omega |\langle Du, \boldsymbol{w}\rangle|^2 \mathrm{d}x \quad \text{for all } \boldsymbol{w} \in \mathbb{S}^{n-1},\ u \in W_0^{1,2}(\Omega)$$

for the component functions of $\boldsymbol{u}$.

Proof of iii) This follows from statement ii) in Theorem 4.1.13.

Proof of iv) Let $\chi_\alpha(A) = \chi_{\{\Pi(A)\in \mathrm{SO}(n)U_\alpha\}}$. By definition of $\widehat{\lambda}$,

$$\widehat{\lambda} = \int_{\operatorname{supp}\nu} \chi_\alpha(A)\mathrm{d}\nu(A) = \frac{1}{|\Omega|}\int_\Omega \chi_\alpha(D\boldsymbol{u}(\boldsymbol{x}))\mathrm{d}\boldsymbol{x} = \frac{|\Omega_\alpha|}{|\Omega|},$$

and similarly $\widehat{\lambda}_{\alpha,\varrho} = |\Omega_{\alpha,\varrho}|/|\Omega|$. The assertion is therefore a consequence of statement iii) in Theorem 4.1.13.

Proof of v) This is a consequence of the inequality

$$\Big|\int_\omega Du\,\mathrm{d}\boldsymbol{x}\Big| \le c(\omega)\big\{\|u\|_{L^2(\omega)} + (\|u\|_{L^2(\omega)}\|Du\|_{L^2(\omega)})^{1/2}\big\}, \quad u \in W^{1,2}(\omega),$$

which follows from embedding theorems, see also (4.42) below.

Proof of vi) This follows from statement iv) in Theorem 4.1.13. □

The foregoing general theorem implies the following explicit error estimate for finite element minimizers.

Corollary 4.5.2. *Let Ω be a polygonal domain and $\mathcal{T}_h$ a regular triangulation in the sense of Ciarlet. Assume that F satisfies the condition* ($\mathrm{C_b}$) *and that $\boldsymbol{u}_h$ is a minimizer of $\mathcal{J}$ in $\mathcal{A}\cap\mathcal{S}_h$. Then there exists a constant c which depends only on F, n, and the shape of the triangles in $\mathcal{T}_h$, but not on h, such that*

$$\frac{1}{|\Omega|}\int_\Omega |\boldsymbol{u}_h(\boldsymbol{x}) - F\boldsymbol{x}|^2 \mathrm{d}\boldsymbol{x} \le ch^{1/2}.$$

Moreover, the L^2 norm of the error of the gradient is bounded in directions $\boldsymbol{w}\in\mathbb{S}^{n-1}$ orthogonal to the layering direction $\boldsymbol{b}$ by

$$\frac{1}{|\Omega|}\int_\Omega \big|(D\boldsymbol{u}_h - F)\boldsymbol{w}\big|^2 \mathrm{d}\boldsymbol{x} \le ch^{1/2}.$$

Finally,

$$d\big(\operatorname{Av}\delta_{D\boldsymbol{u}_h(\boldsymbol{x})}\big) \le ch^{1/4}.$$

Proof. It is easy to see that there exists a finite element function $\boldsymbol{u}_h$ with

$$\frac{1}{|\Omega|}\int_\Omega W(D\boldsymbol{u}_h)\mathrm{d}\boldsymbol{x} \le ch^{1/2}.$$

The informal idea of this construction is the following. Consider for $\delta > 0$ the function

$$\boldsymbol{u}_\delta(\boldsymbol{x}) = Q_\alpha U_\alpha \boldsymbol{x} - \varphi_\delta(\langle \boldsymbol{x}, \boldsymbol{b}\rangle)\boldsymbol{a}.$$

Here $\varphi_\delta(\boldsymbol{x}) = \varphi(\boldsymbol{x}/\delta)$ and $\varphi : \mathbb{R} \to \mathbb{R}$ is the continuous function with $\varphi(0) = 0$ and

$$\varphi'(z+s) = \begin{cases} 0 & \text{if } s \in (0,\lambda) \\ 1 & \text{if } s \in (\lambda, 1) \end{cases} \qquad \text{for all } z \in \mathbb{Z}.$$

Then $D\boldsymbol{u}_\delta \in \{Q_\alpha U_\alpha, Q_\beta U_\beta\}$ and $\boldsymbol{u}_\delta$ is affine on layers separated by affine hyperplanes with normal $\boldsymbol{b}$. The goal is now to choose for $h > 0$ fixed δ in such a way that the nodal interpolation of $\boldsymbol{u}_\delta$ onto $\mathcal{T}_h$ has minimal energy. This requires to balance the following two contributions to the energy.

(a) For each interface across which $D\boldsymbol{u}_\delta$ changes from $Q_\alpha U_\alpha$ to $Q_\beta U_\beta$ one needs a neighborhood of diameter $\mathcal{O}(h)$ which contributes a term of order $\mathcal{O}(h/\delta)$ to the energy.

(b) The function $\boldsymbol{u}_\delta$ satisfies $|\boldsymbol{u}_\delta(\boldsymbol{x}) - F\boldsymbol{x}| = \mathcal{O}(\delta)$ and thus a boundary layer along $\partial\Omega$ of width $\mathcal{O}(\delta)$ is required to interpolate $\boldsymbol{u}_\delta$ and the correct boundary values $F\boldsymbol{x}$.

This shows that one can construct a function $\boldsymbol{u}_{\delta,h}$ with energy

$$\frac{1}{|\Omega|}\int_\Omega W(D\boldsymbol{u}_{\delta,h}) = \mathcal{O}(h/\delta) + \mathcal{O}(\delta).$$

The choice $\delta = \mathcal{O}(\sqrt{h})$ proves the assertion.

The estimates in the corollary follow now immediately from the assertions in Theorem 4.5.1. □

Remark 4.5.3. In the three-dimensional situation, it is natural to assume quadratic growth of the energy W, i.e., to assume that

$$W(X) \geq \kappa \operatorname{dist}^2(X,K), \quad \kappa > 0.$$

In this case, we have to replace the estimate (4.36) by

$$\begin{aligned}
\mathcal{E}(\nu) &= \frac{1}{|\Omega|}\int_\Omega \big(\operatorname{dist}(D\boldsymbol{u},K) + \operatorname{dist}^2(D\boldsymbol{u},K)\big)\mathrm{d}\boldsymbol{x} \\
&\leq \Big(\frac{1}{|\Omega|}\int_\Omega \operatorname{dist}^2(D\boldsymbol{u},K)\mathrm{d}\boldsymbol{x}\Big)^{1/2} + \frac{1}{|\Omega|}\int_\Omega \operatorname{dist}^2(D\boldsymbol{u},K)\mathrm{d}\boldsymbol{x} \\
&\leq \frac{1}{\sqrt{\kappa}}\Big(\frac{1}{|\Omega|}\int_\Omega W(D\boldsymbol{u})\mathrm{d}\boldsymbol{x}\Big)^{1/2} + \frac{1}{\kappa}\frac{1}{|\Omega|}\int_\Omega W(D\boldsymbol{u})\mathrm{d}\boldsymbol{x} \\
&\leq c(h^{1/2} + h^{1/4}).
\end{aligned}$$

This leads to the estimates

$$\frac{1}{|\Omega|}\int_\Omega |u_h(x) - Fx|^2 \mathrm{d}x \le c(h^{1/2} + h^{1/4}),$$
$$\frac{1}{|\Omega|}\int_\Omega \big|(Du_h - F)w\big|^2 \mathrm{d}x \le c(h^{1/2} + h^{1/4}),$$
$$d\big(\operatorname{Av}\delta_{Du_h(\cdot)}, \overline{\nu}\big) \le c(h^{1/4} + h^{1/8}),$$

which follow from Corollary 4.5.2. □

4.6 Extensions to Higher Order Laminates

The theory developed in Section 4.1 provides a sufficient condition for uniqueness and stability of microstructure which is essentially an algebraic condition on the center of mass F and the set K. However, it only applies to simple laminates and Corollary 4.1.15, which asserts that stability implies uniqueness, already indicates that extensions of the theory to higher order laminates are subtle. It is an open question whether uniqueness is sufficient for stability. Theorem 4.2.1 shows that this is true for two-dimensional problems.

The advantage of the approach described here is that it clearly distinguishes between those arguments that rely on uniqueness and those that do not. Therefore some of the estimates can be extended to higher order laminates and we demonstrate this for the case of the cubic to tetragonal phase transformation in three dimensions. Suppose that $\eta_2 > \eta_1$ and define

$$U_1 = \operatorname{diag}(\eta_2, \eta_1, \eta_1), \quad U_2 = \operatorname{diag}(\eta_1, \eta_2, \eta_1), \quad U_3 = \operatorname{diag}(\eta_1, \eta_1, \eta_2).$$

For the rest of this section we consider the set K given by

$$K = \mathrm{SO}(3)U_1 \cup \mathrm{SO}(3)U_2 \cup \mathrm{SO}(3)U_3. \tag{4.37}$$

The polyconvex hull of two of the wells, say $\mathrm{SO}(3)U_1 \cup \mathrm{SO}(3)U_2$, is equal to the second lamination convex hull and a polyconvex measure ν underlying a global affine deformation F is unique if and only if F is a first order laminate (or a matrix in K). Iqbal's results show that even the mass of ν on the two wells is not uniquely determined from its center of mass.

Suppose now that $\nu \in \mathcal{M}^{\mathrm{pc}}(\mathrm{SO}(3)U_1 \cup \mathrm{SO}(3)U_2)$ (in particular, ν could be a second order laminate). It follows from the example following Theorem 2.5.1 that $F = \langle \nu, id \rangle$ satisfies

$$|\operatorname{cof} F e_3| = |\operatorname{cof} U_1 e_3| = |\operatorname{cof} U_2 e_3| > |\operatorname{cof} U_3 e_3|. \tag{4.38}$$

Moreover, if $\nu \in \mathcal{M}^{\mathrm{pc}}(K; F)$ and if F satisfies (4.38), then F is the center of mass of a second order laminate supported on $\mathrm{SO}(3)U_1 \cup \mathrm{SO}(3)U_2$. The strict inequality in (4.38) implies estimates analogously to those in Section 4.1. As before, we define

$$\mathcal{E}(\nu) = \int_{\operatorname{supp}\nu} \big(\operatorname{dist}(A,K) + \operatorname{dist}^2(A,K)\big)\mathrm{d}\nu(A).$$

Theorem 4.6.1. *Suppose that F satisfies (4.38) and that $\nu \in \mathcal{M}^{\mathrm{pc}}(\mathbb{M}^{3\times 3})$. Then*

$$\int_{\operatorname{supp}\nu} |A\boldsymbol{e}_3 - F\boldsymbol{e}_3|^2 \mathrm{d}\nu(A) \leq c\mathcal{E}(\nu).$$

Proof. We conclude as in the proof of Proposition 4.1.9 that

$$\sum_{\ell=1}^{3} \tau_\ell\big(|F\boldsymbol{e}_3|^2 - |U_\ell \boldsymbol{e}_3|^2\big) + \int_{\operatorname{supp}\nu} |\pi(A)\boldsymbol{e}_3 - F\boldsymbol{e}_3|^2 \mathrm{d}\nu(A) \leq c\mathcal{E}(\nu)$$

and

$$\sum_{\ell=1}^{3} \tau_\ell\big(|\operatorname{cof} F\boldsymbol{e}_3|^2 - |\operatorname{cof} U_\ell \boldsymbol{e}_3|^2\big) \leq c\mathcal{E}(\nu).$$

The second inequality implies with (4.38) that $\tau_3 \leq c\mathcal{E}(\nu)$. The first estimate yields in view of this bound and $|F\boldsymbol{e}_3| = |U_1\boldsymbol{e}_3| = |U_2\boldsymbol{e}_2|$ that

$$\int_{\operatorname{supp}\nu} |(\pi(A) - F)\boldsymbol{e}_3|^2 \mathrm{d}\nu(A) \leq c\mathcal{E}(\nu). \tag{4.39}$$

The assertion follows now by the triangle inequality from the two foregoing estimates since $|\pi(A) - A| = \operatorname{dist}(A,K)$. □

As a corollary, we obtain estimates similar to those in Section 4.1 for first order laminates.

Corollary 4.6.2. *Let K be given by (4.37) and suppose that $F \in \mathcal{M}^{\mathrm{pc}}(K)$ satisfies (4.38). Assume that $\boldsymbol{u} \in \mathcal{A}_F$ with $\mathcal{J}(\boldsymbol{u}) \leq \delta$ with $\delta > 0$ where $\mathcal{J}$ has been defined in (4.34). Then there exists a constant c that depends only on F and κ such that the following assertions hold:*

i) (estimates for directional derivatives tangential to the layering direction)

$$\frac{1}{|\Omega|}\int_\Omega \big|(D\boldsymbol{u}(\boldsymbol{x}) - F)\boldsymbol{e}_3\big|^2 \mathrm{d}\boldsymbol{x} \leq c\delta,$$

ii) (estimates for the deformation)

$$\frac{1}{|\Omega|}\int_\Omega |\boldsymbol{u}(\boldsymbol{x}) - F\boldsymbol{x}| \mathrm{d}\boldsymbol{x} \leq c\delta,$$

iii) (weak convergence, averages): the following estimate holds for all $\omega \subset \Omega$ with Lipschitz boundary,

$$\Big|\int_\omega (D\boldsymbol{u} - F)d\boldsymbol{x}\Big| \leq c(\omega,\Omega)\delta^{1/2}.$$

Proof. The first two assertions follow from Theorem 4.6.1 and Poincaré's inequality. The proof of last result is identical to the proof of v) in Theorem 4.5.1.

4.7 Numerical Analysis of Microstructure – A Review

In order to put our results in a more general context, and to trace back the origins of the ideas presented here, we include in this section a brief discussion of related work. At the same time, this allows us to sketch some aspects in the development of the numerical analysis of microstructures that began only about ten years ago.

The first numerical simulations of microstructures in elastic materials undergoing solid to solid phase transformations were reported in [Si89] for a hypothetic hyperelastic material and in [CL89, CLR93] for a three-dimensional model of the cubic to tetragonal transformation with a stored energy function proposed in [Er86] and a choice of parameters suggested by James.

The numerical analysis of finite element schemes for the minimization of nonconvex problems was initiated in [CKL91] in the scalar, one-dimensional setting. Let

$$\mathcal{E}(v) = \int_0^1 [\Phi(v'(y)) + (v(y) - f(y))^2]\,dy$$

where $\Phi \geq 0$ satisfies $\Phi(s) = 0$ if and only if $s \in \{s_L, s_U\}$ and f is affine with $f' \equiv \overline{s} \in (s_L, s_U)$. It is easy to see that the infimum of the energy in the space of all Lipschitz continuous functions is zero, but that there does not exist a minimizer for the functional in this class. In fact, v has to approximate the affine function f with slope $\overline{s}$, but the potential Φ vanishes only in the two points s_L and s_U. Minimizing sequences $\{u_k\}$ develop increasingly finer oscillations, converge weakly to the affine function f, and the sequence $\{u_k'\}$ generates the unique gradient Young measure $\{\nu_x\}_{x\in(0,1)}$ with

$$\nu_x = \nu = \gamma\delta_{s_L} + (1-\gamma)\delta_{s_U}, \quad \overline{s} = \gamma s_L + (1-\gamma)s_U.$$

The same behavior is expected for finite element minimizers, and the authors note: "... we shall show that $u_h'(x)$ and nonlinear functions of $u_h'(x)$ converge weakly. We show below that the topology of this convergence is metrizable since it is convergence in the weak-∗ topology of a suitable Banach space, and we give an error estimate for this convergence in an appropriate metric." ([CKL91], page 322). Indeed, the fundamental estimate in their paper is

$$d(\mathrm{Av}\,\delta_{u_h'(\cdot)}, \nu) \leq ch^{1/4}, \tag{4.40}$$

where d denotes a metric that metrizes the weak-∗ convergence. The crucial estimate in the proof of this convergence result states that the distribution of the values of u_h' is determined by the unique Young measure. The values of u_h' lie in a neighborhood of s_L and s_U on a set of measure close to γ and $1-\gamma$, respectively. Surprisingly, estimates in the metric d were not further pursued in the literature. Our notion of stability in Definition 4.1.4 is based on exactly the same ideas.

The analysis of higher dimensional scalar problems started in [C91, CC92], where properties of minimizers in finite element spaces of the functional

$$\mathcal{E}(v) = \int_\Omega \left[\varphi(Dv(\boldsymbol{x})) + \psi\big(v(\boldsymbol{x}) - \langle \boldsymbol{a}, \boldsymbol{x}\rangle\big)\right] \mathrm{d}\boldsymbol{x}$$

were investigated. Here Ω is a bounded domain in $\mathbb{R}^n$ and Dirichlet conditions $u(\boldsymbol{x}) = \langle \boldsymbol{a}, \boldsymbol{x}\rangle$ are imposed on a subset Γ_0 of $\partial\Omega$ (the lower order term can be omitted if $\Gamma_0 = \partial\Omega$). The function $\varphi : \mathbb{R}^n \to \mathbb{R}$ and $\psi : \mathbb{R} \to \mathbb{R}$ are assumed to be nonnegative with $\varphi(\boldsymbol{w}) = 0$ if and only if $\boldsymbol{w} \in K = \{\boldsymbol{w}_1, \ldots, \boldsymbol{w}_k\} \subset \mathbb{R}^n$, $k \geq 2$, and $\psi(s) = 0$ only for $s = 0$ unless $\psi \equiv 0$. The existence of finite element minimizers follows if φ and ψ satisfy suitable coercivity assumptions. Moreover, based on an explicit construction, one can show that the energy in the finite element space is bounded by $c_0 h^\gamma$ if φ is bounded on bounded sets, where $\gamma > 0$ is related to the growth of ψ. For example, $\gamma = \frac{1}{2}$ if $\psi \equiv 0$.

Also in this case the functional $\mathcal{E}$ does not have a Lipschitz minimizer, but minimizing sequences generate a unique gradient Young measure under the condition that $\boldsymbol{a}$ has a unique representation,

$$\boldsymbol{a} = \sum_{i=1}^k \alpha_i \boldsymbol{w}_i.$$

In fact, $\{\nu_{\boldsymbol{x}}\}_{\boldsymbol{x}\in\Omega}$ is homogeneous with

$$\nu_{\boldsymbol{x}} = \sum_{i=1}^k \alpha_i \delta_{\boldsymbol{w}_i} \quad \text{for a.e. } \boldsymbol{x} \in \Omega.$$

Thus all minimizing sequences generate the same Young measure and consequently the distribution of the gradients of functions with small energy should be determined from the Young measure. This intuitive statement is made precise by what the authors in [C91, CC92] call the probabilistic analysis of the oscillations.

Define the projection $\pi : \mathbb{R}^n \to K$ such that $\pi(\boldsymbol{\xi}) = \boldsymbol{w}_i$ where i is the smallest index such that $|\boldsymbol{\xi} - \boldsymbol{w}_i| = \min_{j=1,\ldots,k} |\boldsymbol{\xi} - \boldsymbol{w}_j|$. Suppose that φ and ψ satisfy the growth conditions

$$\varphi(\boldsymbol{\xi}) \geq \lambda_1 |\boldsymbol{\xi} - \pi(\boldsymbol{\xi})|^p \quad \text{for all } \boldsymbol{\xi} \in \mathbb{R}^n, p > 1, \lambda_1 > 0,$$

and

$$\psi(t) \geq \lambda_2 |t|^q \quad \text{for all } t \in \mathbb{R}, q > 1,$$

and that ω is any subdomain of Ω with Lipschitz boundary. To simplify the statements we assume in the sequel that $p = q = 2$. It is an immediate consequence of the foregoing hypotheses that

$$\int_\omega |Dv - \pi(Dv)|^2 \mathrm{d}\boldsymbol{x} \leq c\mathcal{E}(v) \tag{4.41}$$

(see Lemma 1 in [CC92]), and the estimate

$$\Big|\int_\omega Du\,\mathrm{d}\boldsymbol{x}\Big| \leq c(\omega)\{\|u\|_{L^2(\omega)} + (\|u\|_{L^2(\omega)}\|Du\|_{L^2(\omega)})^{1/2}\}, \quad u \in W^{1,2}(\omega) \tag{4.42}$$

implies that

$$\Big|\int_\omega (Dv - \boldsymbol{a})\mathrm{d}\boldsymbol{x}\Big| \leq c(\mathcal{E}^{1/4}(v) + \mathcal{E}^{1/8}(v)) \tag{4.43}$$

(see Lemma 3 in [CC92]). These estimates allow one to prove estimates for the distribution of the values of Dv. For $\varrho > 0$ with $\varrho < \frac{1}{2}\min_{i\neq j}|\boldsymbol{w}_i - \boldsymbol{w}_j|$ we define

$$\omega_{i,\varrho}(v) = \{\boldsymbol{x} \in \omega : Dv(\boldsymbol{x}) \in B(\boldsymbol{w}_i, \varrho)\},$$

where $B(\boldsymbol{w}_i, \varrho)$ is the ball with center $\boldsymbol{w}_i$ and radius ϱ. Then we have for all $v \in W^{1,2}(\Omega)$ with $u(\boldsymbol{x}) = \langle \boldsymbol{a}, \boldsymbol{x}\rangle$ on Γ_0

$$\Big||\omega_{i,\varrho}(v)| - \alpha_i|\omega|\Big| \leq c(\mathcal{E}^{1/2}(v) + \mathcal{E}^{1/16}(v)) \quad \text{for } i = 1, \ldots, k. \tag{4.44}$$

One of the many consequences of this precise control of the volume fractions is the error estimate in Theorem 6 in [CC92]. Let $f : \Omega \times \mathbb{R}^n \to \mathbb{R}$ be Lipschitz continuous in its second argument, i.e.

$$|f(\boldsymbol{x}, \boldsymbol{\xi}) - f(\boldsymbol{x}, \boldsymbol{\eta})| \leq L|\boldsymbol{\xi} - \boldsymbol{\eta}| \quad \text{for all } \boldsymbol{\xi}, \boldsymbol{\eta} \in \mathbb{R}^n.$$

Then

$$\Big|\int_\Omega [f(\boldsymbol{x}, Dv(\boldsymbol{x})) - \sum_{i=1}^k \alpha_i f(\boldsymbol{x}, \boldsymbol{w}_i)]\mathrm{d}\boldsymbol{x}\Big| \leq c(L)(\mathcal{E}^{1/2}(v) + \mathcal{E}^{1/16}(v)),$$

where the constant c depends on f only via the Lipschitz constant L.

These results were then extended in [CCK95] to the case of point wells in $\mathbb{M}^{m\times n}$ under the assumption that the wells are pairwise compatible, i.e. $\operatorname{rank}(W_i - W_j) = 1$ for $i \neq j$. Then the matrices $X_i = W_1 - W_i = \boldsymbol{a}_i \otimes \boldsymbol{b}_i$ are pairwise compatible, and this is only possible if either all the vectors $\boldsymbol{a}_i$ or the vectors $\boldsymbol{b}_i$ are parallel (see Lemma 2.1 in [CCK95]). A suitable change of coordinates allows one to reduce the problem essentially to the scalar situation and the estimates for the volume fractions can be obtained as in the scalar case.

The paper [CCK95] also presents an adaption of the methods used for point wells to the physically relevant case of energies with potential wells in the exemplary case of the two-well problem in two dimensions (see also [Gd94] for results with nonconforming elements on grids with specific, problem adapted orientations). Assume that $\varphi \geq 0$ satisfies $\varphi(F) = \varphi(RF)$ for all $R \in \mathrm{SO}(2)$ and that $\varphi(X) = 0$ if and only if $X \in K = \mathrm{SO}(2)U_1 \cup \mathrm{SO}(2)U_2$ where Q_1U_1 and Q_2U_2 with $Q_1, Q_2 \in \mathrm{SO}(2)$ are rank-one connected. Let

$$A = \lambda Q_1 U_1 + (1-\lambda) Q_2 U_2 \text{ with } \lambda \in (0,1) \tag{4.45}$$

and

$$Q_1U_1 - Q_2U_2 = \boldsymbol{a} \otimes \boldsymbol{n}, \quad \boldsymbol{a}, \boldsymbol{n} \in \mathbb{R}^2,$$

and consider the minimization problem

$$\inf_{\substack{\boldsymbol{u} \in W^{1,\infty}(\Omega;\mathbb{R}^2) \\ \boldsymbol{u}(\boldsymbol{x}) = A\boldsymbol{x} \text{ on } \partial\Omega}} \int_\Omega \big[\varphi(D\boldsymbol{v}) + \psi(\boldsymbol{v}(\boldsymbol{x}) - A\boldsymbol{x})\big] \mathrm{d}\boldsymbol{x}.$$

Due to the choice of the boundary data, there does not exist a minimizer for the energy. However, minimizing sequences show exactly the expected behavior: the estimates (4.41) and (4.43) follow as in the case of point wells (with slightly modified exponents) and only the proof for the estimates (4.44) of the volume fractions requires some modifications. Define

$$\pi(F) = QU_i \text{ where } QU_i \text{ satisfies } \operatorname{dist}(F, K) = |QU_i - F|,$$

and

$$\widetilde{\omega}_{i,\varrho}(\boldsymbol{v}) = \{\boldsymbol{x} \in \omega : \pi(D\boldsymbol{v}(\boldsymbol{x})) \in \mathrm{SO}(2)U_i,\ |D\boldsymbol{v}(\boldsymbol{x}) - \pi(D\boldsymbol{v}(\boldsymbol{x}))| \leq \varrho\}$$

(if F has the same distance to more than one well, then we choose the one with the smallest index). In this situation the following estimates hold:

$$\big||\widetilde{\omega}_{1,\varrho}| - \lambda|\omega|\big| \leq c\mathcal{E}^{2/5}(\boldsymbol{v}), \quad \big||\widetilde{\omega}_{2,\varrho}| - (1-\lambda)|\omega|\big| \leq c\mathcal{E}^{2/5}(\boldsymbol{v}).$$

The analysis of three-dimensional models begins in [L96a] for a two-well problem with SO(3) invariance described by the energy

$$\mathcal{E}(\boldsymbol{v}) = \int_\Omega \varphi(D\boldsymbol{v}) \mathrm{d}\boldsymbol{x}.$$

Here $\varphi \geq 0$ and $\varphi(F) = 0$ if and only if $F \in K = \mathrm{SO}(3)U_1 \cup \mathrm{SO}(3)U_2$ and φ satisfies the growth condition $\varphi(F) \geq \kappa|F - \pi(F)|^2$. The wells are again assumed to be compatible and boundary conditions that correspond to the center of mass of a simple laminate (4.45) are imposed. This paper improves the foregoing estimates in several aspects. Define $\Pi : \mathbb{M}^{3\times 3} \to \{Q_1U_1, Q_2U_2\}$ by

$$|\Pi(X) - X| = \operatorname{dist}(X, \mathrm{SO}(3)U_1 \cup \mathrm{SO}(3)U_2),$$

and let

$$\omega_{i,\varrho}(\boldsymbol{v}) = \{\boldsymbol{x} \in \omega : \Pi(D\boldsymbol{v}(\boldsymbol{x})) = Q_iU_i,\ |D\boldsymbol{v}(\boldsymbol{x}) - \pi(D\boldsymbol{v}(\boldsymbol{x}))| \le \varrho\}.$$

The estimate

$$\int_\Omega |(D\boldsymbol{v} - A)\boldsymbol{w}|^2 \le c(\mathcal{E}(\boldsymbol{v}) + \mathcal{E}^{1/2}(\boldsymbol{v})), \quad \text{for all } \boldsymbol{w} \in \mathbb{R}^3,\ \langle \boldsymbol{w}, \boldsymbol{n}\rangle = 0,$$

and (4.42) imply immediately

$$\int_\Omega |\boldsymbol{v}(\boldsymbol{x}) - F\boldsymbol{x}|^2 \mathrm{d}\boldsymbol{x} \le c(\mathcal{E}(\boldsymbol{v}) + \mathcal{E}^{1/2}(\boldsymbol{v})),$$
$$\left| \int_\omega (D\boldsymbol{v} - A)\mathrm{d}\boldsymbol{x} \right| \le c(\mathcal{E}^{1/2}(\boldsymbol{v}) + \mathcal{E}^{1/8}(\boldsymbol{v})).$$

Bounds for the volume fractions $\omega_{i,R}$,

$$\big||\omega_{1,\varrho}| - \lambda|\omega|\big| \le c(\mathcal{E}^{1/2}(\boldsymbol{v}) + \mathcal{E}^{1/8}(\boldsymbol{v})),$$
$$\big||\omega_{2,\varrho}| - (1-\lambda)|\omega|\big| \le c(\mathcal{E}^{1/2}(\boldsymbol{v}) + \mathcal{E}^{1/8}(\boldsymbol{v}))$$

follow from the sharper inequality

$$\int_\Omega |D\boldsymbol{v} - \Pi(D\boldsymbol{v})|^2 \mathrm{d}\boldsymbol{x} \le c(\mathcal{E}(\boldsymbol{v}) + \mathcal{E}^{1/2}(\boldsymbol{v})),$$

which is based on an L^2 estimate for the excess rotation $R(D\boldsymbol{u})$ which we define by $\pi(D\boldsymbol{u}(\boldsymbol{x})) = R(D\boldsymbol{u}(\boldsymbol{x}))\Pi(D\boldsymbol{u}(\boldsymbol{x}))$, see e.g. inequality (4.18). These estimates were subsequently applied to various phase transformations using conforming and nonconforming finite element methods, see e.g. [L96b, LL98a, LL98b, BLL99, BL00, LL00] or [GP99, GP00, Pr00] for models including penalizations.

5. Applications to Martensitic Transformations

Many technological applications of shape memory materials are based on alloys that undergo cubic to tetragonal, cubic to orthorhombic, or tetragonal to monoclinic transformations. In this chapter we apply the general theory developed in Chapter 4 to these transformation and investigate the question of uniqueness and stability via the validity of condition ($\mathrm{C_b}$). We restrict our attention to three-dimensional bulk materials and we focus therefore on simple laminates. The application of the stability theory and the resulting error estimates for finite element minimizers developed in Section 4.5 are obvious and therefore not stated explicitly.

Suppose that the set K describing one of the transformations is given by $K = \mathrm{SO}(3)U_1 \cup \ldots \cup \mathrm{SO}(3)U_k$ with symmetric and positive definite matrices U_i that satisfy $\det U_i = \Delta > 0$ for $i = 1, \ldots, k$. Simple laminates are obtained by solving the twinning relation

$$QU_i - U_j = \boldsymbol{a} \otimes \boldsymbol{n}, \quad Q \in \mathrm{SO}(3),\ \boldsymbol{a},\ \boldsymbol{n} \in \mathbb{R}^3, \tag{5.1}$$

for which the general solutions have been given in Proposition A.2.1. Since all the transformations considered here lead to symmetry related wells, that is,

$$U_i = RU_jR, \qquad R \text{ a } 180^\circ \text{ degree rotation,} \tag{5.2}$$

we may apply the result in Proposition A.2.4 which provides explicit formulae for the two solutions of (5.1). We refer to the solutions as type-I and type-II twinning systems, respectively. In the special case that there are two distinct rotations that satisfy (5.2) we call the twinning systems a compound twinning system. We summarize the information for the various transformations in Tables 5.1-5.5, in which we omit occasionally lengthy expressions that are not needed in the text. In these tables we write $R^\pi_{i\pm j}$ and R^π_i for the 180° rotations with axes $\boldsymbol{e}_i \pm \boldsymbol{e}_j$ and $\boldsymbol{e}_i$, respectively.

Suppose now that

$$\nu = \lambda\delta_{Q_iU_i} + (1-\lambda)\delta_{U_j}, \quad Q_iU_i - U_j = \boldsymbol{a} \otimes \boldsymbol{n},\ \lambda \in [0,1],$$

is a simple laminate supported on K, and let $F = \langle \nu, id \rangle$ be its center of mass. In order to apply the general theory based on condition ($\mathrm{C_b}$) in Definition 4.1.4, we need to find so-called test vectors $\boldsymbol{w} \in \mathbb{R}^3$ with

$$|U_i w|^2 = |U_j w|^2 > \max_{k \neq i,j} |U_k w|^2, \quad \text{and} \quad \langle w, n \rangle = 0$$

or

$$|\operatorname{cof} U_i w|^2 = |\operatorname{cof} U_j w|^2 > \max_{k \neq i,j} |\operatorname{cof} U_k w|^2, \quad \text{and} \quad \langle w, U_j^{-1} a \rangle = 0.$$

Then automatically $|Fw|^2 = |U_i w|^2$ and $|\operatorname{cof} F w|^2 = |\operatorname{cof} U_i w|^2$, respectively, and the assumptions ii) in condition (C_b) are satisfied. The first equality is immediate and the second follows from formula (C.2) for the cofactor since $\det U_i = \det U_j$ implies $\langle U_j^{-1} a, n \rangle = 0$. We frequently take advantage of the fact that by Proposition A.2.4 the vectors n and $U_j^{-1} a$ are parallel if the wells are symmetry related. The examples below illustrate that a proof of uniqueness for type-I twins using a vector w as a test vector on F typically also proves uniqueness for the corresponding type-II twin by testing the cofactor matrices by the same vector w.

Carrying out this program in the subsequent sections, we rederive some of Luskin's results as a simple application of our general approach. In particular we find that for cubic to orthorhombic and tetragonal to orthorhombic transformations simple laminates are uniquely determined by their center of mass F unless the lattice parameters describing the energy wells satisfy a certain algebraic condition. The new results are explicit characterizations of the sets $\mathcal{M}^{pc}(K; F)$ in these special cases, and this answers a question raised by James.

5.1 The Cubic to Tetragonal Transformation

The prototype of a solid to solid phase transformation is the cubic to tetragonal transformation with three martensitic wells. An example is the phase transformation in Indium rich InTl alloys which undergo a transformation from face-centered cubic phase into face-centered tetragonal phase. For this transformation we have the following result.

Theorem 5.1.1. *Assume that $\eta_1, \eta_2 > 0$, $\eta_1 \neq \eta_2$, and that*

$$U_1 = \begin{pmatrix} \eta_2 & 0 & 0 \\ 0 & \eta_1 & 0 \\ 0 & 0 & \eta_1 \end{pmatrix}, U_2 = \begin{pmatrix} \eta_1 & 0 & 0 \\ 0 & \eta_2 & 0 \\ 0 & 0 & \eta_1 \end{pmatrix}, U_3 = \begin{pmatrix} \eta_1 & 0 & 0 \\ 0 & \eta_1 & 0 \\ 0 & 0 & \eta_2 \end{pmatrix} \tag{5.3}$$

Let

$$K = SO(3)U_1 \cup SO(3)U_2 \cup SO(3)U_3$$

and suppose that $\nu \in \mathcal{M}^{pc}(K)$ is a simple laminate with $F = \langle \nu, id \rangle$. Then ν is unique and $\mathcal{M}^{pc}(K; F) = \{\nu\}$.

Table 5.1. Rank-one connections (twins) in the tetragonal variants (we abbreviate in this table type by tp and compound by cp).

(ij)	R	tp	$\boldsymbol{n}$	$\boldsymbol{a}$	$U_j^{-1}\boldsymbol{a}$
(12)	R^{π}_{1+2}	cp	$\frac{1}{\sqrt{2}}(1,1,0)$	$\frac{\sqrt{2}(\eta_2^2-\eta_1^2)}{\eta_2^2+\eta_1^2}(\eta_1,-\eta_2,0)$	$\frac{\sqrt{2}(\eta_2^2-\eta_1^2)}{\eta_2^2+\eta_1^2}(1,-1,0)$
(12)	R^{π}_{1-2}	cp	$\frac{1}{\sqrt{2}}(1,-1,0)$	$\frac{\sqrt{2}(\eta_2^2-\eta_1^2)}{\eta_2^2+\eta_1^2}(\eta_1,\eta_2,0)$	$\frac{\sqrt{2}(\eta_2^2-\eta_1^2)}{\eta_2^2+\eta_1^2}(1,1,0)$

Proof. In view of the symmetry relations in Section 5.5, is suffices to consider simple laminates supported on $\mathrm{SO}(3)U_1 \cup \mathrm{SO}(3)U_2$. There exist two rank-one connections between the wells $\mathrm{SO}(3)U_1$ and $\mathrm{SO}(3)U_2$ which generate compound twins, see Table 5.1. In order to simplify the notation in the following statements, we call a matrix F a type-I twin if it is generated from the twinning system with normal $\boldsymbol{n} = \frac{1}{\sqrt{2}}(1,1,0)$, and a type-II twin otherwise.

In view of Theorem 4.1.11 we only need to prove the existence of vectors $\boldsymbol{w}$ with

$$|F\boldsymbol{w}|^2 = |U_1\boldsymbol{w}|^2 = |U_2\boldsymbol{w}|^2 > |U_3\boldsymbol{w}|^2.$$

In order to accomplish this, we assume first that $\eta_2 > \eta_1$. If F is generated from the type-I twinning system, then

$$|F\boldsymbol{w}|^2 = |U_1\boldsymbol{w}|^2 = |U_2\boldsymbol{w}|^2 > |U_3\boldsymbol{w}|^2 \quad \text{for } \boldsymbol{w} = \frac{1}{\sqrt{2}}(1,-1,0),$$

and if F is generated from the type-II twinning system, then

$$|F\boldsymbol{w}|^2 = |U_1\boldsymbol{w}|^2 = |U_2\boldsymbol{w}|^2 > |U_3\boldsymbol{w}|^2 \quad \text{for } \boldsymbol{w} = \frac{1}{\sqrt{2}}(1,1,0).$$

Assume now that $\eta_2 < \eta_1$. In this case we have for both twinning systems that

$$|F\boldsymbol{w}|^2 = |U_1\boldsymbol{w}|^2 = |U_2\boldsymbol{w}|^2 > |U_3\boldsymbol{w}|^2 \quad \text{for } \boldsymbol{w} = (0,0,1),$$

and uniqueness of ν is a consequence of the foregoing inequalities. □

The following theorem complements the results in Theorem 5.1.1 by identifying the implications of assumption ii) in condition ($\mathrm{C_b}$) without the hypothesis that ν be a simple laminate.

Theorem 5.1.2. *Let* $K = \mathrm{SO}(3)U_1 \cup \mathrm{SO}(3)U_2 \cup \mathrm{SO}(3)U_3$ *and* $F \in K^{\mathrm{pc}}$. *Suppose that there exists a* $\boldsymbol{w} \in \mathbb{S}^2$ *with*

$$|Fw|^2 = |U_1 w|^2 = |U_2 w|^2 > |U_3 w|^2. \tag{5.4}$$

Then

$$F \in (\mathrm{SO}(3)U_1 \cup \mathrm{SO}(3)U_2)^{(1)} \quad \text{if} \quad \eta_2 > \eta_1,$$

and

$$F \in (\mathrm{SO}(3)U_1 \cup \mathrm{SO}(3)U_2)^{(2)} = (\mathrm{SO}(3)U_1 \cup \mathrm{SO}(3)U_2)^{\mathrm{pc}} \quad \text{if} \quad \eta_2 < \eta_1.$$

Remark 5.1.3. Assumption (5.4) for $\eta_2 < \eta_1$ can be interpreted as an assumption for $\operatorname{cof} F$ in the case $\eta_2 > \eta_1$. Thus there is a surprising difference in the implications of assumption on F and $\operatorname{cof} F$ for matrices $F \in K^{\mathrm{pc}}$.

Proof. In view of Proposition 4.1.9, assumption (5.4) implies that any polyconvex measure with barycenter F has to be supported on the two wells $\mathrm{SO}(3)U_1 \cup \mathrm{SO}(3)U_2$. Suppose first that $\eta_2 > \eta_1$. In order to conclude from Step 2 in the proof of Proposition 4.1.8 that F is the center of mass of a simple laminate, we have to prove the existence of vectors v, $e \in \mathbb{R}^3$ such that $\langle v, e\rangle = 0$, v is a common eigenvector of U_1 and U_2, and

$$|Fe|^2 = |U_1 e|^2 = |U_2 e|^2.$$

In view of (5.3) we choose $v = (0,0,1)$ and deduce from Theorem 2.5.1 that $F^T F v = \eta_1^2 v$. Moreover, by assumption the vector $w = (w_1, w_2, w_3)$ satisfies

$$|U_1 w|^2 = |U_2 w|^2 \quad \Leftrightarrow \quad (\eta_2^2 - \eta_1^2) w_1^2 = (\eta_2^2 - \eta_1^2) w_2^2$$

and

$$|U_1 w|^2 > |U_3 w|^2 \quad \Leftrightarrow \quad (\eta_2^2 - \eta_1^2) w_1^2 > (\eta_2^2 - \eta_1^2) w_3^2,$$

and therefore $|w_1| = |w_2| > |w_3|$. We define $e = (w_1, w_2, 0) = (w_1, \pm w_1, 0)$ and observe by the foregoing estimate that $e \neq 0$ and that $\langle e, v\rangle = 0$. We now have

$$|Fw|^2 = |F(e + w_3 v)|^2 = |Fe|^2 + 2w_3 \langle Fe, Fv\rangle + w_3^2 |Fv|^2 = |Fe|^2 + w_3^2 \eta_1^2.$$

Similarly,

$$|U_i w|^2 = |U_i e|^2 + w_3^2 \eta_1^2,$$

and hence

$$|Fe|^2 = |U_1 e|^2 = |U_2 e|^2.$$

The assertion of the theorem follows now easily from Proposition 4.1.8. Note that this argument does not work if $\eta_1 > \eta_2$, since the assumptions are satisfied with $w = (0,0,1)$ for all F in the polyconvex hull of the two wells $\mathrm{SO}(3)U_1$ and $\mathrm{SO}(3)U_2$. □

Remark 5.1.4. Suppose that $\eta_2 > \eta_1$. The proof of Theorem 5.1.2 shows in fact the following implication. Assume that F is any matrix in the polyconvex hull of the *three* martensitic wells (5.3). Suppose that there exists a $\boldsymbol{w} \in \mathbb{S}^2$ with $|w_1| \geq |w_2| > |w_3|$ and

$$|F\boldsymbol{w}|^2 = \max_{U \in K} |U\boldsymbol{w}|^2.$$

Then $F \in (\mathrm{SO}(3)U_1 \cup \mathrm{SO}(3)U_2)^{(1)}$. Moreover, if $|w_1| > |w_2|$ then $F \in K$. Thus equality in just one of the necessary conditions

$$|F\boldsymbol{w}|^2 \leq \max_{U \in K} |U\boldsymbol{w}|^2 \quad \forall \boldsymbol{w} \in \mathbb{S}^2$$

already implies that F has to be the center of mass of a simple laminate unless $\boldsymbol{w} = (w_1, w_2, w_3)$ satisfies $|w_1| = |w_2| = |w_3|$, i.e., $\boldsymbol{w}$ is one of the eight vectors $\frac{1}{\sqrt{3}}(\pm 1, \pm 1, \pm 1)$. Therefore it is not surprising that the polyconvex hull of three wells cannot be characterized by

$$\{F : \det F = \eta_1^2 \eta_2,\, |F\boldsymbol{e}|^2 \leq \max_{U \in K} |U\boldsymbol{e}|^2,\, |\operatorname{cof} F\boldsymbol{e}|^2 \leq \max_{U \in K} |\operatorname{cof} U\boldsymbol{e}|^2\, \forall \boldsymbol{e} \in \mathbb{S}^2\}.$$

As a matter of fact, this formula misses a crucial compensation effect between the different wells. The correct representation has therefore to include at least the additional inequalities

$$\begin{aligned} \{F :\ & \det F = \eta_1^2 \eta_2, \\ & |F\boldsymbol{v}|^2 + c(\boldsymbol{v}, \boldsymbol{w}) |\operatorname{cof} F\boldsymbol{w}|^2 \leq \max_{U \in K} \left(|U\boldsymbol{v}|^2 + c(\boldsymbol{v}, \boldsymbol{w}) |\operatorname{cof} U\boldsymbol{w}|^2 \right) \qquad (5.5) \\ & \text{for all } \boldsymbol{v},\, \boldsymbol{w} \in \mathbb{S}^2\}. \end{aligned}$$

The foregoing arguments imply in particular that one needs to understand matrices that realize equality in conditions involving pairs of vectors $(\boldsymbol{v}, \boldsymbol{w})$ with $|U_i\boldsymbol{v}|^2 + c(\boldsymbol{v}, \boldsymbol{w}) |\operatorname{cof} U_i\boldsymbol{w}|^2 = \max_{U \in K} \left(|U\boldsymbol{v}|^2 + c(\boldsymbol{v}, \boldsymbol{w}) |\operatorname{cof} U\boldsymbol{w}|^2 \right)$ for $i = 1, 2, 3$.

The following proposition shows that it is also difficult to construct explicitly laminates that are supported on all three wells. In fact, any laminate supported on three wells must contain at least four Dirac masses.

Proposition 5.1.5. *Suppose that* $K = \mathrm{SO}(3)U_1 \cup \mathrm{SO}(3)U_2 \cup \mathrm{SO}(3)U_3$, *where the matrices* U_i *are given by (5.3) with* $\eta_2 > \eta_1$. *Suppose that* $\nu \in \mathcal{M}^{\mathrm{qc}}(K)$ *is supported on all three wells, i.e., that* ν *has positive mass on* $\mathrm{SO}(3)U_i$ *for* $i = 1, 2, 3$. *Then the support of* ν *must contain at least four points.*

Proof. Assume that ν is supported on three points, i.e., that ν is given by

$$\nu = \lambda_1 X_1 + \lambda_2 X_2 + \lambda_3 X_3, \quad X_i \in \mathrm{SO}(3)U_i,\, \lambda_i > 0,\, \lambda_1 + \lambda_2 + \lambda_3 = 1,$$

and let $F = \langle \nu, id \rangle$. We divide the proof into three steps.

Step 1: Without loss of generality we may assume that

$$\operatorname{rank}(X_1 - X_2) = 1, \qquad \operatorname{rank}\Big(\big(\frac{\lambda_1}{1-\lambda_3}X_1 + \frac{\lambda_2}{1-\lambda_3}X_2\big) - X_3\Big) = 1.$$

Indeed, it follows from Šverák's results that any gradient Young measure supported on three incompatible matrices is a single Dirac mass that at least two of the three matrices X_i must be rank-one connected. We may therefore assume that $X_1 - X_2 = \boldsymbol{a} \otimes \boldsymbol{n}$ with $\boldsymbol{a}$, $\boldsymbol{n} \in \mathbb{R}^3$, $\boldsymbol{a}$, $\boldsymbol{n} \neq 0$. In this situation, we follow arguments by James and Kinderlehrer which imply that

$$\operatorname{rank}\Big(\big(\frac{\lambda_1}{1-\lambda_3}X_1 + \frac{\lambda_2}{1-\lambda_3}X_2\big) - X_3\Big) = 1. \tag{5.6}$$

To show this, we use the identity

$$\begin{aligned}\operatorname{cof} F = \sum_{i=1}^{3} \lambda_i \operatorname{cof} X_i - &\frac{\lambda_1\lambda_2}{1-\lambda_3}\operatorname{cof}(X_2 - X_1)\\ &-\lambda_3(1-\lambda_3)\operatorname{cof}\Big(\big(\frac{\lambda_1}{1-\lambda_3}X_1 + \frac{\lambda_2}{1-\lambda_3}X_2\big) - X_3\Big)\end{aligned}$$

and insert the minors relations

$$\begin{aligned} F &= \lambda_1 X_1 + \lambda_2 X_2 + \lambda_3 X_3,\\ \operatorname{cof} F &= \lambda_1 \operatorname{cof} X_1 + \lambda_2 \operatorname{cof} X_2 + \lambda_3 \operatorname{cof} X_3,\end{aligned}$$

to obtain that

$$\operatorname{cof}\Big(\big(\frac{\lambda_1}{1-\lambda_3}X_1 + \frac{\lambda_2}{1-\lambda_3}X_2\big) - X_3\Big) = 0.$$

Step 2: We have

$$\lambda_{\max}(\widetilde{X}) = \sup_{\boldsymbol{e}\in\mathbb{S}^2} |\widetilde{X}\boldsymbol{e}| < \eta_2 \quad \text{where} \quad \widetilde{X} = \frac{\lambda_1}{1-\lambda_3}X_1 + \frac{\lambda_2}{1-\lambda_3}X_2.$$

Assume that $\lambda_{\max}(\widetilde{X}) = \eta_2$. Then there exists an $\widetilde{\boldsymbol{e}} \in \mathbb{S}^2$ such that $|\widetilde{X}\widetilde{\boldsymbol{e}}| = \eta_2$ and thus

$$\eta_2 = |\widetilde{X}\widetilde{\boldsymbol{e}}| \le \widetilde{\lambda}|U_1\widetilde{\boldsymbol{e}}| + (1-\widetilde{\lambda})|U_2\widetilde{\boldsymbol{e}}| \le \eta_2 \text{ with } \widetilde{\lambda} = \frac{\lambda_1}{1-\lambda_3}.$$

In particular, $|U_1\widetilde{\boldsymbol{e}}| = \eta_2$ and $|U_2\widetilde{\boldsymbol{e}}| = \eta_2$ which implies that $\widetilde{\boldsymbol{e}}$ has to be parallel to $\boldsymbol{e}_1$ and $\boldsymbol{e}_2$, respectively. This is only possible if $\widetilde{\boldsymbol{e}} = 0$, a contradiction.

Step 3: The wells $\mathrm{SO}(3)\widetilde{X}$ *and* $\mathrm{SO}(3)U_3$ *are incompatible.*

Without loss of generality we may assume that $X_2 = U_2$. Since there are exactly two rank-one connections between the wells $\mathrm{SO}(3)U_1$ and $\mathrm{SO}(3)U_2$, the matrix X_1 has to be equal to either $Q_1 U_1$ or $Q_2 U_1$. A short calculation shows that

$$Q_1 = \begin{pmatrix} \frac{2\eta_1\eta_2}{\eta_2^2+\eta_1^2} & \frac{\eta_2^2-\eta_1^2}{\eta_2^2+\eta_1^2} & 0 \\ -\frac{\eta_2^2-\eta_1^2}{\eta_2^2+\eta_1^2} & \frac{2\eta_1\eta_2}{\eta_2^2+\eta_1^2} & 0 \\ 0 & 0 & 1 \end{pmatrix}, \qquad Q_2 = \begin{pmatrix} \frac{2\eta_1\eta_2}{\eta_2^2+\eta_1^2} & -\frac{\eta_2^2-\eta_1^2}{\eta_2^2+\eta_1^2} & 0 \\ \frac{\eta_2^2-\eta_1^2}{\eta_2^2+\eta_1^2} & \frac{2\eta_1\eta_2}{\eta_2^2+\eta_1^2} & 0 \\ 0 & 0 & 1 \end{pmatrix}.$$

We may suppose that $X_1 = Q_1 U_1$, and thus

$$\widetilde{X} = \operatorname{diag}(\widehat{X}, \eta_1) = \begin{pmatrix} \eta_1 + \frac{\eta_1\widetilde{\lambda}(\eta_2^2-\eta_1^2)}{\eta_2^2+\eta_1^2} & \frac{\eta_1\widetilde{\lambda}(\eta_2^2-\eta_1^2)}{\eta_2^2+\eta_1^2} & 0 \\ -\frac{\eta_1\widetilde{\lambda}(\eta_2^2-\eta_1^2)}{\eta_2^2+\eta_1^2} & \eta_2 - \frac{\eta_2\widetilde{\lambda}(\eta_2^2-\eta_1^2)}{\eta_2^2+\eta_1^2} & 0 \\ 0 & 0 & \eta_1 \end{pmatrix}. \tag{5.7}$$

By the polar decomposition theorem, there exists a $\widehat{Q} \in SO(2)$ such that

$$Y = \operatorname{diag}(\widehat{Q}, 1)\widetilde{X} = (\widetilde{X}^T\widetilde{X})^{1/2} = \operatorname{diag}(\widehat{Y}, \eta_1),$$

and $\widehat{Y} \in \mathbb{M}^{2\times 2}$ is symmetric. There exists a rank-one connection between $\widetilde{X}$ and the well $SO(3)U_3$ if and only if there exists a rank-one connection between Y and $SO(3)U_3$. In view of Proposition A.2.1, this rank-one connection exists if and only if the middle eigenvalue λ_{mid} of

$$Z = U_3^{-1}Y^2U_3^{-1} = U_3^{-1}\widetilde{X}^T\widetilde{X}U_3^{-1}$$

is equal to one. By (5.7),

$$Z = \operatorname{diag}(\widehat{Z}, \frac{\eta_1^2}{\eta_2^2}), \quad \widehat{Z} \in \mathbb{M}^{2\times 2}$$

with

$$\widehat{Z} = \frac{1}{\eta_1^2}\widehat{Y}^T\widehat{Y} = \frac{1}{\eta_1^2}\widehat{X}^T\widehat{X}, \quad \det \widehat{Z} = \frac{\eta_2^2}{\eta_1^2}.$$

By Step 2,

$$\lambda_{\max}(\widehat{Z}) < \frac{\eta_2^2}{\eta_1^2} \text{ for } \widetilde{\lambda} \notin \{0, 1\}.$$

This implies

$$\lambda_{\text{mid}}\frac{\eta_2^2}{\eta_1^2} > \lambda_{\max}\lambda_{\text{mid}} = \det \widehat{Z} = \frac{\eta_2^2}{\eta_1^2}$$

and hence the middle eigenvalue of Z is strictly bigger than one for $\widetilde{\lambda} \notin \{0, 1\}$. The wells are therefore incompatible.

This contradicts (5.6) and the assertion of the proposition follows. □

Table 5.2. Rank-one connections (twins) in the cubic to trigonal transformation.

(ij)	R	type	n	a	$U_j^{-1}a$
(12)	R_2^π	compound	$(0,1,0)$	*	$\frac{2\beta(2\alpha+\beta)}{\alpha^2+2\alpha\beta+3\beta^2}(1,0,1)$
(12)	R_{13}^π	compound	$\frac{1}{\sqrt{2}}(1,0,1)$	*	$\frac{2\sqrt{2}\beta(2\alpha+\beta)}{\alpha^2+2\beta^2}(0,1,0)$

5.2 The Cubic to Trigonal Transformation

If the lengths of the sides of the unit cell in the cubic and the trigonal phase are the same, then this transformation is in a suitable basis characterized by the four strains

$$U_1 = \begin{pmatrix} \alpha & \beta & \beta \\ \beta & \alpha & \beta \\ \beta & \beta & \alpha \end{pmatrix}, \quad U_2 = \begin{pmatrix} \alpha & -\beta & \beta \\ -\beta & \alpha & -\beta \\ \beta & -\beta & \alpha \end{pmatrix},$$
$$U_3 = \begin{pmatrix} \alpha & \beta & -\beta \\ \beta & \alpha & -\beta \\ -\beta & -\beta & \alpha \end{pmatrix}, \quad U_4 = \begin{pmatrix} \alpha & -\beta & -\beta \\ -\beta & \alpha & \beta \\ -\beta & \beta & \alpha \end{pmatrix}.$$

In this situation, we have the following result.

Corollary 5.2.1. *Assume that* $\alpha > \beta$, $\alpha > 0$ *and that the set* K *is given by* $K = SO(3)U_1 \cup \ldots \cup SO(3)U_4$. *Suppose that* $\nu \in \mathcal{M}^{pc}(K)$ *is a simple laminate and let* $F = \langle \nu, id \rangle$. *Then* $\mathcal{M}^{pc}(K; F) = \{\nu\}$.

Proof. By symmetry and invariance under the point group it suffices to prove the statement of the corollary for $i = 1$ and $j = 2$. If F is generated from the compound twinning system with $n = e_2$ (cf. Table 5.2), then

$$|U_1 w|^2 = |U_2 w|^2 = (\alpha+\beta)^2 + 2\beta^2 > (\alpha-\beta)^2 = |U_3 w|^2 = |U_4 w|^2$$

with $w = (e_1 + e_3)/\sqrt{2}$. We conclude from Proposition 4.1.9 that ν must be supported on $SO(3)U_1 \cup SO(3)U_2$ and since $-e_1 + e_3$ is a common eigenvector of U_1 and U_2 (with eigenvalue $\alpha - \beta$) we obtain from Proposition 4.1.8 the uniqueness of ν. Similarly, if F is generated using the second normal parallel to $e_1 + e_3$, then

$$|\operatorname{cof} U_1 w|^2 = |\operatorname{cof} U_2 w|^2 = (\alpha^2 + \alpha\beta - 2\beta^2)^2$$
$$> (\alpha-\beta)^2(\alpha^2 + 2\beta^2) = |\operatorname{cof} U_3 w|^2 = |\operatorname{cof} U_4 w|^2,$$

where $w = \frac{1}{\sqrt{2}}(e_1 - e_3)$. We conclude as before. □

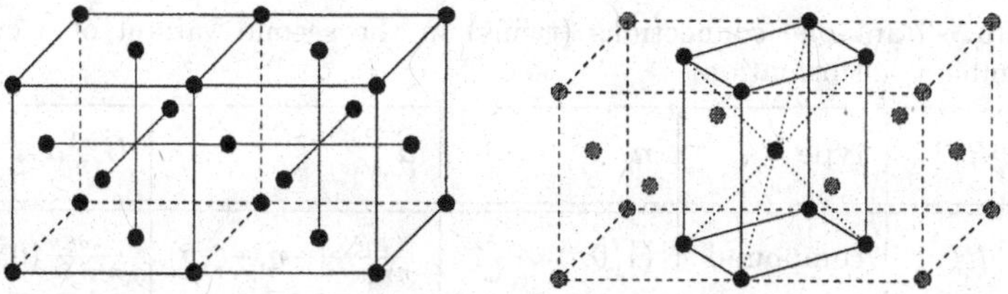

Fig. 5.1. The second type of the cubic to orthorhombic phase transformation. The tetragonal unit cell within the cubic lattice is stretched.

5.3 The Cubic to Orthorhombic Transformation

The orthorhombic phase is characterized by three mutually perpendicular axes of twofold symmetry. If these three axes are parallel to the edges of the cubic cell of the parent phase, then the transformation strains are described by the six matrices

$$U_1 = \begin{pmatrix} \alpha & 0 & 0 \\ 0 & \beta & 0 \\ 0 & 0 & \gamma \end{pmatrix}, \quad U_2 = \begin{pmatrix} \beta & 0 & 0 \\ 0 & \alpha & 0 \\ 0 & 0 & \gamma \end{pmatrix}, \quad U_3 = \begin{pmatrix} \alpha & 0 & 0 \\ 0 & \gamma & 0 \\ 0 & 0 & \beta \end{pmatrix},$$
$$U_4 = \begin{pmatrix} \beta & 0 & 0 \\ 0 & \gamma & 0 \\ 0 & 0 & \alpha \end{pmatrix}, \quad U_5 = \begin{pmatrix} \gamma & 0 & 0 \\ 0 & \alpha & 0 \\ 0 & 0 & \beta \end{pmatrix}, \quad U_6 = \begin{pmatrix} \gamma & 0 & 0 \\ 0 & \beta & 0 \\ 0 & 0 & \alpha \end{pmatrix}.$$

There seems to be no material known with this symmetry in the martensitic phase. A second type of cubic to orthorhombic transformations occurs in materials with a fcc parent phase, see Figure 5.1. It is characterized by two axes of symmetry along face diagonals of the cubic cell and one along the edge orthogonal to this face. The six variants are usually described by the matrices

$$U_1 = \begin{pmatrix} \xi & \eta & 0 \\ \eta & \xi & 0 \\ 0 & 0 & \zeta \end{pmatrix}, \quad U_2 = \begin{pmatrix} \xi & -\eta & 0 \\ -\eta & \xi & 0 \\ 0 & 0 & \zeta \end{pmatrix}, \quad U_3 = \begin{pmatrix} \xi & 0 & \eta \\ 0 & \zeta & 0 \\ \eta & 0 & \xi \end{pmatrix}, \tag{5.8}$$
$$U_4 = \begin{pmatrix} \xi & 0 & -\eta \\ 0 & \zeta & 0 \\ -\eta & 0 & \xi \end{pmatrix}, \quad U_5 = \begin{pmatrix} \zeta & 0 & 0 \\ 0 & \xi & \eta \\ 0 & \eta & \xi \end{pmatrix}, \quad U_6 = \begin{pmatrix} \zeta & 0 & 0 \\ 0 & \xi & -\eta \\ 0 & -\eta & \xi \end{pmatrix}$$

with

$$\xi = \frac{\alpha + \gamma}{2} > \eta = \frac{\alpha - \gamma}{2} > 0, \quad \zeta = \beta > 0. \tag{5.9}$$

This transformation is for example found in CuAlNi alloys, the high temperature phase is body centered cubic, the low temperature phase has orthorhombic symmetry.

Table 5.3. Rank-one connections (twins) in the second variant of a cubic to orthorhombic transformation.

(ij)	R	type	$\boldsymbol{n}$	$\boldsymbol{a}$	$U_j^{-1}\boldsymbol{a}$
(12)	R_1^π	compound	$(1,0,0)$	$\frac{4\eta\xi}{\eta^2+\xi^2}(-\eta,-\xi,0)$	$\frac{4\eta\xi}{\eta^2+\xi^2}(0,1,0)$
(12)	R_2^π	compound	$(0,1,0)$	$\frac{4\eta\xi}{\eta^2+\xi^2}(\xi,-\eta,0)$	$\frac{4\eta\xi}{\eta^2+\xi^2}(1,0,0)$
(13)	R_{2-3}^π	type-I	$\frac{1}{\sqrt{2}}(0,1,-1)$	*	*
(13)	R_{2-3}^π	type-II	*	$\frac{1}{\sqrt{2}}(-\eta,\zeta,-\xi)$	$\frac{1}{\sqrt{2}}(0,1,-1)$

Theorem 5.3.1. *Assume that $K = \mathrm{SO}(3)U_1 \cup \ldots \cup \mathrm{SO}(3)U_6$ where the matrices U_i are defined in (5.8). Suppose that $\nu \in \mathcal{M}^{\mathrm{pc}}(K)$ is a simple laminate supported on K and let $F = \langle \nu, id \rangle$.*

1. *Suppose the rank-one connection used in the definition of F determines a compound twin. Then $\mathcal{M}^{\mathrm{pc}}(K;F) = \{\nu\}$ for all the parameters ξ, η, and ζ satisfying (5.9), except those such that*
$$\zeta^2 = \xi^2 + \eta^2, \quad \text{or} \quad \zeta^2 = \frac{(\xi^2-\eta^2)^2}{\xi^2+\eta^2}.$$
If these identities hold, then $\mathcal{M}^{\mathrm{pc}}(K;F)$ consists of a one-parameter family of simple laminates supported on at most four matrices. Moreover, F has two distinct representations as a simple laminate if and only if $\lambda = \frac{1}{2}$.
2. *If F is generated by a type-I twinning system, then $\mathcal{M}^{\mathrm{pc}}(K;F) = \{\nu\}$ for all the parameters ξ, η, and ζ satisfying (5.9), except those such that*
$$\zeta = \xi + \eta.$$
If this identity holds, then $\mathcal{M}^{\mathrm{pc}}(K;F)$ can be obtained from a three-well problem in two dimensions. Moreover, F has two distinct representations as a simple laminate with mass λ on one well and mass $1-\lambda$ on the other well in the type-I twinning system if and only if λ is equal to one of the two values λ_1 and λ_2 defined by
$$\lambda_1 = \frac{1}{3}\Big(4 - \frac{\xi}{\eta} - \frac{\eta}{\xi}\Big) \quad \text{and} \quad \lambda_2 = \frac{1}{3}\Big(\frac{\xi}{\eta} + \frac{\eta}{\xi} - 1\Big)$$
and λ_1 and λ_2 lie in the open interval $(0,1)$.
3. *If F is generated by a type-II twinning system, then $\mathcal{M}^{\mathrm{pc}}(K;F) = \{\nu\}$ for all the parameters ξ, η, and ζ satisfying (5.9), except those such that*

$$\zeta = \xi - \eta.$$

If this identity holds, then $\mathcal{M}^{\mathrm{pc}}(K;F)$ can be obtained from a three-well problem in two dimensions. Moreover, F has two distinct representations as a simple laminate if and only if the mass λ on one of the two wells is equal to one of the two values λ_1 and λ_2 defined in 2.

Remark 5.3.2. It follows from the results in Chapter 2 that the semiconvex hulls are all equal for the three-well problems. Therefore the characterizations of the polyconvex hulls in the exceptional cases are also characterizations for the quasiconvex hulls.

Proof. We divide the proof into two parts. We first prove the uniqueness assertions and then the characterizations for the sets $\mathcal{M}^{\mathrm{pc}}(K;F)$ for the exceptional cases.

In order to prove uniqueness, it suffices to consider laminates for which one of the matrices is in $\mathrm{SO}(3)U_1$. Assume first that F is a compound twin with normal $\boldsymbol{n} = \boldsymbol{e}_1$.

We have

$$|U_1\boldsymbol{e}_3|^2 = |U_2\boldsymbol{e}_3|^2 > \max_{i=3,\dots,6} |U_i\boldsymbol{e}_3|^2, \qquad \zeta^2 > \eta^2 + \xi^2,$$

$$|U_1\boldsymbol{e}_2|^2 = |U_2\boldsymbol{e}_2|^2 = |U_5\boldsymbol{e}_3|^2 = |U_6\boldsymbol{e}_3|^2 > |U_3\boldsymbol{e}_2|^2 = |U_4\boldsymbol{e}_2|^2, \quad \zeta^2 < \eta^2 + \xi^2.$$

Moreover,

$$|\operatorname{cof} U_1\boldsymbol{e}_1|^2 = \ldots = |\operatorname{cof} U_4\boldsymbol{e}_1|^2 > |\operatorname{cof} U_5\boldsymbol{e}_1|^2 = |\operatorname{cof} U_6\boldsymbol{e}_1|^2$$

for $(\eta^2 + \xi^2)\zeta^2 > (\eta^2 - \xi^2)^2$ and

$$|\operatorname{cof} U_1\boldsymbol{e}_3|^2 = |\operatorname{cof} U_2\boldsymbol{e}_3|^2 > \max_{i=3,\dots,6} |\operatorname{cof} U_i\boldsymbol{e}_3|^2$$

for $(\eta^2 + \xi^2)\zeta^2 < (\eta^2 - \xi^2)^2$, and we can therefore choose appropriate vectors $\boldsymbol{w}_i$ if the assumptions in the theorem hold. The nonuniqueness part is discussed in detail below.

Suppose now that F is obtained from the type-I twinning system between $\mathrm{SO}(3)U_1$ and $\mathrm{SO}(3)U_3$. Let $\boldsymbol{w} = (\xi, \xi, \xi)$. Then

$$|U_1\boldsymbol{w}|^2 = |U_3\boldsymbol{w}|^2 = |U_5\boldsymbol{w}|^2 > |U_2\boldsymbol{w}|^2 = |U_4\boldsymbol{w}|^2 = |U_6\boldsymbol{w}|^2. \tag{5.10}$$

Therefore it only remains to show that U_5 cannot be part of the microstructure. We have to prove the existence of a vector $\boldsymbol{v} = (t, s, s)$ such that

$$|U_1\boldsymbol{v}|^2 = |U_3\boldsymbol{v}|^2 > |U_5\boldsymbol{v}|^2$$

(since $\boldsymbol{n}$ is parallel to $(0, 1, -1)$ we have automatically that $|F\boldsymbol{v}|^2 = |U_1\boldsymbol{v}|^2$). This is equivalent to

$$g(s,t) = \{s^2\zeta^2 + (t\eta + s\xi)^2 + (s\eta + t\xi)^2\} - \{t^2\zeta^2 + 2s^2(\eta+\xi)^2\} > 0.$$

Since $g(t,t) = 0$ and

$$\partial_s g(t,t) = 2t(\zeta^2 - 2(\eta+\xi)^2),$$

this is possible if $\zeta^2 \neq (\eta+\xi)^2$, i.e., the hypothesis of the theorem holds. Similarly, if F is a type-II twin, then we have for $\boldsymbol{w} = (-\xi, \xi, \xi)$ that

$$\begin{aligned} &|\operatorname{cof} U_1 \boldsymbol{w}|^2 = |\operatorname{cof} U_3 \boldsymbol{w}|^2 = |\operatorname{cof} U_6 \boldsymbol{w}|^2 \\ &> |\operatorname{cof} U_2 \boldsymbol{w}|^2 = |\operatorname{cof} U_4 \boldsymbol{w}|^2 = |\operatorname{cof} U_5 \boldsymbol{w}|^2 \end{aligned}$$

and we only have to exclude U_6 from the microstructure. Let $\boldsymbol{v} = (s,t,t)$. Then

$$|\operatorname{cof} U_1 \boldsymbol{v}|^2 = |\operatorname{cof} U_3 \boldsymbol{v}|^2$$

and we have to determine $s, t \in \mathbb{R}$ such that

$$g(s,t) = |\operatorname{cof} U_1 \boldsymbol{v}|^2 - |\operatorname{cof} U_6 \boldsymbol{v}|^2 > 0.$$

Since $g(-t,t) = 0$ and

$$\partial_s g(-t,t) = -2t(\eta+\xi)^2(\zeta^2 - (\eta-\xi)^2)$$

this is possible if the assumptions in ii) hold. Finally, the remaining cases follow by symmetry.

We now turn towards the characterizations in the case that the polyconvex measure is not uniquely determined. We discuss the three different cases separately.

1) We assume first that $\zeta^2 = \xi^2 + \eta^2$ and that

$$F = tQU_1 + (1-t)U_2 = U_2 + t\boldsymbol{a} \otimes \boldsymbol{n},$$

where Q and $\boldsymbol{a}$ are given by Proposition A.2.4 with $\boldsymbol{n} = \boldsymbol{e}_1$, i.e.

$$F = \begin{pmatrix} \xi(1 - \frac{4t\eta^2}{\xi^2+\eta^2}) & -\eta & 0 \\ \eta(-1 + \frac{4t\xi^2}{\xi^2+\eta^2}) & \xi & 0 \\ 0 & 0 & \zeta \end{pmatrix}. \tag{5.11}$$

Assume that $\nu \in \mathcal{M}^{\mathrm{pc}}(K)$ satisfies $\langle \nu, id \rangle = F$. Since

$$|F\boldsymbol{e}_2|^2 = |U_i \boldsymbol{e}_2|^2, \quad |F\boldsymbol{e}_3|^2 = |U_i \boldsymbol{e}_3|^2, \quad i = 1, \ldots, 6,$$

and

$$|\operatorname{cof} F\boldsymbol{e}_1|^2 = \cdots = |\operatorname{cof} U_4 \boldsymbol{e}_1|^2 > |\operatorname{cof} U_5 \boldsymbol{e}_1|^2 > |\operatorname{cof} U_6 \boldsymbol{e}_1|^2,$$

ν must be supported on $\mathrm{SO}(3)U_1\cup\ldots\cup\mathrm{SO}(3)U_4$ and thus ν can be represented as

$$\nu = \sum_{i=1}^{4} \lambda_i \nu_i U_i, \quad \nu_i \in \mathcal{P}(\mathrm{SO}(3)), \quad \lambda_i \geq 0, \quad \lambda_1 + \lambda_2 + \lambda_3 + \lambda_4 = 1.$$

Let $X_i = \langle \nu_i, id \rangle \in \operatorname{conv}\mathrm{SO}(3)$. Thus $\langle \nu, id \rangle = \sum_{i=1}^{4} \lambda_i X_i U_i$ and

$$|F\boldsymbol{e}_2| = \Big|\sum_{i=1}^{4} \lambda_i X_i U_i \boldsymbol{e}_2\Big| \leq \sum_{i=1}^{4} \lambda_i |X_i U_i \boldsymbol{e}_2| \leq |F\boldsymbol{e}_2|.$$

Therefore we have equality everywhere in this chain of inequalities and the strict convexity of the norm implies that all the vectors $X_i U_i \boldsymbol{e}_2$, $i = 1, \ldots, 4$, must have the same direction and maximal length $\xi^2 + \eta^2$. The same assertion follows for $X_i U_i \boldsymbol{e}_3$ and $X_i \operatorname{cof} U_i \boldsymbol{e}_1$, $i = 1, \ldots, 4$. We conclude in view of (5.8) that

$$X_1 \begin{pmatrix} \eta \\ \xi \\ 0 \end{pmatrix} = X_2 \begin{pmatrix} -\eta \\ \xi \\ 0 \end{pmatrix} = X_3 \begin{pmatrix} 0 \\ \zeta \\ 0 \end{pmatrix} = X_4 \begin{pmatrix} 0 \\ \zeta \\ 0 \end{pmatrix},$$

and

$$X_1 \begin{pmatrix} 0 \\ 0 \\ \zeta \end{pmatrix} = X_2 \begin{pmatrix} 0 \\ 0 \\ \zeta \end{pmatrix} = X_3 \begin{pmatrix} \eta \\ 0 \\ \xi \end{pmatrix} = X_4 \begin{pmatrix} -\eta \\ 0 \\ \xi \end{pmatrix},$$

and since $\operatorname{cof}(AB) = \operatorname{cof}(A)\operatorname{cof}(B)$ we have additionally that

$$X_1 \begin{pmatrix} \xi \\ -\eta \\ 0 \end{pmatrix} = X_2 \begin{pmatrix} \xi \\ \eta \\ 0 \end{pmatrix} = X_3 \begin{pmatrix} \xi \\ 0 \\ -\eta \end{pmatrix} = X_4 \begin{pmatrix} \xi \\ 0 \\ \eta \end{pmatrix}.$$

Since

$$\left| X_1 \begin{pmatrix} \xi \\ \pm\eta \\ 0 \end{pmatrix} \right|^2 = \left| X_1 \begin{pmatrix} 0 \\ 0 \\ \zeta \end{pmatrix} \right|^2 = \xi^2 + \eta^2 = \zeta^2,$$

the matrices X_i are length preserving on an orthogonal frame and have singular values bounded by one, since they are a convex combination of matrices in $\mathrm{SO}(3)$. This implies by Lemma 5.3.3 below that $X_i \in \mathrm{O}(3)$. On the other hand, $X_i = \langle \nu_i, id \rangle$ with $\nu_i \in \mathcal{P}(\mathrm{SO}(3))$ and thus $X_i = Q_i \in \mathrm{SO}(3)$. The proper rotation mapping one orthogonal frame $\{\boldsymbol{v}_1, \boldsymbol{v}_2, \boldsymbol{v}_3\}$ onto another orthogonal frame $\{\boldsymbol{w}_1, \boldsymbol{w}_2, \boldsymbol{w}_3\}$ is given by $Q = \boldsymbol{w}_1 \otimes \boldsymbol{v}_1 + \boldsymbol{w}_2 \otimes \boldsymbol{v}_2 + \boldsymbol{w}_3 \otimes \boldsymbol{v}_3$, and we obtain

$$X_2^T X_1 U_1 = \begin{pmatrix} \xi\big(1-\frac{4\eta^2}{\xi^2+\eta^2}\big) & -\eta & 0 \\ \eta\big(-1+\frac{4\xi^2}{\xi^2+\eta^2}\big) & \xi & 0 \\ 0 & 0 & \zeta \end{pmatrix},$$

$$X_2^T X_3 U_3 = \begin{pmatrix} \xi\big(1-\frac{2\eta^2}{\xi^2+\eta^2}\big) & -\eta & 0 \\ \eta\big(-1+\frac{2\xi^2}{\xi^2+\eta^2}\big) & \xi & 0 \\ \frac{2\xi\eta}{\sqrt{\xi^2+\eta^2}} & 0 & \zeta \end{pmatrix},$$

$$X_2^T X_4 U_4 = \begin{pmatrix} \xi\big(1-\frac{2\eta^2}{\xi^2+\eta^2}\big) & -\eta & 0 \\ \eta\big(-1+\frac{2\xi^2}{\xi^2+\eta^2}\big) & \xi & 0 \\ \frac{-2\xi\eta}{\sqrt{\xi^2+\eta^2}} & 0 & \zeta \end{pmatrix}.$$

We conclude

$$X_2^T F = \begin{pmatrix} \widetilde{F}_{11} & -\eta & 0 \\ \widetilde{F}_{21} & \xi & 0 \\ \widetilde{F}_{31} & 0 & \zeta \end{pmatrix} \tag{5.12}$$

with

$$\widetilde{F}_{11} = \lambda_1\xi\big(1-\frac{4\eta^2}{\xi^2+\eta^2}\big) + \lambda_2\xi + (\lambda_3+\lambda_4)\xi\big(1-\frac{2\eta^2}{\xi^2+\eta^2}\big),$$
$$\widetilde{F}_{21} = \lambda_1\eta\big(-1+\frac{4\xi^2}{\xi^2+\eta^2}\big) - \lambda_2\eta + (\lambda_3+\lambda_4)\eta\big(-1+\frac{2\xi^2}{\xi^2+\eta^2}\big),$$
$$\widetilde{F}_{31} = (\lambda_3-\lambda_4)\frac{2\xi\eta}{\sqrt{\xi^2+\eta^2}}.$$

We obtain from (5.11) and (5.12) that $X_2 = \mathbb{I}$ and

$$\lambda_3 = \lambda_4, \quad \lambda_1+\lambda_3 = t, \quad \lambda_1+\lambda_2+\lambda_3+\lambda_4 = 1, \quad \lambda_i \geq 0.$$

If we consider $\lambda_3 = s$ as a free parameter, we can represent all solutions to these equations by

$$\lambda_1(s) = t-s, \quad \lambda_2(s) = 1-t-s, \quad \lambda_3(s) = s, \quad \lambda_4(s) = s$$

for $s \in [0, \min\{t, 1-t\}]$. In the special case $t = \frac{1}{2}$ we may choose $s = 0$ and $s = \frac{1}{2}$, and obtain two representations of F as a simple laminate with $\lambda_1 = \lambda_2 = \frac{1}{2}$ and $\lambda_3 = \lambda_4 = \frac{1}{2}$, respectively. On the other hand, if $t \neq \frac{1}{2}$, then it is easy to see that the only representation of F as a simple laminate corresponds to $\lambda_1 = t$, $\lambda_2 = 1-t$. If we define $\mu = 1-2s$, $s \in [0, \min\{t, 1-t\}]$, and

$$G_1 = \frac{\lambda_1}{1-2\lambda_3} X_1 U_1 + \frac{\lambda_2}{1-2\lambda_3} U_2 = \frac{t-s}{1-2s} X_1 U_1 + \frac{1-t-s}{1-2s} U_2,$$

$$G_2 = \frac{1}{2}(X_3 U_3 + X_4 U_4),$$

then G_1 and G_2 are rank-one connected and

$$F = \mu G_1 + (1-\mu) G_2, \quad \mu \in [\max\{1-2t, 2t-1\}, 1].$$

This proves the assertion for the case that $\boldsymbol{n} = \boldsymbol{e}_1$ in the definition of the compound twins. The other case with $\boldsymbol{n} = \boldsymbol{e}_2$ follows by symmetry. Assume now that $\zeta^2 = (\xi^2 - \eta^2)^2/(\xi^2 + \eta^2)$, i.e. $\zeta^2 < \xi^2 + \eta^2$. If F is generated from the compound twin system with $\boldsymbol{n} = \boldsymbol{e}_1$, then $|U_i \boldsymbol{e}_2|^2 = |F\boldsymbol{e}_2|^2 > |U_3 \boldsymbol{e}_2|^2 = |U_4 \boldsymbol{e}_2|^2$ for $i = 1,2,5,6$, and thus the polyconvex measure must be supported on $SO(3)U_1 \cup SO(3)U_2 \cup SO(3)U_5 \cup SO(3)U_6$. Since $|\operatorname{cof} U_i \boldsymbol{e}_1|^2 = |\operatorname{cof} F\boldsymbol{e}_1|^2$ and $|\operatorname{cof} U_i \boldsymbol{e}_3|^2 = |\operatorname{cof} F\boldsymbol{e}_3|^2$ for $i = 1,2,5,6$, we conclude as before. The last case, the compound twin with $\boldsymbol{n} = \boldsymbol{e}_2$, follows again by symmetry.

2) Assume now that $\zeta = \xi + \eta$. We consider the case that F is obtained from the type-I twinning system between $SO(3)U_1$ and $SO(3)U_3$, i.e.

$$F = U_3 + t\boldsymbol{a} \otimes \boldsymbol{n}, \quad \boldsymbol{n} = \frac{1}{\sqrt{2}} \begin{pmatrix} 0 \\ 1 \\ -1 \end{pmatrix}$$

and $\boldsymbol{a}$ is given by Proposition A.2.4. It follows from (5.10) that every measure $\nu \in \mathcal{M}^{\mathrm{pc}}(K)$ with $\langle \nu, id \rangle$ is supported on $SO(3)U_1 \cup SO(3)U_3 \cup SO(3)U_5$. Since $\frac{1}{\sqrt{3}}(1,1,1)$ is a common eigenvector of U_1, U_3 and U_5 under the assumption that $\zeta = \xi + \eta$ we are in a situation in which Theorem 2.5.1 applies. In order to simplify the calculations, we choose a new basis $\mathcal{B}$ of $\mathbb{R}^3$ by

$$\mathcal{B} = \left\{ \frac{1}{\sqrt{3}} \begin{pmatrix} 1 \\ 1 \\ 1 \end{pmatrix}, \ \frac{1}{\sqrt{6}} \begin{pmatrix} 1 \\ 1 \\ -2 \end{pmatrix}, \ \frac{1}{\sqrt{2}} \begin{pmatrix} 1 \\ -1 \\ 0 \end{pmatrix} \right\},$$

and obtain the representations $\widetilde{U}_i$ of U_i in this basis as

$$\widetilde{U}_1 = \operatorname{diag}(\xi + \eta, \widehat{U}_1), \quad \widetilde{U}_3 = \operatorname{diag}(\xi + \eta, \widehat{U}_3), \quad \widetilde{U}_5 = \operatorname{diag}(\xi + \eta, \widehat{U}_5)$$

with

$$\widehat{U}_1 = \begin{pmatrix} \xi + \eta & 0 \\ 0 & \xi - \eta \end{pmatrix}, \quad \widehat{U}_3 = \begin{pmatrix} \xi - \frac{\eta}{2} & -\frac{\sqrt{3}\eta}{2} \\ -\frac{\sqrt{3}\eta}{2} & \xi + \frac{\eta}{2} \end{pmatrix}, \quad \widehat{U}_5 = \begin{pmatrix} \xi - \frac{\eta}{2} & \frac{\sqrt{3}\eta}{2} \\ \frac{\sqrt{3}\eta}{2} & \xi + \frac{\eta}{2} \end{pmatrix}.$$

The matrix $\widehat{F}$ satisfies $\widehat{F} = tQ_1\widehat{U}_1 + (1-t)\widehat{U}_3$ and in order to prove nonuniqueness of the polyconvex measure we first solve the system $\widehat{F} = R(sQ_5^T \widehat{U}_5 - (1-s)\widehat{U}_3)$ for s and t with $R = R(s) \in SO(2)$. The solutions are given by $(s,t) = (0,0)$ (corresponding to $F = \widehat{U}_3$) and $(s,t) = (\bar{s}, \bar{t})$ with

$$\bar{s} = \bar{t} = \frac{1}{3}\Big(4 - \frac{\xi}{\eta} - \frac{\eta}{\xi}\Big).$$

This implies that we have two representations of F as a simple laminate for parameters $\bar{t} \in (0,1)$. If $\bar{t} \notin (0,1)$, then the two curves on the surface of constant determinant $\det F = \xi^2 - \eta^2$ corresponding to the (symmetrized) rank-one lines between $(\widehat{U}_1)^T\widehat{U}_1$ and $U_3^T U_3$ and $(\widehat{U}_5)^T\widehat{U}_5$ and $U_3^T U_3$, respectively, do not intersect. However, there is a one parameter family of second order laminates that generate F which can for example be generated by fixing any matrix of the form $G_1 = \mu Q_2 \widehat{U}_1 + (1-\mu)\widehat{U}_2$ and defining G_2 to be the intersection point of the rank-one line through G_1 and F with the boundary of the quasiconvex hull of $\mathrm{SO}(2)\widehat{U}_1 \cup \mathrm{SO}(2)\widehat{U}_3 \cup \mathrm{SO}(2)\widehat{U}_5$

3) We assume finally that $\zeta = \xi - \eta$ and that F is the barycenter of a simple laminate given by the type-II twins between $\mathrm{SO}(3)U_1$ and $\mathrm{SO}(3)U_3$. For completeness, we summarize the arguments in this case. As in the proof of Theorem 5.3.1 we deduce with $\boldsymbol{w} = (-1,1,1)$ and $\boldsymbol{v} = (s,t,t)$ as test vectors that any $\nu \in \mathcal{M}^{\mathrm{pc}}(K;F)$ is supported on $\mathrm{SO}(3)U_1 \cup \mathrm{SO}(3)U_3 \cup \mathrm{SO}(3)U_6$. Since $\frac{1}{\sqrt{3}}(1,1,1)$ is a common eigenvector of U_1, U_3 and U_6 under the assumption that $\zeta = \xi - \eta$ we are in a situation in which Theorem 2.5.1 applies. In order to simplify the calculations, we choose a new basis $\mathcal{B}$ of $\mathbb{R}^3$ by

$$\mathcal{B} = \left\{ \frac{1}{\sqrt{3}}\begin{pmatrix} -1 \\ 1 \\ 1 \end{pmatrix},\ \frac{1}{\sqrt{6}}\begin{pmatrix} -1 \\ 1 \\ -2 \end{pmatrix},\ \frac{1}{\sqrt{2}}\begin{pmatrix} 1 \\ 1 \\ 0 \end{pmatrix} \right\}.$$

The transformed matrices are given by

$$\widetilde{U}_1 = \mathrm{diag}(\xi - \eta, \widehat{U}_1), \quad \widetilde{U}_3 = \mathrm{diag}(\xi - \eta, \widehat{U}_3), \quad \widetilde{U}_6 = \mathrm{diag}(\xi - \eta, \widehat{U}_6)$$

with

$$\widehat{U}_1 = \begin{pmatrix} \xi - \eta & 0 \\ 0 & \xi + \eta \end{pmatrix}, \quad \widehat{U}_3 = \begin{pmatrix} \xi + \frac{\eta}{2} & -\frac{\sqrt{3}\eta}{2} \\ -\frac{\sqrt{3}\eta}{2} & \xi - \frac{\eta}{2} \end{pmatrix}, \quad \widehat{U}_6 = \begin{pmatrix} \xi + \frac{\eta}{2} & \frac{\sqrt{3}\eta}{2} \\ \frac{\sqrt{3}\eta}{2} & \xi - \frac{\eta}{2} \end{pmatrix}.$$

The analysis has thus been reduced to the three-well problem in Section 2.2, since the matrices $J\widehat{U}_j J^T$ are identical to the matrices considered there. The proof of the theorem is now complete. □

The proof of the foregoing theorem used the following fact.

Lemma 5.3.3. *Assume that $A \in \mathbb{M}^{n\times n}$ satisfies $|A\boldsymbol{v}_i| = 1$ on an orthonormal basis $\{\boldsymbol{v}_1, \ldots, \boldsymbol{v}_n\}$ and that the singular values of A are bounded by one. Then $A \in \mathrm{O}(n)$.*

Proof. We may assume that A is symmetric. Let $\{\lambda_i, \boldsymbol{e}_i\}$ be an eigensystem for A, $A = \sum \lambda_i \boldsymbol{e}_i \otimes \boldsymbol{e}_i$, and define $\alpha_i^{(j)}$ by $v_j = \sum \alpha_i^{(j)} \boldsymbol{e}_i$. Then

$$1 = |Av_j|^2 = |\sum_{i=1}^{n} \lambda_i \alpha_i^{(j)} e_i|^2 = \sum_{i=1}^{n} \lambda_i^2 (\alpha_i^{(j)})^2 \leq 1 \quad \text{for all } j,$$

since $|\lambda_i| \leq 1$. If $|\lambda_i| < 1$, then $\alpha_i^{(j)} = 0$ for all j and hence $v_j = 0$, contradicting the assumption that $|Av_j| = 1$. Hence $\lambda_i \in \{\pm 1\}$, and we conclude that $A^T A = \mathbb{I}$ with $\det A \in \{\pm 1\}$. □

5.4 The Tetragonal to Monoclinic Transformations

It is an open problem to characterize uniqueness of microstructure in the cubic to monoclinic phase transformation. However, the ideas described in the analysis of the cubic to tetragonal transformation can be applied to three possible tetragonal to monoclinic transformations, which we will refer to as type-I through type-III. The matrices for the type-I transformation are given by

$$U_1 = \begin{pmatrix} \theta_1 & 0 & \theta_4 \\ 0 & \theta_2 & 0 \\ \theta_4 & 0 & \theta_3 \end{pmatrix}, \quad U_2 = \begin{pmatrix} \theta_1 & 0 & -\theta_4 \\ 0 & \theta_2 & 0 \\ -\theta_4 & 0 & \theta_3 \end{pmatrix},$$
$$U_3 = \begin{pmatrix} \theta_2 & 0 & 0 \\ 0 & \theta_1 & \theta_4 \\ 0 & \theta_4 & \theta_3 \end{pmatrix}, \quad U_4 = \begin{pmatrix} \theta_2 & 0 & 0 \\ 0 & \theta_1 & -\theta_4 \\ 0 & -\theta_4 & \theta_3 \end{pmatrix}.$$

The matrices U_i are positive definite for $\theta_1, \theta_2, \theta_3 > 0$ and $\theta_1\theta_3 - \theta_4^2 > 0$, and we may assume that $\theta_4 > 0$. We summarize the relevant rank-one connections in Table 5.4.

The matrices for the type-II transformation are given by

$$U_1 = \begin{pmatrix} \eta_1 & \eta_3 & \eta_4 \\ \eta_3 & \eta_1 & -\eta_4 \\ \eta_4 & -\eta_4 & \eta_2 \end{pmatrix}, \quad U_2 = \begin{pmatrix} \eta_1 & -\eta_3 & \eta_4 \\ -\eta_3 & \eta_1 & \eta_4 \\ \eta_4 & \eta_4 & \eta_2 \end{pmatrix},$$
$$U_3 = \begin{pmatrix} \eta_1 & \eta_3 & -\eta_4 \\ \eta_3 & \eta_1 & \eta_4 \\ -\eta_4 & \eta_4 & \eta_2 \end{pmatrix}, \quad U_4 = \begin{pmatrix} \eta_1 & -\eta_3 & -\eta_4 \\ -\eta_3 & \eta_1 & -\eta_4 \\ -\eta_4 & -\eta_4 & \eta_2 \end{pmatrix},$$

and we assume that

$$\eta_1, \eta_2 > 0, \quad \eta_1^2 > \eta_3^2, \quad \det U_i = (\eta_1 + \eta_3)(\eta_2(\eta_1 - \eta_3) - 2\eta_4^2) > 0. \quad (5.13)$$

These conditions imply that the matrices U_i are positive definite. The relevant rank-one connections are summarized in Table 5.5. Finally, the transformation of type-III is described by the following matrices:

Table 5.4. *Rank-one connections for the type-I tetragonal to monoclinic transformation.*

(ij)	R	type	$\boldsymbol{n}$	$\boldsymbol{a}$	$U_j^{-1}\boldsymbol{a}$ parallel to
(12)	R_1^π	compound	$(1,0,0)$	*	$(0,0,1)$
(12)	R_3^π	compound	$(0,0,1)$	*	$(1,0,0)$
(13)	R_{1-2}^π	type-I	$\frac{1}{\sqrt{2}}(1,-1,0)$	*	*
(13)	R_{1-2}^π	type-II	*	*	$(1,-1,0)$

Table 5.5. *Rank-one connections for the type-II tetragonal to monoclinic transformation.*

(ij)	R	type	$\boldsymbol{n}$	$\boldsymbol{a}$	$U_j^{-1}\boldsymbol{a}$ parallel to
(12)	R_2^π	type-I	$(0,1,0)$	*	*
(12)	R_2^π	type-II	*	*	$(0,1,0)$
(13)	R_3^π	compound	$(0,0,1)$	*	$(1,-1,0)$
(13)	R_{1-2}^π	compound	$\frac{1}{\sqrt{2}}(1,-1,0)$	*	$(0,0,1)$

$$U_1 = \begin{pmatrix} \delta_1 & \delta_4 & 0 \\ \delta_4 & \delta_2 & 0 \\ 0 & 0 & \delta_3 \end{pmatrix}, \quad U_2 = \begin{pmatrix} \delta_1 & -\delta_4 & 0 \\ -\delta_4 & \delta_2 & 0 \\ 0 & 0 & \delta_3 \end{pmatrix},$$
$$U_3 = \begin{pmatrix} \delta_2 & \delta_4 & 0 \\ \delta_4 & \delta_1 & 0 \\ 0 & 0 & \delta_3 \end{pmatrix}, \quad U_4 = \begin{pmatrix} \delta_2 & -\delta_4 & 0 \\ -\delta_4 & \delta_1 & 0 \\ 0 & 0 & \delta_3 \end{pmatrix}$$

with $\delta_1, \delta_2, \delta_3 > 0$, $\delta_4 \neq 0$ and $\delta_1\delta_2 - \delta_4^2 > 0$. This case is an example for a three dimensional situation which can be reduced to a two-dimensional one, see Theorem 2.5.1, and we do not discuss it in the following theorem.

Theorem 5.4.1. *Assume that the matrices U_i, $i = 1\ldots,4$, describing the tetragonal to monoclinic transformations of type I and II and the corresponding vectors $\boldsymbol{a}$ and $\boldsymbol{n}$ in the representation of the rank-one connections between* *the wells are given by the expressions in Tables 5.4–5.5. Let*

$$K = \mathrm{SO}(3)U_1 \cup \ldots \cup \mathrm{SO}(3)U_4$$

and suppose that F describes a global deformation corresponding to the center of mass of a simple laminate, i.e.

$$F = \lambda Q U_i + (1-\lambda)U_j, \quad QU_i - U_j = \boldsymbol{a} \otimes \boldsymbol{n},$$

$i \neq j$, $\lambda \in [0,1]$, *and* $Q \in \mathrm{SO}(3)$. *Then the microstructure underlying F is unique, and in fact* $\mathcal{M}^{\mathrm{pc}}(K;F) = \{\nu\}$ *with* $\nu = \lambda\delta_{QU_i} + (1-\lambda)\delta_{U_j}$, *unless the situation is symmetry related to one of the following exceptional choices of lattice parameters and twinning systems:*

1. Tetragonal to monoclinic transformation of type I: *If* $i = 1$ *and* $j = 2$, *i.e., F is generated from the compound twinning system between* $\mathrm{SO}(3)U_1$ *and* $\mathrm{SO}(3)U_2$, *then F is not unique if*

$$\boldsymbol{n} = (1,0,0) \quad \text{and} \quad \theta_2^2(\theta_3^2+\theta_4^2) = (\theta_1\theta_3 - \theta_4)^2$$

or

$$\boldsymbol{n} = (0,0,1) \quad \text{and} \quad \theta_2^2 = \theta_1^2 + \theta_4^2.$$

2. Tetragonal to monoclinic transformation of type II: *If* $i = 1$ *and* $j = 3$, *i.e., F is generated from the compound twinning system between* $\mathrm{SO}(3)U_1$ *and* $\mathrm{SO}(3)U_3$, *then F is not unique if*

$$\boldsymbol{n} = (0,0,1) \text{ and } 2\eta_1\eta_3 = \eta_4^2.$$

In all the exceptional cases, $\mathcal{M}^{\mathrm{pc}}(K;F)$ *consists of a one-parameter family of laminates and the matrix F has two different representations as a simple laminate if and only if* $\lambda = \frac{1}{2}$.

Remarks 5.4.2. 1) The proof of the theorem shows that uniqueness of the microstructure can be obtained based on condition ($\mathrm{C_b}$) with $\boldsymbol{n} = \boldsymbol{e}_3$ in Definition 4.1.4. Therefore stability of microstructure is an immediate consequence of the results in Chapter 4.

2) The analysis of the transformation of type-III can be reduced in view of Theorem 2.5.1 and the examples following Proposition 2.2.4 to the two-dimensional situation, and the uniqueness results follow from the corresponding analysis in Chapter 4.

Proof. We sketch the proof of the theorem in the setting of our general framework for uniqueness based on Definition 4.1.4 and Theorem 4.1.14

Proof for the tetragonal to monoclinic transformation of type-I. We consider separately the different twinning systems and establish the existence of the test vectors $\boldsymbol{w}$ which imply uniqueness of microstructure based on Definition 4.1.4.

Case $i = 1$, $j = 2$, $\boldsymbol{n} = \boldsymbol{e}_1$: Assume first that $\theta_2^2(\theta_3^2 + \theta_4^2) \neq (\theta_1\theta_3 - \theta_4)^2$. Since $U_2^{-1}\boldsymbol{a}$ is parallel to $(0, 0, 1)$, we may use $\boldsymbol{w}_1 = (1, 0, 0)$ and $\boldsymbol{w}_2 = (0, 1, 0)$ as test vectors for the cofactor matrices. We obtain

$$\begin{aligned}|F\boldsymbol{w}_1|^2 =& |\operatorname{cof} U_1\boldsymbol{w}_1|^2 = |\operatorname{cof} U_2\boldsymbol{w}_1|^2 = \theta_2^2(\theta_3^2 + \theta_4^2),\\ & |\operatorname{cof} U_3\boldsymbol{w}_1|^2 = |\operatorname{cof} U_4\boldsymbol{w}_1|^2 = (\theta_1\theta_3 - \theta_4)^2,\end{aligned}$$

and

$$\begin{aligned}|\operatorname{cof} F\boldsymbol{w}_2|^2 =& |\operatorname{cof} U_1\boldsymbol{w}_2|^2 = |\operatorname{cof} U_2\boldsymbol{w}_2|^2 = (\theta_1\theta_3 - \theta_4)^2,\\ & |\operatorname{cof} U_3\boldsymbol{w}_1|^2 = |\operatorname{cof} U_4\boldsymbol{w}_1|^2 = \theta_2^2(\theta_3^2 + \theta_4^2),\end{aligned}$$

and thus we may choose $\boldsymbol{w}$ correspondingly to ensure uniqueness. It remains to establish the characterization of $\mathcal{M}^{\mathrm{pc}}(K; F)$ for $\theta_2^2(\theta_3^2 + \theta_4^2) = (\theta_1\theta_3 - \theta_4)^2$. Suppose that

$$\begin{aligned}F &= \lambda Q U_1 + (1 - \lambda) U_2 = U_2 + \lambda \boldsymbol{a} \otimes \boldsymbol{n}\\ &= \begin{pmatrix} \theta_1 & 0 & -\theta_4 \\ 0 & \theta_2 & 0 \\ -\theta_4 & 0 & \theta_3 \end{pmatrix} + \frac{2\lambda(\theta_1 + \theta_3)\theta_4}{\theta_3^2 + \theta_4^2} \begin{pmatrix} -\theta_4 & 0 & 0 \\ 0 & 0 & 0 \\ \theta_3 & 0 & 0 \end{pmatrix}.\end{aligned}$$

We assume as in the proof of Theorem 5.3.1 that $\nu \in \mathcal{M}^{\mathrm{pc}}(K; F)$ is given by

$$\nu = \lambda_1 \delta_{X_1 U_1} + \lambda_2 \delta_{X_2 U_2} + \lambda_3 \delta_{X_3 U_3} + \lambda_4 \delta_{X_4 U_4}$$

with $X_i \in \operatorname{conv} \mathrm{SO}(3)$. It follows that the matrices X_i map the orthogonal vectors $\{\operatorname{cof} U_i \boldsymbol{e}_1, \operatorname{cof} U_i \boldsymbol{e}_2, U_i \boldsymbol{e}_3\}$, $i = 1, \ldots, 4$, onto the orthogonal vectors $\{\operatorname{cof} F\boldsymbol{e}_1, \operatorname{cof} F\boldsymbol{e}_2, F\boldsymbol{e}_3\}$ of same length. We may apply Lemma 5.3.3 and conclude that $X_i \in \mathrm{SO}(3)$ for $i = 1, \ldots, 4$. Therefore the matrices X_i are uniquely determined and a short calculation shows that

$$X_1 U_1 = \begin{pmatrix} \theta_1 - \frac{2\theta_4^2(\theta_1+\theta_3)}{\theta_3^2+\theta_4^2} & 0 & -\theta_4 \\ 0 & \theta_2 & 0 \\ -\theta_4 + \frac{2\theta_3\theta_4(\theta_1+\theta_3)}{\theta_3^2+\theta_4^2} & 0 & \theta_3 \end{pmatrix},$$

that $X_2 = \mathbb{I}$ and therefore $X_2 U_2 = U_2$, that

$$X_3 U_3 = \begin{pmatrix} \theta_1 - \frac{\theta_4^2(\theta_1+\theta_3)}{\theta_3^2+\theta_4^2} & -\frac{\theta_4^2(\theta_1+\theta_3)}{\theta_3^2+\theta_4^2} & -\theta_4 \\ 0 & \theta_2 & 0 \\ -\theta_4 + \frac{\theta_3\theta_4(\theta_1+\theta_3)}{\theta_3^2+\theta_4^2} & \frac{\theta_3\theta_4(\theta_1+\theta_3)}{\theta_3^2+\theta_4^2} & \theta_3 \end{pmatrix},$$

and finally that

$$X_4U_4 = \begin{pmatrix} \theta_1 - \frac{\theta_4^2(\theta_1+\theta_3)}{\theta_3^2+\theta_4^2} & \frac{\theta_4^2(\theta_1+\theta_3)}{\theta_3^2+\theta_4^2} & -\theta_4 \\ 0 & \theta_2 & 0 \\ -\theta_4 + \frac{\theta_3\theta_4(\theta_1+\theta_3)}{\theta_3^2+\theta_4^2} & -\frac{\theta_3\theta_4(\theta_1+\theta_3)}{\theta_3^2+\theta_4^2} & \theta_3 \end{pmatrix}.$$

We may now solve for the unknown volume fractions λ_i, and if we consider $\lambda_4 = s \in [0,1]$ as a free parameter, then we find the solutions

$$\lambda_1 = \lambda - s, \quad \lambda_2 = 1 - \lambda - s, \quad \lambda_3 = \lambda_4 = s.$$

Let

$$Z_1 = \frac{\lambda - s}{1-2s} X_1U_1 + \frac{1-\lambda-s}{1-2s} U_2, \quad Z_2 = \frac{1}{2} X_3U_3 + \frac{1}{2} X_4U_4.$$

With these definitions, it is easy to see that the polyconvex measures ν is in fact a laminates since

$$\operatorname{rank}(X_1U_1 - U_2) = 1, \quad \operatorname{rank}(X_3U_3 - X_4U_4) = 1, \quad \operatorname{rank}(Z_1 - Z_2) = 1,$$

and hence

$$\begin{aligned} \mathcal{M}^{\mathrm{pc}}(K;F) = \{\nu =& (\lambda - s)\delta_{X_1U_1} + (1-\lambda-s)\delta_{U_2} + s\delta_{X_3U_3} + s\delta_{X_4U_4}, \\ & s \in [0, \min\{\lambda, 1-\lambda\}]\}. \end{aligned}$$

A short calculation shows that there exist two different representations of F as a simple laminate if and only if $\lambda = \frac{1}{2}$.

Case $i = 1$, $j = 2$, $\boldsymbol{n} = \boldsymbol{e}_3 = (0,0,1)$: We may use $\boldsymbol{w}_1 = \boldsymbol{e}_1$ and $\boldsymbol{w}_2 = \boldsymbol{e}_2$ as test vectors for the matrices and we deduce

$$|F\boldsymbol{w}_1|^2 = |U_1\boldsymbol{w}_1|^2 = |U_2\boldsymbol{w}_1|^2 = \theta_1^2 + \theta_4^2, \quad |U_3\boldsymbol{w}_1|^2 = |U_4\boldsymbol{w}_1|^2 = \theta_2^2,$$

and

$$|F\boldsymbol{w}_2|^2 = |U_1\boldsymbol{w}_2|^2 = |U_2\boldsymbol{w}_2|^2 = \theta_2^2, \quad |U_3\boldsymbol{w}_2|^2 = |U_4\boldsymbol{w}_2|^2 = \theta_1^2 + \theta_4^2,$$

respectively. This proves the uniqueness of the microstructure unless we have $\theta_1^2 + \theta_4^2 = \theta_2^2$. Assume now that this identity holds and that F is a simple laminate generated from this twinning system,

$$\begin{aligned} F &= \lambda QU_1 + (1-\lambda)U_2 = U_2 + \lambda \boldsymbol{a} \otimes \boldsymbol{n} \\ &= \begin{pmatrix} \theta_1 & 0 & -\theta_4 \\ 0 & \theta_2 & 0 \\ -\theta_4 & 0 & \theta_3 \end{pmatrix} + \frac{2\lambda(\theta_1+\theta_3)\theta_4}{\theta_2^2} \begin{pmatrix} 0 & 0 & \theta_1 \\ 0 & 0 & 0 \\ 0 & 0 & -\theta_4 \end{pmatrix}. \end{aligned}$$

We suppose again that the polyconvex measure $\nu \in \mathcal{M}^{\mathrm{pc}}(K;F)$ is given by

$$\nu = \lambda_1 \delta_{X_1 U_1} + \lambda_2 \delta_{X_2 U_2} + \lambda_3 \delta_{X_3 U_3} + \lambda_4 \delta_{X_4 U_4}$$

with $X_i \in \operatorname{conv} \mathrm{SO}(3)$. Now the matrices X_i, $i = 1, \ldots, 4$, map the orthogonal vectors $\{U_i \boldsymbol{e}_1, U_i \boldsymbol{e}_2, \operatorname{cof} U_i \boldsymbol{e}_3\}$, $i = 1, \ldots, 4$, onto the orthogonal vectors $\{F\boldsymbol{e}_1, F\boldsymbol{e}_2, \operatorname{cof} F\boldsymbol{e}_3\}$ of same length. We may apply Lemma 5.3.3 and conclude that $X_i \in \mathrm{SO}(3)$ for $i = 1, \ldots, 4$. Therefore the matrices X_i are uniquely determined and a short calculation shows that

$$X_1 U_1 = \begin{pmatrix} \theta_1 & 0 & -\theta_4 + \frac{2\theta_1\theta_4(\theta_1+\theta_3)}{\theta_2^2} \\ 0 & \theta_2 & 0 \\ -\theta_4 & 0 & \theta_3 - \frac{2\theta_4^2(\theta_1+\theta_3)}{\theta_2^2} \end{pmatrix},$$

that $X_2 = \mathbb{I}$ and thus $X_2 U_2 = U_2$, that

$$X_3 U_3 = \begin{pmatrix} \theta_1 & 0 & -\theta_4 + \frac{\theta_1\theta_4(\theta_1+\theta_3)}{\theta_2^2} \\ 0 & \theta_2 & \frac{\theta_4(\theta_1+\theta_3}{\theta_2^2} \\ -\theta_4 & 0 & \theta_3 - \frac{\theta_4^2(\theta_1+\theta_3)}{\theta_2^2} \end{pmatrix},$$

and finally that

$$X_4 U_4 = \begin{pmatrix} \theta_1 & 0 & -\theta_4 + \frac{\theta_1\theta_4(\theta_1+\theta_3)}{\theta_2^2} \\ 0 & \theta_2 & -\frac{\theta_4(\theta_1+\theta_3}{\theta_2^2} \\ -\theta_4 & 0 & \theta_3 - \frac{\theta_4^2(\theta_1+\theta_3)}{\theta_2^2} \end{pmatrix}.$$

As before, we can solve for the unknowns λ_i and we obtain with $\lambda_4 = s$ as a free parameter the solutions

$$\lambda_1 = \lambda - s, \quad \lambda_2 = 1 - \lambda - s, \quad \lambda_3 = \lambda_4 = s.$$

It is easy to check that ν corresponds again to a second order laminate.

Case $i = 1$, $j = 3$: If F is a type-I twin, then the interface normal is parallel to $\boldsymbol{e}_1 - \boldsymbol{e}_2$ and we may choose $\boldsymbol{w} = \boldsymbol{e}_1 + \boldsymbol{e}_2 + \boldsymbol{e}_3$ as a test vector. We find

$$\begin{aligned} |F\boldsymbol{w}|^2 = |U_1\boldsymbol{w}|^2 = |U_3\boldsymbol{w}|^2 &= \theta_2^2 + (\theta_1 + \theta_4)^2 + (\theta_3 + \theta_4)^2, \\ |U_2\boldsymbol{w}|^2 = |U_4\boldsymbol{w}|^2 &= \theta_2^2 + (\theta_1 - \theta_4)^2 + (\theta_3 - \theta_4)^2. \end{aligned}$$

In the type-II twinning system, $U_3^{-1}\boldsymbol{a}$ is parallel to $\boldsymbol{e}_1 - \boldsymbol{e}_2$, and the choice of $\boldsymbol{w} = \boldsymbol{e}_1 + \boldsymbol{e}_2 - \boldsymbol{e}_3$ as a test vector for the cofactor matrices shows that

$$\begin{aligned}|\operatorname{cof} F\boldsymbol{w}|^2 &= |\operatorname{cof} U_1\boldsymbol{w}|^2 = |\operatorname{cof} U_3\boldsymbol{w}|^2\\ &= \theta_2^2(\theta_1+\theta_4)^2+\theta_2^2(\theta_3+\theta_4)^2+(\theta_1\theta_3-\theta_4^2)^2,\end{aligned}$$

and that

$$|\operatorname{cof} U_2\boldsymbol{w}|^2 = |\operatorname{cof} U_4\boldsymbol{w}|^2 = \theta_2^2(\theta_1-\theta_4)^2+\theta_2^2(\theta_3-\theta_4)^2+(\theta_1\theta_3-\theta_4^2)^2;$$

these inequalities prove the uniqueness of ν for all choices of the lattice parameters.

Proof for the tetragonal to monoclinic transformation of type-II. We begin our analysis with the uniqueness results for the type-I/II twinning system, then we turn towards the characterization for the compound twins between $SO(3)U_1$ and $SO(3)U_3$.

Case $i=1$, $j=2$: If F is generated form a type-I twinning system, then we choose $\boldsymbol{w}=(\pm1,0,1)$ as a test vector for F and obtain

$$\begin{aligned}|F\boldsymbol{w}|^2 = |U_1\boldsymbol{w}|^2 = |U_2\boldsymbol{w}|^2 &= (\pm\eta_1+\eta_4)^2+(\eta_2\pm\eta_4)^2+(\pm\eta_3-\eta_4)^2,\\ |U_3\boldsymbol{w}|^2 = |U_4\boldsymbol{w}|^2 &= (\pm\eta_1-\eta_4)^2+(\eta_2\mp\eta_4)^2+(\pm\eta_3+\eta_4)^2.\end{aligned}$$

We conclude

$$|U_1\boldsymbol{w}|^2-|U_3\boldsymbol{w}|^2 = \pm4(\eta_1+\eta_2-\eta_3)\eta_4$$

and we obtain uniqueness with the appropriate choice of the sign since (5.13) implies that $\eta_1+\eta_2-\eta_3>0$. For the type-II twinning system we use the same vector $\boldsymbol{w}$ for the cofactor matrices, and we get

$$\begin{aligned}|\operatorname{cof} F\boldsymbol{w}|^2 &= |\operatorname{cof} U_1\boldsymbol{w}|^2 = |\operatorname{cof} U_2\boldsymbol{w}|^2\\ &= (\eta_1+\eta_3)^2(\eta_1-\eta_3\mp\eta_4)^2+\big((\eta_1+\eta_3)\eta_4\mp(\eta_1\eta_2-\eta_4^2)\big)^2\\ &\quad+\big((\eta_1+\eta_3)\eta_4\mp(\eta_2\eta_3+\eta_4^2)\big)^2\end{aligned}$$

and

$$\begin{aligned}|\operatorname{cof} U_3\boldsymbol{w}|^2 &= |\operatorname{cof} U_4\boldsymbol{w}|^2\\ &= (\eta_1+\eta_3)^2(\eta_1-\eta_3\pm\eta_4)^2+\big((\eta_1+\eta_3)\eta_4\pm(\eta_1\eta_2-\eta_4^2)\big)^2\\ &\quad+\big((\eta_1+\eta_3)\eta_4\pm(\eta_2\eta_3+\eta_4^2)\big)^2,\end{aligned}$$

and hence

$$|\operatorname{cof} U_1\boldsymbol{w}|^2-|\operatorname{cof} U_3\boldsymbol{w}|^2 = \mp4(\eta_1+\eta_2-\eta_3)(\eta_1+\eta_3)^2\eta_4.$$

We conclude uniqueness as before.

Case $i=1$, $j=3$, $\boldsymbol{n}=\boldsymbol{e}_3=(0,0,1)$: Assume first that $2\eta_1\eta_3\neq\eta_4^2$. We choose $\boldsymbol{w}=(\pm1,1,0)$ and obtain

$$|Fw|^2 = |U_1 w|^2 = |U_3 w|^2 = 2(\eta_1 \pm \eta_3)^2 + (\eta_4 \mp \eta_4)^2,$$
$$|U_2 w|^2 = |U_4 w|^2 = 2(\eta_1 \mp \eta_3)^2 + (\eta_4 \pm \eta_4)^2,$$

and thus

$$|U_1 w|^2 - |U_2 w|^2 = \pm 4(2\eta_1\eta_3 \mp \eta_4^2),$$

and uniqueness of microstructure is an immediate consequence. Assume now that $2\eta_1\eta_3 = \eta_4^2$, and that F is a simple laminate generated from this twinning system,

$$F = U_3 + \lambda a \otimes n = \begin{pmatrix} \eta_1 & \eta_3 & -\eta_4 \\ \eta_3 & \eta_1 & \eta_4 \\ -\eta_4 & \eta_4 & \eta_2 \end{pmatrix} + \frac{2\lambda\eta_4(\eta_1 + \eta_2 - \eta_3)}{(\eta_1 + \eta_3)^2} \begin{pmatrix} 0 & 0 & \eta_1 - \eta_3 \\ 0 & 0 & -(\eta_1 - \eta_3) \\ 0 & 0 & -2\eta_4 \end{pmatrix}.$$

To simplify notation, we define

$$\eta = \frac{2\eta_4(\eta_1 + \eta_2 - \eta_3)}{(\eta_1 + \eta_3)^2}.$$

We assume again that the polyconvex measure $\nu \in \mathcal{M}^{\mathrm{pc}}(K; F)$ is given by

$$\nu = \lambda_1 \delta_{X_1 U_1} + \lambda_2 \delta_{X_2 U_2} + \lambda_3 \delta_{X_3 U_3} + \lambda_4 \delta_{X_4 U_4}$$

with $X_i \in \operatorname{conv} SO(3)$. We conclude as before that the matrices X_i map the orthogonal vectors $\{U_i e_1, U_i e_2, \operatorname{cof} U_i e_3\}$, $i = 1, \ldots, 4$, onto the orthogonal vectors $\{Fe_1, Fe_2, \operatorname{cof} Fe_3\}$ of same length. We may apply Lemma 5.3.3 and obtain that $X_i \in SO(3)$ for $i = 1, \ldots, 4$. Therefore the matrices X_i are uniquely determined and a short calculation shows that

$$X_1 U_1 = \begin{pmatrix} \eta_1 & \eta_3 & -\eta_4 + \eta(\eta_1 - \eta_3) \\ \eta_3 & \eta_1 & \eta_4 - \eta(\eta_1 - \eta_3) \\ -\eta_4 & \eta_4 & \eta_2 - 2\eta\eta_4 \end{pmatrix},$$

that

$$X_2 U_2 = \begin{pmatrix} \eta_1 & \eta_3 & -\eta_4 + \eta\eta_1 \\ \eta_3 & \eta_1 & \eta_4 + \eta\eta_3 \\ -\eta_4 & \eta_4 & \eta_2 - \eta\eta_4 \end{pmatrix},$$

that $X_3 = \mathbb{I}$ and therefore $X_3 U_3 = U_3$, and finally that

$$X_4 U_4 = \begin{pmatrix} \eta_1 & \eta_3 & -\eta_4 - \eta\eta_3 \\ \eta_3 & \eta_1 & \eta_4 - \eta\eta_1 \\ -\eta_4 & \eta_4 & \eta_2 - \eta\eta_4 \end{pmatrix}.$$

We solve for the volume fractions λ_i and find the solutions

$$\lambda_1 = \lambda - s, \quad \lambda_2 = s, \quad \lambda_3 = 1 - \lambda - s, \quad \lambda_4 = s.$$

As before, we obtain a one-parameter family of second order laminates with center of mass equal to F.

Case $i = 1$, $j = 3$, $\boldsymbol{n} = \boldsymbol{e}_3 = (0,0,1)$: We use $\boldsymbol{w} = (-1,1,0)$ as a test vector for F and obtain

$$\begin{aligned} |\operatorname{cof} F\boldsymbol{w}|^2 = |\operatorname{cof} U_1\boldsymbol{w}|^2 = |\operatorname{cof} U_3\boldsymbol{w}|^2 &= 2(\eta_1 + \eta_3)^2(\eta_2^2 + 2\eta_4^2) \\ |\operatorname{cof} U_2\boldsymbol{w}|^2 = |\operatorname{cof} U_4\boldsymbol{w}|^2 &= 2\big((\eta_2(\eta_3 - \eta_1) + 2\eta_4^2\big)^2, \end{aligned}$$

and since $\eta_1\eta_2 - \eta_4^2 > 0$ we deduce

$$|\operatorname{cof} U_1\boldsymbol{w}|^2 - |\operatorname{cof} U_2\boldsymbol{w}|^2 = 4(\eta_1\eta_2 - \eta_4^2)(\eta_2\eta_3 + \eta_4^2) + 2\eta_2^2(\eta_1 + \eta_3)^2 > 0.$$

This implies uniqueness of the microstructure and concludes the proof of the theorem. □

5.5 Reduction by Symmetry Operations

In the foregoing sections, we discussed the uniqueness of simple laminates for representative twinning systems. The analysis for the remaining ones can be reduced to the presented cases by symmetry operations. To simplify the notation, we define

$$R\{U_i, U_j\}R^T = \{RU_iR^T, RU_jR^T\}.$$

Suppose now that we have analyzed the twins generated by the system $Q_{1,2}U_i - U_j = \boldsymbol{a}_{1,2} \otimes \boldsymbol{n}_{1,2}$ and that $\{U_k, U_\ell\} = R\{U_i, U_j\}R^T$. Then

$$\begin{aligned} & Q_{1,2}U_i - U_j = \boldsymbol{a}_{1,2} \otimes \boldsymbol{n}_{1,2} \\ \Leftrightarrow \quad & RQ_{1,2}R^T RU_iR^T - RU_jR^T = R\boldsymbol{a}_{1,2} \otimes R\boldsymbol{n}_{1,2} \\ \Leftrightarrow \quad & \widetilde{Q}_{1,2}U_k - U_\ell = \widetilde{\boldsymbol{a}}_{1,2} \otimes \widetilde{\boldsymbol{n}}_{1,2} \end{aligned}$$

with $\widetilde{Q}_{1,2} = RQ_{1,2}R^T$, $\widetilde{\boldsymbol{a}} = R\boldsymbol{a}$, $\widetilde{\boldsymbol{n}} = R\boldsymbol{n}$. The results for $\{U_k, U_\ell\}$ follow therefore from those for $\{U_i, U_j\}$.

The Cubic to Tetragonal Transformation. In this case it suffices to analyze the simple laminates supported on the two wells $\mathrm{SO}(3)U_1 \cup \mathrm{SO}(3)U_2$ since

$$R_{23}^\pi\{U_1, U_3\}R_{23}^\pi = R_{13}^\pi\{U_2, U_3\}R_{13}^\pi = \{U_1, U_2\}.$$

The Cubic to Trigonal Transformation. In this case it suffices to analyze the simple laminates supported on the two wells $SO(3)U_1 \cup SO(3)U_2$ and $SO(3)U_1 \cup SO(3)U_3$ since

$$\begin{aligned} R^{\pi}_{1-2}\{U_1, U_4\}R^{\pi}_{1-2} &= \{U_1, U_2\}, & R^{\pi}_{23}\{U_2, U_4\}R^{\pi}_{23} &= \{U_2, U_1\}, \\ R^{\pi}_{3}\{U_3, U_4\}R^{\pi}_{3} &= \{U_1, U_2\}, & R^{\pi}_{12}\{U_2, U_3\}R^{\pi}_{12} &= \{U_2, U_1\}. \end{aligned}$$

The Cubic to Orthorhombic Transformation. In this case it suffices to analyze the simple laminates supported on $SO(3)U_1 \cup SO(3)U_2$ and on $SO(3)U_1 \cup SO(3)U_3$ since

$$\begin{aligned} R^{\pi}_{3}\{U_1, U_4\}R^{\pi}_{3} &= \{U_1, U_3\}, & R^{\pi}_{1-2}\{U_1, U_5\}R^{\pi}_{1-2} &= \{U_1, U_3\}, \\ R^{\pi}_{12}\{U_1, U_6\}R^{\pi}_{12} &= \{U_1, U_3\}, & R^{\pi}_{2}\{U_2, U_3\}R^{\pi}_{2} &= \{U_1, U_3\}, \\ R^{\pi}_{1}\{U_2, U_4\}R^{\pi}_{1} &= \{U_1, U_3\}, & R^{\pi}_{1-2}\{U_2, U_5\}R^{\pi}_{1-2} &= \{U_2, U_3\}, \\ R^{\pi}_{12}\{U_2, U_6\}R^{\pi}_{12} &= \{U_2, U_3\}, & R^{\pi}_{23}\{U_3, U_4\}R^{\pi}_{23} &= \{U_1, U_2\}, \\ R^{\pi}_{13}\{U_3, U_5\}R^{\pi}_{13} &= \{U_2, U_3\}, & R^{\pi}_{1-3}\{U_3, U_6\}R^{\pi}_{1-3} &= \{U_2, U_3\}, \\ R^{\pi}_{1-3}\{U_4, U_5\}R^{\pi}_{1-3} &= \{U_1, U_4\}, & R^{\pi}_{13}\{U_4, U_6\}R^{\pi}_{13} &= \{U_1, U_4\}, \\ R^{\pi}_{13}\{U_5, U_6\}R^{\pi}_{13} &= \{U_1, U_2\}. \end{aligned}$$

The Tetragonal to Monoclinic Transformation. It suffices for both the tetragonal to monoclinic transformations analyzed in Section 5.4 to analyze the simple laminates supported on the pairs of wells $SO(3)U_1 \cup SO(3)U_2$ and on $SO(3)U_1 \cup SO(3)U_3$. In fact, we have for the type-I transformation

$$\begin{aligned} R^{\pi}_{2}\{U_1, U_4\}R^{\pi}_{2} &= \{U_1, U_3\}, & R^{\pi}_{1}\{U_2, U_3\}R^{\pi}_{1} &= \{U_1, U_3\}, \\ R^{\pi}_{3}\{U_2, U_4\}R^{\pi}_{3} &= \{U_1, U_3\}, & R^{\pi}_{1-2}\{U_2, U_4\}R^{\pi}_{1-2} &= \{U_1, U_2\}, \end{aligned}$$

and for the type-II transformation

$$\begin{aligned} R^{\pi}_{12}\{U_1, U_4\}R^{\pi}_{12} &= \{U_1, U_2\}, & R^{\pi}_{1-2}\{U_2, U_3\}R^{\pi}_{1-2} &= \{U_2, U_1\}, \\ R^{\pi}_{1}\{U_2, U_4\}R^{\pi}_{1} &= \{U_3, U_1\}, & R^{\pi}_{3}\{U_2, U_4\}R^{\pi}_{3} &= \{U_1, U_2\}. \end{aligned}$$

6. Algorithmic Aspects

The numerical analysis of nonconvex variational problems, microstructures, and Young measures poses a wealth of challenging problems, both from the analytical and the computational point of view. One of the fundamental issues in this context is the question of what the appropriate objects are that one wants to obtain as the output of a computation: a (highly oscillating) minimizer of the nonconvex energy in a finite element space, a minimizer of the relaxed problem, or a discretization of the Young measure.

It turns out that each of these choices has its own advantages. We discussed in Chapter 4 properties of minimizers of nonconvex variational problems in finite element spaces, and we described the information they contain about the minimizing microstructure for affine boundary conditions that are the barycenter of a simple laminate. In this chapter we focus on the other two aspects - computation of relaxed functionals and of Young measures.

The idea to replace the original variational principle by the relaxed one is motivated by the observation that the implementation of a direct energy minimization in a finite element space faces two serious obstacles. First, there is no class of steepest descent type algorithms known that produces reliable results for nonconvex energies without ingenious guesses for the initialization of the routine. Secondly, the rigidity results for Lipschitz functions whose gradients are functions of bounded variation show that minimizers of variational problems modeling solid to solid phase transformations must have an intrinsically complicated structure that renders them inaccessible to naïve approximation schemes.

The relaxed energy has under suitable growth conditions on the density W an integral representation given by

$$\mathcal{J}^{\mathrm{qc}}(\boldsymbol{u}) = \int_\Omega W^{\mathrm{qc}}(D\boldsymbol{u})\mathrm{d}\boldsymbol{x}.$$

Here W^{qc} denotes the quasiconvex envelope of W defined by

$$W^{\mathrm{qc}}(F) = \inf_{\substack{\boldsymbol{u}\in W^{1,\infty}(\Omega;\mathbb{R}^3)\\ \boldsymbol{u}(\boldsymbol{x})=F\boldsymbol{x} \text{ on } \partial\Omega}} \frac{1}{|\Omega|}\int_\Omega W(D\boldsymbol{u})\mathrm{d}\boldsymbol{x}. \tag{6.1}$$

Equivalently, W^{qc} can be defined as the largest quasiconvex function less than or equal to W. If we define the rank-one convex envelope W^{rc} and the polyconvex envelope W^{pc} analogously, then

$$W^{\mathrm{rc}} \geq W^{\mathrm{qc}} \geq W^{\mathrm{pc}}.$$

We therefore obtain an upper bound for the relaxed functional by replacing W^{qc} by (an approximation of) W^{rc}. There are no general, structural conditions known that ensure the equality of W^{rc} and W^{qc}, but these envelopes coincide in all examples for which explicit characterizations have been obtained. The infimum of the nonconvex energy is equal to the minimum of the relaxed energy, and it is therefore expected that a numerical minimization of the relaxed functional does not generate mesh sensitive oscillations. The drawback of this approach is that exactly these oscillations contain a lot of information about the underlying microstructure which is difficult to recover from the minimizers of the relaxed energy. One approach to recover this information is the algorithm for the computation of laminates that we describe in Section 6.2.

Another motivation for the computation of rank-one convex envelopes of functions is the computation of rank-one convex hulls of sets. The rank-one convex hull of a compact set K can be characterized by

$$K^{\mathrm{rc}} = \{X : \mathrm{dist}^{\mathrm{rc}}(X, K) = 0\},$$

and is thus equal to the zero set of the rank-one convexification of the distance function to K.

6.1 Computation of Envelopes of Functions

Our approach to the approximation of rank-one convex functions is rather in the spirit of finite difference schemes than finite element methods. The algorithm computes a function f^h defined on a uniform grid

$$\mathcal{G}_h = \{hF : F \in \mathbb{Z}^{m \times n}\}$$

in the space of all matrices. The reason for this choice is a surprising rigidity of finite element spaces. To illustrate this, consider for $n = 2$, $m = 1$, and $N \in \mathbb{N}$ a uniform triangulation $\mathcal{T}_h$ of the square $Q = [0,1]^2$ with nodes $n_{ij} = (i/N, j/N)$ and triangles, the edges of which are parallel to the vectors $\{(1,0), (0,1)\ (1,1)\}$, see Figure 6.1. Assume that we want to approximate the convex function

$$f(x_1, x_2) = |x_1 + x_2 - 1|$$

in the space $\mathcal{S}_h(\mathcal{T}_h)$ of all continuous functions which are affine on the elements in the triangulation. At a first glance, it seems natural to seek

$$\tilde{f}_h = \sup\{g_h \leq f : g_h \in \mathcal{S}_1(\Omega_h),\ g_h \text{ is } \mathcal{D}\text{-convex}\}.$$

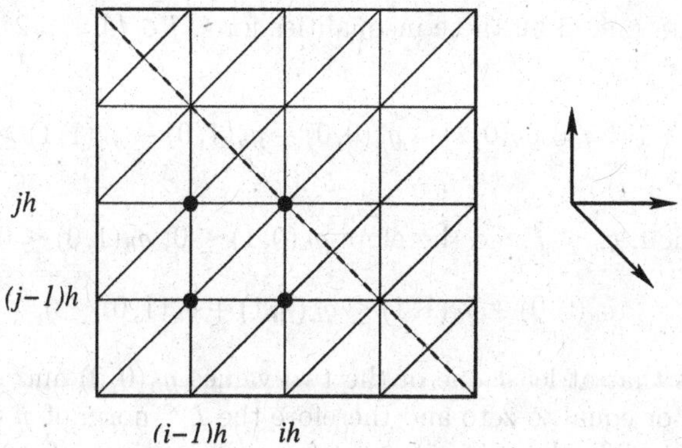

Fig. 6.1. *The function* $f(x_1, x_2) = |x_1 + x_2 - 1|$ *cannot be approximated by* $\mathcal{D}$-*convex functions* $g_h \leq f$ *in* $\mathcal{S}_1(\Omega_h)$ *with* $\mathcal{D} = \{(1,0),(0,1),(1,-1)\}$. *The dashed line corresponds to the zero set of* f, *and the arrows on the right hand side of the figure indicate the directions of convexity.*

However, there are two fundamental difficulties with this approach. First, the maximum of two functions in $\mathcal{S}_1(\Omega_h)$ is not necessarily in $\mathcal{S}_1(\Omega_h)$. Secondly, finite element spaces are too rigid to approximate convex functions from below. Indeed, we assert that

$$|g_h - f|_{\infty;\Omega} \geq 1 \quad \text{for all } g_h \in \mathcal{S}_h,\ g_h \leq f \text{ and } g_h \text{ is } \mathcal{D}\text{-convex.}$$

Consequently the scheme does not converge for $h \to 0$.

In order to prove this estimate, we fix a $\mathcal{D}$-convex function $g_h \in \mathcal{S}_1(\Omega_h)$ such that $g_h \leq f$ and consider g_h on the cube

$$C_{i,j} = [\,(i-1)h, ih\,] \times [\,(j-1)h, jh\,],$$

see Figure 6.1. Let $M_{i,j} = \big((i-\frac{1}{2})h, (j-\frac{1}{2})h\big)$ be its center. Since g_h is affine on the two triangles in $C_{i,j}$, we have

$$g_h(M_{i,j}) = \frac{1}{2}\big(g_h(n_{i-1,j-1}) + g_h(n_{i,j})\big),$$

while the convexity of g_h in direction $(1,-1)$ implies that

$$g_h(M_{i,j}) \leq \frac{1}{2}\big(g_h(n_{i-1,j}) + g_h(n_{i,j-1})\big).$$

We define

$$d_{i,j} = g_h(n_{i-1,j}) - g_h(n_{i-1,j-1}) + g_h(n_{i,j-1}) - g_h(n_{i,j})$$

and we deduce from the two foregoing estimates that

$$d_{i,j} \geq 0.$$

If we take the sum of all these inequalities for $i, j \in \{1, \dots, 2^k\}$, we obtain

$$\sum_{i,j=1}^{2^k} d_{i,j} = g_h(0,1) - g_h(0,0) + g_h(1,0) - g_h(1,1) \geq 0.$$

By assumption, $g_h \leq f$ and therefore $g_h(0,1) \leq 0$, $g_h(1,0) \leq 0$, and thus

$$g_h(0,0) + g_h(1,1) \leq g_h(0,1) + g_h(1,0) \leq 0.$$

This implies that at least one of the two values $g_h(0,0)$ and $g_h(1,1)$ has to be less than or equal to zero and therefore the L^∞ norm of $f - g_h$ is at least one. The same difficulty arises if one defines alternatively (for smooth enough functions)

$$\widehat{f}_h = \sup \{ g_h \leq \Pi_1 f : g_h \in \mathcal{S}_1(\Omega_h), \, g_h \text{ is } \mathcal{D}\text{-convex} \},$$

where Π_1 given by $\Pi_1 f(p_\alpha) = f(p_\alpha)$ for all nodes p_α in the triangulation denotes the interpolation operator onto $\mathcal{S}_1(\Omega_h)$.

The Algorithm for the Computation of Rank-one Convex Envelopes. Our approach to the computation of an approximation f^h of the $\mathcal{D}$-convex envelope of a given function f uses a discretization for both the function and the set of directions $\mathcal{D}$. For $h > 0$ fixed we let

$$\mathcal{G}_h = \{ hF : F \in \mathbb{Z}^{m \times n} \},$$

and we choose a set of directions $\mathcal{D}_h \subset \mathcal{G}_h$ as an approximation of $\mathcal{D}$. A typical choice which allows us to obtain explicit estimates is

$$\mathcal{D}_h = \{ h(\boldsymbol{a} \otimes \boldsymbol{b}) : \boldsymbol{a} \in \mathbb{Z}^m, \, \boldsymbol{b} \in \mathbb{Z}^n, \, |\boldsymbol{a}|_{\ell^\infty}, \, |\boldsymbol{b}|_{\ell^\infty} \leq h^{-1/3} \},$$

see the statements of the theorems below. The advantage of this choice is that it leads naturally to a robust definition of a discretely $\mathcal{D}_h$-convex approximation of f.

The idea behind the algorithm is to perform convexifications along rank-one lines until the function is stable under this operations. In view of our definitions, the line

$$\ell(F, R) = \{ F + tR, \, t \in \mathbb{R} \}$$

intersects the grid $\mathcal{G}_h$ in infinitely many, evenly spaced points for all matrices $F \in \mathcal{G}_h$ and all directions $R \in \mathcal{D}_h$. We consider f along this line as a function of one variable defined in a set of nodes and denote its piecewise affine interpolation by $\Pi f\big|_{\ell(F,R)}$, see Figure 6.3. We now define the notion of discrete $\mathcal{D}$-convexity for functions defined on $\mathcal{G}_h$.

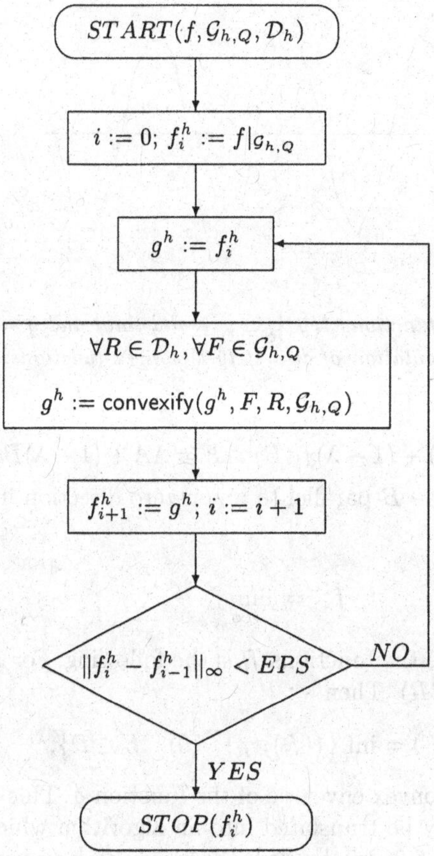

Fig. 6.2. *The algorithm for the computation of the rank-one convex envelope.*

Definition 6.1.1. *Let $\mathcal{D}_h \subset \mathcal{G}_h$. We say that a function $f : \mathcal{G}_h \to \mathbb{R}$ is discretely $\mathcal{D}_h$-convex on $\mathcal{G}_h$ if the functions $\Pi f\big|_{\ell(F,R)}$ are convex (as functions of one variable) for all $F \in \mathcal{G}_h$ and all $R \in \mathcal{D}_h$. The discretely $\mathcal{D}_h$-convex envelope $f^{h,D}$ of f is the largest discretely $\mathcal{D}_h$-convex function $g^h \leq f$.*

Remark 6.1.2. The pointwise maximum of two discretely $\mathcal{D}_h$-convex functions is a discretely $\mathcal{D}_h$-convex function. Thus $f^{h,D}$ is well-defined.

Remark 6.1.3. For computations it is necessary to use a finite domain, and we denote by $\mathcal{G}_{h,Q}$ the intersection of the grid with a cube $Q \in \mathbb{M}^{m\times n}$. We then define discretely $\mathcal{D}_h$-convex functions on $\mathcal{G}_{h,Q}$ analogously.

The algorithm we are now going to describe is based on an idea by Kohn and Strang for the case of rank-one convexity, i.e., for $\mathcal{D}$ being the space of all rank-one matrices. Let $f_0 = f$ and define iteratively

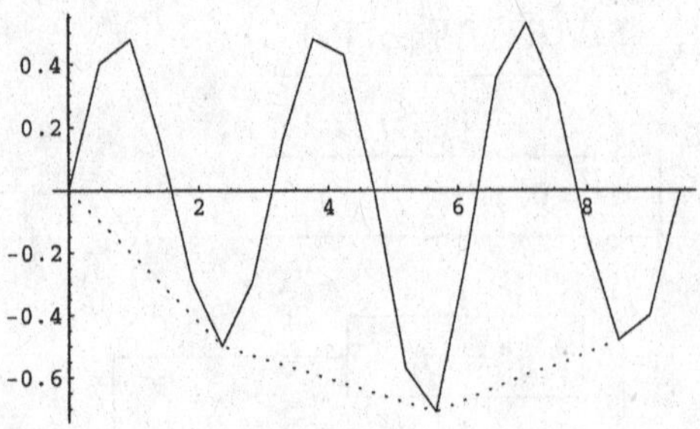

Fig. 6.3. *Sketch of the functions* $\Pi g^h|_{\ell(F,R)}$ *(solid line) and* $\tilde{g}^h$ *(dotted line) in the algorithm for the computation of discretely* $\mathcal{D}$*-convex functions.*

$$f_{i+1}(F) = \inf\{\lambda f_i(A) + (1-\lambda)f_i(B) :\ F = \lambda A + (1-\lambda)B,\ \lambda \in [0,1],\ \text{and } A - B \text{ parallel to a non-zero direction in } \mathcal{D}\}.$$

Then

$$f^D = \lim_{i\to\infty} f_i.$$

A different way to define the functions f_i is the following. For F, $R \in \mathbb{M}^{m\times n}$ define $f_{F,R}(t) = f(F + tR)$. Then

$$f_{i+1}(F) = \inf\{((f_i)_{F,R})^{**}(0) : R \in \mathcal{D}\},$$

where g^{**} denotes the convex envelope of the function g. This interpretation of the formula can easily be translated into an algorithm which we describe schematically in Figure 6.2. In the flow chart we use a parameter EPS to control the performance of the algorithm. Since it is a priori not clear whether the discretely rank-one convex function can be determined with finitely many iterations, one needs to introduce a stop criterion. The proposed algorithm terminates if the difference between the functions f_i^h and f_{i+1}^h are small.

The input consists of the restriction of the given function f to a uniform grid of width h and a set $\mathcal{D}_h \subset \mathcal{G}_h$ of directions; the output is (for $EPS > 0$ an approximation to) a discretely $\mathcal{D}_h$ convex function defined on $\mathcal{G}_{h,Q}$. The central part of the algorithm is a subroutine convexify$(g^h, F, R, \mathcal{G}_{h,Q})$ which computes the convexification of the restriction of the function g^h to the intersection of the grid $\mathcal{G}_h$ with a line $\ell(F,R)$. The convexification is defined as the largest convex and piecewise affine function $\tilde{g}^h \leq \Pi g^h|_{\ell(F,R)}$ restricted to the nodes in $\mathcal{G}_h$ on $\ell(F,R)$, see Figure 6.3.

The next theorem describes the properties of the functions f_i^h.

Theorem 6.1.4. *Assume that* $h > 0$ *and that* $f : \mathbb{M}^{m\times n} \to \mathbb{R}$ *is continuous. Then there exists a discretely* $\mathcal{D}_h$*-convex function* $f^h : \mathcal{G}_{h,Q} \to \mathbb{R}$ *such that the functions* f_i^h *defined in the flow-chart of the algorithm in Figure 6.2 with* $EPS = 0$ *converge to* f^h.

The case of the rank-one convex envelope, which corresponds to the choice of all rank-one matrices for $\mathcal{D}$, has attracted a lot of attention because of its close connection to the quasiconvex envelope. In particular, a reliable algorithm for the computation of the rank-one convex envelope can be used to get upper bounds on the quasiconvex envelope. There are two important assumptions one has to make in order to state an explicit convergence result: On the one hand, the $\mathcal{D}$-convexification of any function f depends in general on the values of f on the entire space $\mathbb{M}^{m\times n}$, not only on a finite neighborhood of a given point. Since computations can be performed only on compact sets, one has to assume a condition of the form $(f^D)_{|Q} = (f_{|Q})^D$ (see also the remark following the theorem below). This will ensure convergence of the functions f^h to f^D as $h \to 0$. A sufficient condition is, for example, that $f \geq g^D$ on $\mathbb{M}^{m\times n}$ with $f = g^D$ on $\mathbb{M}^{m\times n} \setminus Q$. On the other hand, the proof relies on the representation

$$f^{\mathrm{rc}}(F) = \inf\Big\{\sum_{i=1}^{N} \lambda_i(F_i) : (\lambda_i, F_i) \in \mathcal{H}_N,\, F = \sum_{i=1}^{N} \lambda_i F_i\Big\}, \tag{6.2}$$

where the infimum is taken over all pairs $(\lambda_i, F_i)_{i=1,\ldots,N}$ satisfying condition $\mathcal{H}_N$ in Definition A.1.4. There is no criterion known that ensures that the envelope for a given function can be obtained with a finite N. Therefore convergence can only be obtained if this is assumed. In this situation the following convergence result holds.

Theorem 6.1.5. *Assume that f is Lipschitz continuous and that there exists a rank-one convex function $g : \mathbb{M}^{m\times n} \to \mathbb{R}$ such that $f \geq g$ on Q and $f = g$ on $\mathbb{M}^{m\times n} \setminus Q$. Suppose in addition that f^{rc} can be computed by formula (6.2) with $N \leq N_0$, and that*

$$\mathcal{D}_h = \big\{h(\boldsymbol{a} \otimes \boldsymbol{b}) : \boldsymbol{a} \in \mathbb{Z}^m,\, \boldsymbol{b} \in \mathbb{Z}^n,\, |\boldsymbol{a}|_{\ell^\infty},\, |\boldsymbol{b}|_{\ell^\infty} \leq h^{-1/3}\big\}.$$

Then there exists a constant C which depends only on m, and n such that

$$\|f^{\mathrm{rc}} - f^h\|_{L^\infty(\mathcal{G}_{h,Q})} \leq C|f|_{1,\infty;Q}h^{1/3},$$

where f^h is the rank-one convex function the existence of which is guaranteed in Theorem 6.1.4.

Remark 6.1.6. The condition that f coincides with a rank-one convex function outside a compact set can of course be weakened. For example, to get convergence of the algorithm at the established rate on a compact set Q_1, it is sufficient that the rank-one convex envelope can be obtained by using only points in some compact set Q_2. Then one runs the algorithm on Q_2.

The proof of the theorem relies on the following approximation result.

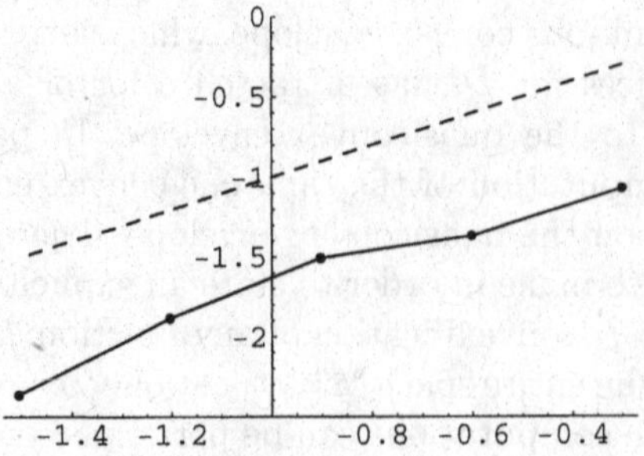

Fig. 6.4. *The logarithm (base 10) of the L^∞-norm versus the logarithm of the width of the grid for the modified Kohn-Strang example. The dashed line has slope one.*

Lemma 6.1.7. *Assume that $h \in (0,1)$, that f and $\mathcal{D}_h$ are as in Theorem 6.1.5. and that the pairs (λ_i, F_i), $i = 1,\dots,N$ satisfy condition $\mathcal{H}_N$ with $F = \sum_{i=1}^N \lambda_i F_i$. Suppose that $F^h \in \mathcal{G}_h$ satisfies $|F - F^h| \le c_0 h^{1/3}$. Then there exist pairs (λ_i^h, F_i^h), $i = 1,\dots,N$, which satisfy condition $\mathcal{H}_N$ and a constant c_1 which depends only on m, n, and $\max_{i=1,\dots,N}|F_i|$ such that*

i) $F_i^h \in \mathcal{G}_h$ *and* $F^h = \sum_{i=1}^N \lambda_i^h F_i^h$;
ii) $|F_i^h - F_i| \le \big(c_0 + (N-1)c_1\big)h^{1/3}$ *for* $i = 1,\dots,N$;
iii) we have the estimate

$$\Big|\sum_{i=1}^N \big(\lambda_i f(F_i) - \lambda_i^h f(F_i^h)\big)\Big| \le \big(c_0 + (N-1)c_1\big)|f|_{1,\infty} h^{1/3}.$$

This lemma is also the key ingredient in the convergence proof for the computation of laminates in Section 6.2 below.

We now present the results of some numerical experiments for the computation of envelopes of functions and sets. In our examples we use the sets $\mathcal{D}_h$ of rank-one directions defined by

$$\mathcal{D}_{h,k} = \{hR :\ R = \boldsymbol{a} \otimes \boldsymbol{b}, \boldsymbol{a} \in \mathbb{Z}^m, \boldsymbol{b} \in \mathbb{Z}^n, |\boldsymbol{a}|_{\ell^\infty}, |\boldsymbol{b}|_{\ell^\infty} \le k\}.$$

The Kohn-Strang Example. One classical example in this context is the function

$$f(F) = \begin{cases} 1 + |F|^2 & \text{if } F \ne 0, \\ 0 & \text{else}, \end{cases}$$

which was originally studied in the context of optimal design problems by Kohn and Strang. This example is very appealing because the semiconvex envelopes can be calculated explicitly. In fact, $f^{rc} = f^{pc}$ where

Table 6.1. *Numerical results for the modified Kohn-Strang function in (6.4), see also Figure 6.4. The formula for the minimizing laminate (6.3) shows that the full rank-one cone is used in the constructions. This is reflected in the fact that the error decreases as the parameter k in the sets* $\mathcal{D}_k$ *increases. In addition, a third iteration of the algorithm further reduces the error.*

h	$\mathcal{D}_k$	1st iteration	2nd iteration	3rd iteration
0.125	1	0.070 718	0.067 708	0.067 188
	2	0.035 938	0.031 250	0.031 250
	3	0.034 636	0.031 250	0.031 250
	4	0.034 636	0.031 250	0.031 250
0.0625	1	0.077 139	0.076 384	0.076 384
	2	0.022 042	0.021 856	0.021 354
	3	0.019 142	0.013 951	0.013 238
	4	0.019 142	0.013 951	0.013 238
0.03125	1	0.076 660	0.076 489	0.076 488
	2	0.027 269	0.026 795	0.026 795
	3	0.012 500	0.009 863	0.009 732
	4	0.010 326	0.004 561	0.004 390

$$f^{\mathrm{rc}}(F) = \begin{cases} 1 + |F|^2 & \text{if } \varrho(F) \geq 1, \\ 2(\varrho - D) & \text{if } \varrho(F) \leq 1, \end{cases}$$

where $D = |\det F|$ and $\varrho(F) = \sqrt{|F|^2 + 2D}$. We sketch the proof of this formula following Kohn and Strang for the convenience of the reader and in order to emphasize two important consequences of the calculation.

Since f is convex on all rank-one lines that do not pass through zero, it is clear that

$$f_1(F) = \begin{cases} 0 & \text{if } F = 0, \\ 2|F| & \text{if } \operatorname{rank}(F) = 1,\ |F| \leq 1, \\ 1 + |F|^2 & \text{otherwise.} \end{cases}$$

The calculation of f_2 requires an 'inspired guess'. We use polar decomposition to write F as

$$F = QU = \lambda_1 \boldsymbol{u}_1 \otimes \boldsymbol{v}_1 + \lambda_2 \boldsymbol{u}_2 \otimes \boldsymbol{v}_2,$$

where λ_1, λ_2 are the singular values of F and $\{\boldsymbol{u}_1, \boldsymbol{u}_2\}$ and $\{\boldsymbol{v}_1, \boldsymbol{v}_2\}$ are orthonormal vectors in $\mathbb{R}^2$. Then F can be rewritten with $\theta \in (0,1)$ as

$$F = \theta\lambda_1 \boldsymbol{u}_1 \otimes \boldsymbol{v}_1 + (1-\theta)\big(\lambda_1 \boldsymbol{u}_1 \otimes \boldsymbol{v}_1 + \frac{\lambda_2}{1-\theta} \boldsymbol{u}_2 \otimes \boldsymbol{v}_2\big), \tag{6.3}$$

and thus

$$f_2(F) \leq \theta f_1(\lambda_1 u_1 \otimes v_1) + (1-\theta) f_2\big(\lambda_1 u_1 \otimes v_1 + \frac{\lambda_2}{1-\theta} u_2 \otimes v_2\big).$$

For $\lambda_1 \leq 1$ this implies the estimate

$$f_2(F) \leq 2\theta\lambda_1 + (1-\theta)\Big(1 + \lambda_1^2 + \frac{\lambda_2^2}{(1-\theta)^2}\Big).$$

In case that $\lambda_1 + \lambda_2 \leq 1$ this expression is minimized for $1 - \theta = \lambda_2/(1-\lambda_1)$ and this leads to

$$f_2(F) = \begin{cases} 1 + \lambda_1^2 + \lambda_2^2 & \text{if } \lambda_1 + \lambda_2 \geq 1, \\ 2(\lambda_1 + \lambda_2) - 2\lambda_1\lambda_2 & \text{if } \lambda_1 + \lambda_2 \leq 1. \end{cases}$$

Hence $f^{\mathrm{rc}} \leq f_2$ and since f_2 turns out to be polyconvex, $f^{\mathrm{rc}} = f_2$. The derivation of the formula has two important implications.

First, $f^{\mathrm{rc}}(F)$ can be obtained for all F as a laminate which is supported on at most four points, and since f and f^{rc} agree with the convex function $1 + |F|^2$ outside of the compact set $|F|_{\ell^\infty} \leq 1$, all the assumptions in Theorem 6.1.5, except the continuity of f, are satisfied. Replacing f by a continuous function $\widetilde{f}$ with $f^{\mathrm{rc}} \leq \widetilde{f} \leq f$ all the hypotheses are fulfilled and the algorithm converges at least at the established rate. One possible choice for $\widetilde{f}$ is

$$\widetilde{f}(F) = \begin{cases} 1 + |F|^2 & \text{if } |F| \geq \sqrt{2} - 1, \\ 2\sqrt{2}|F| & \text{if } |F| \leq \sqrt{2} - 1. \end{cases} \tag{6.4}$$

We use this function in our numerical experiments and summarize our results in Figure 6.4.

Secondly, the entire rank-one cone is used in the construction of the second order laminates. The impact of this on the computation can be seen in Table 6.1: the approximation error decreases as we pass from $\mathcal{D}_1$ to $\mathcal{D}_4$, and the fact that the error is further reduced in a third iteration of the convexification routine is another manifestation of the fact that not all the directions used in the construction of the relaxation are contained in the discrete sets $\mathcal{D}_h$.

The Eight Point Example. We now consider the eight point set in Section 2.1. Let f be the ℓ^1-distance to the set K given by

$$K = \Big\{ \begin{pmatrix} x & y \\ y & z \end{pmatrix} : |x| = a,\ |y| = b,\ |z| = c \Big\}$$

with a, b, $c > 0$ and $ac - b^2 \geq 0$. For our experiments we chose $a = \frac{3}{4}$, $b = \frac{1}{2}$, and $c = \frac{1}{4}$. In this case,

$$K^{(4)} = K^{\mathrm{lc}} = K^{\mathrm{rc}} = K^{\mathrm{qc}} \subsetneq K^{\mathrm{pc}},$$

see Theorem 2.1.1. The results of our computation with $Q = [-1, 1]^4$ and $h = 1/16$ are summarized in Figure 6.5.

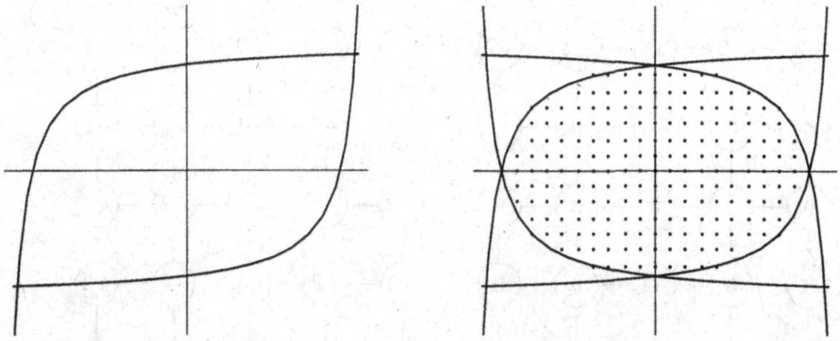

Fig. 6.5. *Computation of the rank-one convex hull for the eight point set. The left figure shows the intersection of* K^{pc} *with the diagonal matrices* ($y = 0$) *which is formed by the intersection of two hyperbolae. The right figure displays the smaller set* K^{rc}, *which is* K^{pc} *intersected with two additional cones. The dots correspond to the grid points in* $\mathcal{G}_h$ *with* $h = 1/16$ *in which the discrete rank-one convex envelope of the* ℓ^1*-distance to* K *is smaller than* 0.0001.

6.2 Computation of Laminates

We now present a variant of the algorithm for the computation of rank-one convex envelopes that is designed for the computation of laminates. At the same time, it provides information about oscillations in minimizing sequences. Assume for example that we choose boundary conditions $\boldsymbol{u}(\boldsymbol{x}) = F\boldsymbol{x}$ which enforce the formation of a unique microstructure in the nonconvex problem. Then we can obtain an approximation of the microstructure from minimizers of the nonconvex energy based on the methods described in Chapter 4. The relaxed problem, however, is also minimized by the affine deformation $\boldsymbol{u}(\boldsymbol{x}) = F\boldsymbol{x}$ which does not provide any information about the underlying microstructure. Suppose now that the rank-one convex and the quasiconvex envelope of W coincide at F. Then there exists a laminate ν with center of mass F such that $W^{\mathrm{qc}}(F) = W^{\mathrm{rc}}(F) = \langle \nu, W \rangle$. This laminate can be approximated by finite laminates ν_N supported on N matrices, and Theorem 6.2.4 ensures that the proposed algorithm computes a laminate ν_h supported on on a discretization $\mathcal{G}_h$ of the space of all matrices with

$$|\langle \nu_h, W \rangle| \leq |\langle \nu_N, W \rangle| + 5(N-1)c_1\sqrt{mn}\,|f|_{1,\infty}h^{1/3}.$$

We therefore obtain for h small enough approximating laminates ν_h for which $\langle \nu_h, W \rangle$ is arbitrarily close to $W^{\mathrm{rc}}(F) = \langle \nu, W \rangle$. These laminates represent microstructures with small energy which constitute a good approximation to the minimizing microstructure.

An algorithm similar in spirit was proposed by Aranda and Pedregal. They focus on laminates realizing a given matrix F as the center of mass and try to construct a generalized solution of the variational problem by splitting F in an optimal way along rank-one lines. One therefore minimizes

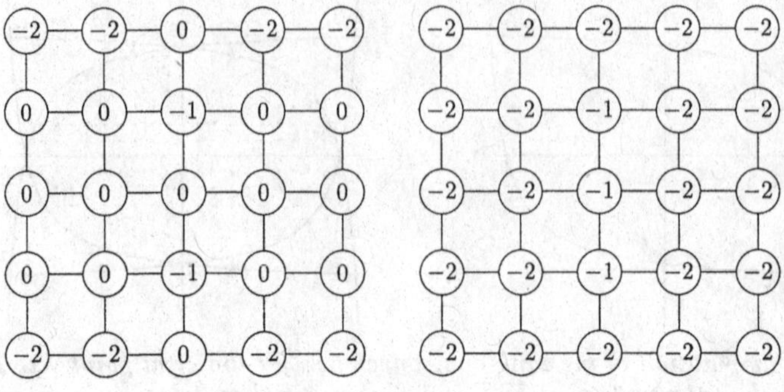

Fig. 6.6. *A typical problem with splitting algorithms if one tries to split a fixed matrix. The computation for the point in the center with directions* $(1,0)$ *and* $(0,1)$ *(corresponding to separate convexity) gets stuck after the first step, in which the value is found to be* -1*. However, the correct value in the center is* -2*, see also Table 6.2. The right figure shows the values of the function* f_1 *in (6.5), which is generated at each node by minimizing among all splittings with direction* $\{(1,0),\ (0,1)\}$.

$$\Psi(R,t^+,t^-) = \frac{t^+}{t^- + t^+} W(F - t^- R) + \frac{t^-}{t^- + t^+} W(F + t^+ R)$$

among all rank-one matrices R and parameters t^- and $t^+ \geq 0$. This step involves a discretization of the set of all rank-one matrices, which can for example be realized by taking $R = s\boldsymbol{a} \otimes \boldsymbol{b}$, where $(s, \boldsymbol{a}, \boldsymbol{b})$ belongs to a suitable discretization of a compact subset of $\mathbb{R} \times \mathbb{S}^m \times \mathbb{S}^n$. The same procedure is then applied to $F^\pm = F \pm t^\pm F$. However, at least any deterministic implementation faces the typical difficulty in the numerical analysis of nonconvex problems: the algorithm finds a local minimum, but not the global one. See Figure 6.6 for a cartoon of this situation.

Here we propose a different approach, namely to compute simultaneously Young measures for all matrices in a given subset of a discretization of $\mathbb{M}^{m\times n}$. In fact, the algorithm for the computation of $\mathcal{D}$-convex envelopes can easily be modified to find certain laminates which consist of finitely many atoms supported on an equidistant grid $\mathcal{G}_h$. In the following we describe first the algorithm and prove its convergence, and then we present results of numerical experiments.

We define by induction classes of laminates $L_k(Q)$ which are supported on at most 2^k matrices in the cube Q and can be generated by successively splitting a single Dirac mass along rank-one lines. Let

$$L_0(Q) = \{\delta_F : F \in Q\}.$$

If

$$\nu_k = \sum_{i=1}^{M} \lambda_i \delta_{F_i} \in L_k(Q),\ M \leq 2^k, \lambda_i > 0, \sum_{i=1}^{M} \lambda_i = 1,$$

then all laminates that can be generated by splitting (up to relabeling) the first M', $M' \leq M$, of the matrices F_i along rank-one lines belong to $L_{k+1}(Q)$:

$$\nu_{k+1} = \sum_{i=1}^{M'} (\mu_i^+ \delta_{F_i^+} + \mu_i^- \delta_{F_i^-}) + \sum_{i=M'+1}^{M} \lambda_i \delta_{F_i} \in L_{k+1}(Q),$$

if $\mu_i^\pm > 0$, $\lambda_i = \mu_i^+ + \mu_i^-$, and

$$F_i^\pm \in Q,\ \mathrm{rank}(F_i^+ - F_i^-) = 1,\ F_i = \frac{\mu_i^+}{\lambda_i} F_i^+ + \frac{\mu_i^-}{\lambda_i} F_i^- \quad \text{for } i = 1, \ldots, M'.$$

For computations one needs to define a subclass of laminates in $L_k(Q)$ which is supported on a finite set of points in $\mathcal{G}_{h,Q}$ and uses only a finite set $\mathcal{D}_h$ of directions.

Definition 6.2.1. *Let* $L_0(\mathcal{G}_{h,Q}, \mathcal{D}_h) = \{\delta_F : F \in \mathcal{G}_{h,Q}\}$. *If*

$$\nu_k = \sum_{i=1}^{M} \lambda_i \delta_{F_i} \in L_k(\mathcal{G}_{h,Q}, \mathcal{D}_h),\ M \leq 2^k, \lambda_i > 0, \sum_{i=1}^{M} \lambda_i = 1,$$

then all laminates that can be generated by splitting (up to relabeling) the first M', $M' \leq M$, *of the matrices* F_i *into matrices* F_i^+, $F_i^- \in \mathcal{G}_{h,Q}$ *along lines parallel to directions in* $\mathcal{D}_h$ *belong to* $L_{k+1}(\mathcal{G}_{h,Q}, \mathcal{D}_h)$:

$$\nu_{k+1} = \sum_{i=1}^{M'} (\mu_i^+ \delta_{F_i^+} + \mu_i^- \delta_{F_i^-}) + \sum_{i=M'+1}^{M} \lambda_i \delta_{F_i} \in L_{k+1}(\mathcal{G}_{h,Q}, \mathcal{D}_h),$$

if $\mu_i^\pm > 0$, $\lambda_i = \mu_i^+ + \mu_i^-$, $F_i^\pm \in \mathcal{G}_{h,Q}$, $F_i^+ - F_i^- \parallel D \in \mathcal{D}_h$, *and*

$$\mathrm{rank}(F_i^+ - F_i^-) = 1,\ F_i = \frac{\mu_i^+}{\lambda_i} F_i^+ + \frac{\mu_i^-}{\lambda_i} F_i^- \quad \textit{for } i = 1, \ldots, M'.$$

Remark 6.2.2. If $\nu \in L_k(\mathcal{G}_{h,Q}, \mathcal{D}_h)$, then ν can be written as a convex combination $\nu = \lambda \nu^+ + (1-\lambda)\nu^-$ with ν^+, $\nu^- \in L_{k-1}(\mathcal{G}_{h,Q}, \mathcal{D}_h)$ such that $F^+ = \langle \nu^+, id \rangle$ and $F^- = \langle \nu^-, id \rangle$ belong to $\mathcal{G}_{h,Q}$, F^+ and F^- are rank-one connected with $F^+ - F^- \parallel D \in \mathcal{D}_h$, and $F = \lambda F^+ + (1-\lambda)F^-$. This follows from the fact that in the first step δ_F is split into two Dirac masses centered in points in $\mathcal{G}_{h,Q}$ along a rank-one line parallel to a direction in $\mathcal{D}_h$, $\delta_F = \lambda \delta_{F^+} + (1-\lambda)\delta_{F^-}$. The subsequent splitting steps generate Young measures ν^+ and ν^- with center of mass equal to F^+ and F^-, if the weights are rescaled by λ and $1-\lambda$, respectively.

Remark 6.2.3. If

$$\nu = \sum_{i=1}^{N} \lambda_i \delta_{F_i} \in L_k(\mathcal{G}_{h,Q}, \mathcal{D}_h),$$

then the pairs $(\lambda_i, F_i)_{i=1,\dots,N}$ satisfy condition $\mathcal{H}_N$ in Definition A.1.4.

The observation here is that the algorithm for the computation of the rank-one convex envelope of a given function f implicitly constructs minimizing Young measures in L_k if the basic convexification routine is executed k times. Assume for simplicity that f is rank-one convex outside $Q = [-1,1]^{m\times n}$ and that $\mathcal{G}_{h,Q} = \mathcal{G}_h \cap Q = \{F_1, \dots, F_\ell\}$. For a given function $f : \mathcal{G}_{h,Q} \to \mathbb{R}$, a given set of directions $\mathcal{D}_h \subset \mathcal{G}_h$, and $F \in \mathcal{G}_{h,Q}$ we define an optimal splitting of F,

$$F = \lambda(f,F)F^+(f,F) + \big(1 - \lambda(f,F)\big)F^-(f,F)$$

with $\lambda \in [0,1]$ and $F^+ - F^-$ parallel to a direction $D \in \mathcal{D}_h$ by requiring that

$$\begin{aligned} &\lambda(f,F)f\big(F^+(f,F)\big) + \big(1-\lambda(f,F)\big)f\big(F^-(f,F)\big) \\ &\quad = \min\big\{\mu f(G^+) + (1-\mu)f(G^-) : G^+, G^- \in \mathcal{G}_{h,Q}, \mu \in [0,1], \\ &\qquad\qquad G^+ - G^- \parallel D \in \mathcal{D}_h, F = \mu G^+ + (1-\mu)G^-\big\}. \end{aligned}$$

The Algorithm for the Computation of Laminates. The algorithm is now defined in the flow-chart in Figure 6.7. The output for the computation of laminates in L_k for all points in $\mathcal{G}_{h,Q}$ consists of an $\ell \times k$ field `fractions` (which contains the volume fractions of F^+ in the optimal splittings) and an $\ell \times (2k)$ field `atoms` (which contains the matrices $F^\pm$ in the optimal splitting). In order to describe the performance of the algorithm and to prove its convergence, we define f_i^h to be the function that is generated in the i-th iteration of the algorithm, i.e., $f_0 = f|_{\mathcal{G}_{h,Q}}$ and f_{i+1}^h is generated from f_i^h by finding an optimal splitting of each matrix F_j with respect to f_i^h,

$$f_{i+1}^h(F_j) = \lambda(f_i^h, F_j) f_i^h\big(F^+(f_i^h, F_j)\big) + \big(1 - \lambda(f_i^h, F_j)\big) f_i^h\big(F^-(f_i^h, F_j)\big).$$

Similarly we construct from the output variables `fractions` and `atoms` laminates $\nu_{j,i} \in L_i(\mathcal{G}_{h,Q}, \mathcal{D}_h)$ with center of mass F_j and i splitting levels by

$$\nu_{j,0} = \delta_{F_j}$$

and

$$\nu_{j,i+1} = \texttt{fraction}[j,i]\, \nu_{\alpha,i} + \big(1 - \texttt{fraction}[j,i]\big)\, \nu_{\beta,i}$$

where

$$F_\alpha = F^+(f_i^h, F_j) = \texttt{atoms}[j,2i], \quad \text{and} \quad F_\beta = F^-(f_i^h, F_j) = \texttt{atoms}[j,2i+1],$$

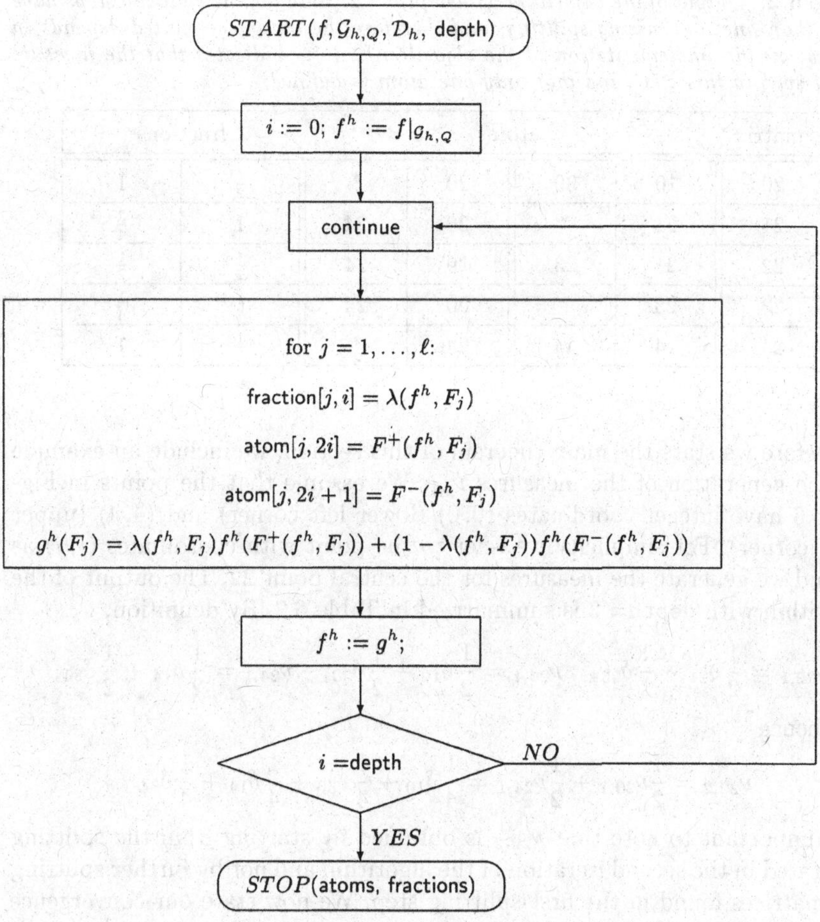

Fig. 6.7. *The algorithm for the computation of laminates.*

are the matrices arising in the optimal splitting of F_j in the $(i+1)^{\text{st}}$ step. It follows from the foregoing definitions that

$$f_i^h(F_j) = \langle \nu_{j,i}, f \rangle. \tag{6.5}$$

Indeed, $f_0^h(F_j) = f(F_j) = \langle \nu_{j,0}, f \rangle$, and if the assertion holds for f_i and $\nu_{j,i}$, then

$$\begin{aligned} f_{i+1}^h(F_j) &= \lambda(f_i^h, F_j) f_i^h\big(F^+(f_i^h, F_j)\big) + \big(1 - \lambda(f_i^h, F_j)\big) f_i^h\big(F^-(f_i^h, F_j)\big) \\ &= \lambda(f_i^h, F_j) \langle \nu_{\alpha,i}, f \rangle + \big(1 - \lambda(f_i^h, F_j)\big) \langle \nu_{\beta,i}, f \rangle \\ &= \langle \nu_{j,i}, f \rangle. \end{aligned}$$

Table 6.2. *Output of the algorithm with depth = 2 (many of the splitting steps have more than one minimizing splitting, and therefore the output generated depends in general on the implementation of the algorithm). A '*' indicates that the measure is not split in this step, and that only one atom is defined.*

matrix	atoms				fractions	
20	10	30	20	*	$\frac{1}{2}$	1
21	21	*	20	24	1	$\frac{3}{4}$
22	21	23	20	24	$\frac{1}{2}$	$\frac{1}{2}$
23	23	*	20	24	1	$\frac{1}{4}$
24	14	34	24	*	$\frac{1}{2}$	1

Before we state the main theorem of this section, we include an example for the generation of the measures $\nu_{j,i}$. We assume that the points in Figure 6.6 have integer coordinates $(0,0)$ (lower left corner) and $(4,4)$ (upper right corner). For simplicity we refer to the point with coordinates (i,j) as ij, and we generate the measures for the central point 22. The output of the algorithm with `depth`= 2 is summarized in Table 6.2. By definition,

$$\nu_{22,1} = \frac{1}{2}\delta_{21} + \frac{1}{2}\delta_{23}, \quad \nu_{20,1} = \frac{1}{2}\delta_{10} + \frac{1}{2}\delta_{30}, \quad \nu_{24,1} = \frac{1}{2}\delta_{14} + \frac{1}{2}\delta_{34},$$

and hence

$$\nu_{22,2} = \frac{1}{2}\nu_{20,1} + \frac{1}{2}\nu_{24,1} = \frac{1}{4}\delta_{10} + \frac{1}{4}\delta_{30} + \frac{1}{4}\delta_{14} + \frac{1}{4}\delta_{34}.$$

It is important to note that $\nu_{22,2}$ is obtained by starting from the splitting generated in the second iteration of the algorithm and not by further splitting the matrices found in the first splitting step. We now state our convergence result for laminates in $L_k(Q)$ which are supported on at most 2^k points. Recall that we assume that the points in $\mathcal{G}_{h,Q}$ have been labeled from 1 to ℓ.

Theorem 6.2.4. *Assume that $Q = [-1,1]^{m\times n}$ and $f \in W^{1,\infty}(Q)$. Suppose that $\nu \in L_k(Q)$ with $\langle \nu, id\rangle = F_j \in [-\frac{1}{2},\frac{1}{2}]^{m\times n}$, $j \in \{1,\dots,\ell\}$ is supported on N matrices and that $h > 0$ is small enough so that γ defined by*

$$\gamma = 1 - 2(N-1)c_1 h^{1/3} \tag{6.6}$$

is greater than zero. Finally, let $\nu_h = \nu_{j,k}$ be the Young measure computed by the algorithm in Figure 6.7 with `depth`= *k and*

$$\mathcal{D}_h = \{h(\boldsymbol{a}\otimes\boldsymbol{b}) : \boldsymbol{a}\in\mathbb{Z}^m,\ \boldsymbol{b}\in\mathbb{Z}^n,\ |\boldsymbol{a}|_{\ell^\infty},\ |\boldsymbol{b}|_{\ell^\infty} \le h^{-1/3}\}.$$

Then

$$|\langle \nu_h, f\rangle| \le |\langle \nu, f\rangle| + 5(N-1)c_1\sqrt{mn}|f|_{1,\infty}h^{1/3}, \tag{6.7}$$

where c_1 is the constant appearing in Lemma 6.1.7.

Proof. We divide the proof in two steps. The first step shows that the measures $\nu_{j,i}$ are minimizing in L_i, and the second one proves the estimate (6.7) based on the construction of discrete laminates in Lemma 6.1.7.

Step 1: Optimality of $\nu_{j,i}$. We have for $i = 0, \ldots, k$, and $j = 1, \ldots, \ell$, that

$$\langle \nu_{j,i}, f \rangle = \min \big\{ \langle \nu, f \rangle : \nu \in L_i(\mathcal{G}_{h,Q}, \mathcal{D}_h),\, \langle \nu, id \rangle = F_j \big\}.$$

This statement is obvious for $i = 0$ and $i = 1$, since the algorithm chooses an optimal splitting. Assume now that the assertion has been established for $i-1$, and suppose that it does not hold for i. Then there exists a $j \in \{1, \ldots, \ell\}$ such that

$$\langle \nu_{j,i}, f \rangle > \min \big\{ \langle \nu, f \rangle : \nu \in L_i(\mathcal{G}_{h,Q}, \mathcal{D}_h),\, \langle \nu, id \rangle = F_j \big\},$$

and let $\widetilde{\nu}$ be an element in $L_i(\mathcal{G}_{h,Q}, \mathcal{D}_h)$ realizing the minimum on the right hand side (this minimum exists since $L_i(\mathcal{G}_{h,Q}, \mathcal{D}_h)$ is a finite set). We may split $\widetilde{\nu}$ as

$$\widetilde{\nu} = \widetilde{\lambda}\, \widetilde{\nu}^+ + (1 - \widetilde{\lambda})\, \widetilde{\nu}^-, \quad \widetilde{\nu}^\pm \in L_{i-1}(\mathcal{G}_{h,Q}, \mathcal{D}_h), \quad \lambda \in [0,1]$$

(see the remark following Definition 6.2.1), such that $\widetilde{F}^+ = \langle \widetilde{\nu}^+, id \rangle$ and $\widetilde{F}^- = \langle \widetilde{\nu}^-, id \rangle$ belong to $\mathcal{G}_{h,Q}$ with

$$F_j = \widetilde{\lambda}\, \widetilde{F}^+ + (1 - \widetilde{\lambda})\, \widetilde{F}^-, \qquad \operatorname{rank}(\widetilde{F}^+ - \widetilde{F}^-) \le 1.$$

Since $\widetilde{F}^\pm \in \mathcal{G}_{h,Q}$, there exist indices α, $\beta \in \{1, \ldots, \ell\}$ such that $\widetilde{F}^+ = F_\alpha$ and $\widetilde{F}^- = F_\beta$. In view of the fact that the measures $\nu_{\alpha,i-1}$ and $\nu_{\beta,i-1}$ are minimizing, we conclude

$$\begin{aligned}
f_i^h(F_j) = \langle \nu_{j,i}, f \rangle > \langle \widetilde{\nu}, f \rangle &\\
&= \widetilde{\lambda}\, \langle \widetilde{\nu}^+, f \rangle + (1 - \widetilde{\lambda})\, \langle \widetilde{\nu}^-, f \rangle \\
&\ge \widetilde{\lambda}\, \langle \nu_{\alpha,i-1}, f \rangle + (1 - \widetilde{\lambda})\, \langle \nu_{\beta,i-1}, f \rangle \\
&= \widetilde{\lambda}\, f_{i-1}^h(F_\alpha) + (1 - \widetilde{\lambda})\, f_{i-1}^h(F_\beta).
\end{aligned}$$

This implies that there exists a better splitting for F_j in the i-th step, and this contradicts the definition of the loop in the algorithm.

Step 2: Proof of the error estimate (6.7). We now construct explicitly a Young measure $\widetilde{\nu}_h \in L_k(\mathcal{G}_{h,Q}, \mathcal{D}_h)$ which is close to ν in the sense of Lemma 6.1.7. The estimate (6.7) is then a consequence of the Lipschitz continuity of f. The only difficulty here is that we cannot apply Lemma 6.1.7 directly, since the generated Young measure might not belong to $L_k(\mathcal{G}_{h,Q}, \mathcal{D}_h)$ (the constructed laminate is supported on matrices which lie in an $\mathcal{O}(h^{1/3})$-neighborhood of the support of ν). The remedy here is to construct from ν by rescaling a Young measure $\widetilde{\nu} \in L_k(Q)$ for which the distance of the atoms in its support to the boundary of Q is sufficiently large. With $\gamma > 0$ given by (6.6) we define $\varphi : \mathbb{M}^{m\times n} \to \mathbb{M}^{m\times n}$ by

$$\varphi(F) = \gamma(F - F_j) + F_j.$$

Suppose that $\nu \in L_k(Q)$ is given by

$$\nu = \sum_{i=1}^{N} \lambda_i \delta_{X_i},$$

and let

$$\widetilde{\nu} = \sum_{i=1}^{N} \lambda_i \delta_{Y_i} \quad \text{with} \quad Y_i = \varphi(X_i) \quad \text{and} \quad F_j = \sum_{i=1}^{N} \lambda_i Y_i.$$

Since φ is affine, it is clear that $\widetilde{\nu} \in L_k(Q)$, and by construction

$$\begin{aligned} |X_i - Y_i| &= \big|X_i - (\gamma(X_i - F_j) + F_j)\big| = \big|X_i - F_j - \gamma(X_i - F_j)\big| \\ &= (1-\gamma)|X_i - F_j| \le 4(N-1)\sqrt{mn}\, c_1 h^{1/3}, \end{aligned}$$

We now assert that the ℓ^∞-distance of the matrices Y_i to ∂Q is at least $(N-1)c_1 h^{1/3}$. By definition

$$\begin{aligned} |Y_{i;k\ell}| &= \big|\gamma(X_{i;k\ell} - F_{i;k\ell}) + F_{i;k\ell}\big| \\ &\le (1-\gamma)|F_{i;k\ell}| + \gamma|X_{i;k\ell}|, \end{aligned}$$

and since $F_j \in [-\frac{1}{2}, \frac{1}{2}]^{m\times n}$, and $X_i \in Q$ we obtain

$$\begin{aligned} |Y_{i;k\ell}| &\le \frac{1}{2}(1-\gamma) + \gamma \\ &= \frac{1}{2} + \frac{1}{2}(1 - 2(N-1)c_1 h^{1/3}) \\ &= 1 - (N-1)c_1 h^{1/3}. \end{aligned}$$

Lemma 6.6 now ensures the existence of a laminate $\widetilde{\nu}_h \in L_k(\mathcal{G}_{h,Q}, \mathcal{D}_h)$ represented by

$$\widetilde{\nu}_h = \sum_{i=1}^{N} \lambda_i^h \delta_{Y_i^h} \quad \text{with} \quad Y_i^h \in \mathcal{G}_{h,Q} \quad \text{and} \quad F_j = \sum_{i=1}^{N} \lambda_i^h Y_i^h.$$

Moreover, the points Y_i^h are close to Y_i in the sense that

$$|Y_i^h - Y_i| \le (N-1)c_1 h^{1/3} \quad \text{for } i = 1, \ldots, N,$$

and we have the estimate

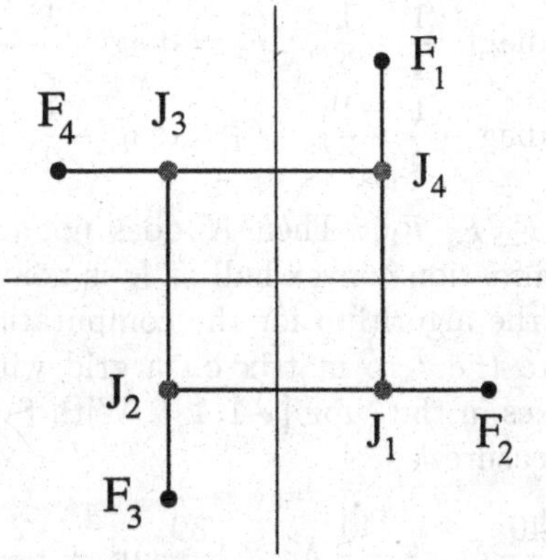

Fig. 6.8. *The four point configuration in the diagonal matrices that supports an infinite laminate. The lamination convex hull is given by the four line segments and the square with corners $J_1, \ldots, J_4$.*

$$\left|\sum_{i=1}^{N} (\lambda_i f(Y_i) - \lambda_i^h f(Y_i^h))\right| \leq (N-1)c_1 |f|_{1,\infty} h^{1/3}.$$

Since $\nu_h = \nu_{j,k}$ is by Step 1 minimizing in $L_k(\mathcal{G}_{h,Q}, \mathcal{D}_h)$ we may estimate

$$\begin{aligned}
|\langle \nu_h, f\rangle| &\leq |\langle \widetilde{\nu}_h, f\rangle| \\
&= \left|\sum_{i=1}^{N} \lambda_i^h f(Y_i^h)\right| \\
&\leq \left|\sum_{i=1}^{N} (\lambda_i^h f(Y_i^h) - \lambda_i f(Y_i))\right| + \left|\sum_{i=1}^{N} (\lambda_i f(Y_i) - \lambda_i f(X_i))\right| \\
&\quad + \left|\sum_{i=1}^{N} \lambda_i f(X_i)\right| \\
&\leq 5(N-1)c_1\sqrt{mn}|f|_{1,\infty} h^{1/3} + |\langle \nu, f\rangle|.
\end{aligned}$$

The proof is now an immediate consequence of Steps 1 and 2. □

Numerical Experiments for the Computation of Laminates. We conclude this section with two numerical experiments for the generation of laminates by the algorithm in Figure 6.7. We first use our scheme to find approximations of the infinite laminate supported on four points. Then we report on computations for eight point set in Theorem 2.1.1.

An Infinite Laminate. A canonical example for the performance of an algorithm is the following four point set. Let

$$F_1 = \operatorname{diag}(\ \frac{1}{4},\ \frac{1}{2}),\quad F_2 = \operatorname{diag}(\ \frac{1}{2}, -\frac{1}{4}),$$
$$F_3 = \operatorname{diag}(-\frac{1}{4}, -\frac{1}{2}),\quad F_4 = \operatorname{diag}(-\frac{1}{2},\ \frac{1}{4}),$$

and define $K = \{F_1, F_2, F_3, F_4\}$. Then K does not contain any rank-one connection, but the lamination convex hull of K is nontrivial, see Figure 6.8 for a sketch. We used the algorithm for the computation of laminates with center of mass equal to the zero matrix on a grid with 17 grid points on the one dimensional axes in the cube $[-1,1]^4$. With five splitting levels, we obtained the Young measure

$$\nu_5 = \frac{39}{162}\delta_{F_1} + \frac{40}{162}\delta_{F_2} + \frac{40}{162}\delta_{F_3} + \frac{39}{162}\delta_{F_4} + \frac{2}{162}\delta_{J_2} + \frac{2}{162}\delta_{J_4}$$

and the rank-one tree shown in Figure 6.9. Here

$$F = 0,\quad F^+ = \operatorname{diag}(\frac{1}{4}, 0),\quad F^- = \operatorname{diag}(-\frac{1}{4}, 0).$$

One obtains the corresponding subtrees if one uses less than five splittings (this is not surprising, since the algorithm is completely deterministic).

The Eight Point Example. Recall that the eight point set K is given by

$$K = \left\{ \begin{pmatrix} x & y \\ y & z \end{pmatrix} : |x| = a,\ |y| = b,\ |z| = c \right\}$$

with a, b, $c > 0$ and $ac - b^2 > 0$. In this case,

$$K^{(4)} = K^{\mathrm{lc}} = K^{\mathrm{rc}} = K^{\mathrm{qc}} \neq K^{\mathrm{pc}},$$

see Theorem 2.1.1 for the precise statement. For our experiments we choose f to be the ℓ^1 distance to K with $a = \frac{3}{4}$, $b = \frac{1}{2}$, and $c = \frac{1}{4}$. In our first experiment we chose $F = 0$ and computed on $\mathcal{G}_{h,Q}$ with $h = 1/8$ and $\mathcal{D}_{h,2}$, a set of 64 directions. The algorithm correctly finds that $F \in K^{(2)}$ and produces the laminate

$$\nu = \frac{7}{24}\delta_{F_1} + \frac{5}{24}\delta_{F_2} + \frac{5}{24}\delta_{F_3} + \frac{7}{24}\delta_{F_4}$$

with

$$F_1 = \begin{pmatrix} -\frac{3}{4} & -\frac{1}{4} \\ -\frac{1}{4} & -\frac{1}{2} \end{pmatrix},\quad F_2 = \begin{pmatrix} \frac{3}{4} & -\frac{1}{4} \\ -\frac{1}{4} & -\frac{1}{2} \end{pmatrix},\quad F_3 = \begin{pmatrix} -\frac{3}{4} & \frac{1}{4} \\ \frac{1}{4} & \frac{1}{2} \end{pmatrix},\quad F_4 = \begin{pmatrix} \frac{3}{4} & \frac{1}{4} \\ \frac{1}{4} & \frac{1}{2} \end{pmatrix}.$$

It is easy to see that ν is indeed a laminate since

$$F_1 - F_2 = -\frac{3}{2} e_1 \otimes e_1,\quad F_3 - F_4 = -\frac{3}{2} e_1 \otimes e_1,$$

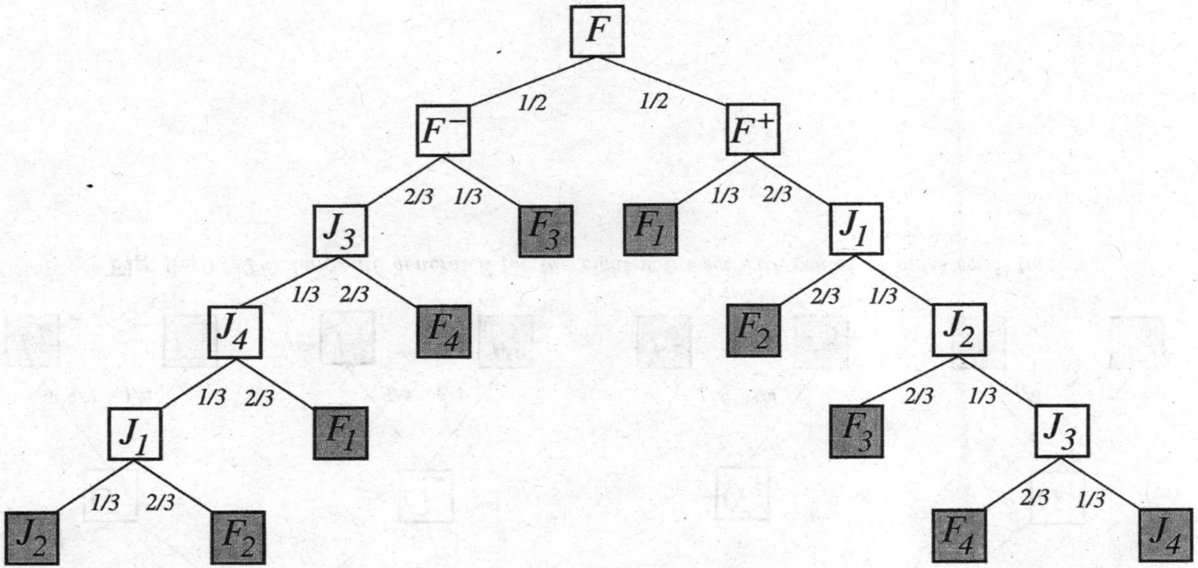

Fig. 6.9. *Approximation of an infinite laminate with finite ones. The laminate computed with the algorithm using five splitting steps. The shaded squares correspond to the points in which the Young measure is supported. The volume fractions in the splittings are indicated along the lines denoting rank-one connections.*

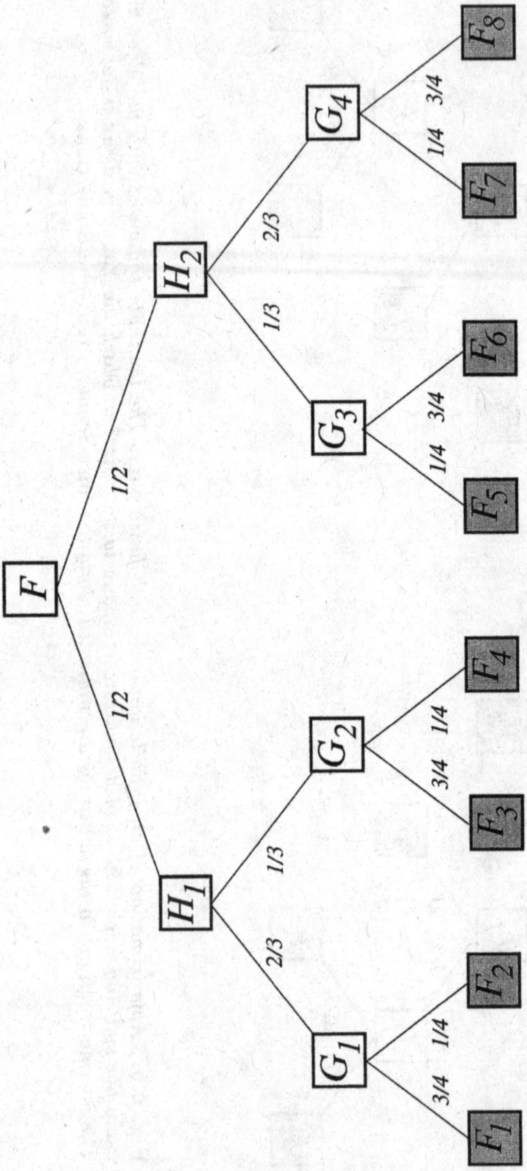

Fig. 6.10. *The laminate generated for the eight point set with center of mass equal to zero.*

and

$$\left(\frac{7}{12}F_1 + \frac{5}{12}F_2\right) - \left(\frac{5}{12}F_3 + \frac{7}{12}F_4\right) = \begin{pmatrix} -\frac{1}{4} & -\frac{1}{2} \\ -\frac{1}{2} & -1 \end{pmatrix}.$$

It is not surprising that the result is exact since all the necessary rank-one directions are contained in the set $\mathcal{D}_h$. Minimizing laminates are not unique, and if we restrict $\mathcal{D}_h$ to $\mathcal{D}_{h,1}$, a set of 16 matrices, then we we obtain the following laminate supported on all eight matrices in K and with center of mass equal to zero:

$$\nu = \frac{6}{24}\delta_{F_1} + \frac{2}{24}\delta_{F_2} + \frac{3}{24}\delta_{F_3} + \frac{1}{24}\delta_{F_4} + \frac{1}{24}\delta_{F_5} + \frac{3}{24}\delta_{F_6} + \frac{2}{24}\delta_{F_7} + \frac{6}{24}\delta_{F_8},$$

where (it is convenient to relabel the matrices)

$$F_1 = \begin{pmatrix} -\frac{3}{4} & -\frac{1}{4} \\ -\frac{1}{4} & -\frac{1}{2} \end{pmatrix}, \quad F_2 = \begin{pmatrix} -\frac{3}{4} & -\frac{1}{4} \\ -\frac{1}{4} & \frac{1}{2} \end{pmatrix}, \quad F_3 = \begin{pmatrix} \frac{3}{4} & \frac{1}{4} \\ -\frac{1}{4} & -\frac{1}{2} \end{pmatrix}, \quad F_4 = \begin{pmatrix} \frac{3}{4} & \frac{1}{4} \\ -\frac{1}{4} & \frac{1}{2} \end{pmatrix},$$

$$F_5 = \begin{pmatrix} -\frac{3}{4} & \frac{1}{4} \\ \frac{1}{4} & -\frac{1}{2} \end{pmatrix}, \quad F_6 = \begin{pmatrix} -\frac{3}{4} & \frac{1}{4} \\ \frac{1}{4} & \frac{1}{2} \end{pmatrix}, \quad F_7 = \begin{pmatrix} -\frac{3}{4} & \frac{1}{4} \\ \frac{1}{4} & -\frac{1}{2} \end{pmatrix}, \quad F_8 = \begin{pmatrix} -\frac{3}{4} & \frac{1}{4} \\ \frac{1}{4} & \frac{1}{2} \end{pmatrix}.$$

In this notation, $F_1 - F_2$, $F_3 - F_4$, $F_5 - F_6$, $F_7 - F_8$ are all parallel to $e_2 \otimes e_2$. If one defines

$$G_1 = \frac{3}{4}F_1 + \frac{1}{4}F_2, \quad G_2 = \frac{3}{4}F_3 + \frac{1}{4}F_4,$$
$$G_3 = \frac{1}{4}F_5 + \frac{3}{4}F_6, \quad G_4 = \frac{1}{4}F_7 + \frac{3}{4}F_8,$$

then $G_1 - G_2$ and $G_3 - G_4$ are parallel to $e_1 \otimes e_1$. Finally, with H_1 and H_2 given by

$$H_1 = \frac{2}{3}G_1 + \frac{1}{3}G_2, \quad H_2 = \frac{1}{3}G_3 + \frac{2}{3}G_4,$$

the matrix $H_1 - H_2$ is parallel to $(e_1 + e_2) \otimes (e_1 + e_2)$ and

$$F = \frac{1}{2}H_1 + \frac{1}{2}H_2.$$

It remains a challenging problem to find a class of reliable algorithms for the minimization of nonconvex problems. Such a scheme would allow one to design an integrated approach to the numerical analysis of microstructures by combining computation of relaxed energy densities with the minimization the relaxed functionals and an approximation of the underlying microstructures in a postprocessing step via the computation of laminates.

7. Bibliographic Remarks

We begin by describing some important contributions that form the background for the mathematical theory in a broader context. Then we provide detailed references for the material presented in the chapters of this text.

Ball&James [BJ87, BJ92] and Chipot&Kinderlehrer [CK88] derived the mathematical description of fundamental mechanisms for the appearance of microstructure in single crystals based on energy minimization, and Bhattacharya&Kohn [BhK96, BhK97] extended the theory to polycrystals. Magnetic effects, and the coupling of magnetic and elastic properties were analyzed by DeSimone [DS93], James&Kinderlehrer [JK93], Tartar [Ta95] and DeSimone&James [DSJ97]. Kohn&Müller [KM92, KM94] presented an analysis of domain branching which was then applied to magnetic domain patterns in Choksi, Kohn&Otto [CKO99]. James&Hane [JH00] and Pitteri&Zanzotto [PZ00] provided a detailed description of the crystallographic aspects. Further applications include studies of formation of blisters by Ortiz&Gioia [OG94] and Ben Belgacem, Conti, DeSimone&Müller [BCDM00] and of dislocation patterns by Ortiz&Repetto [OR99].

This research has led to new, simply stated, but deep questions in the calculus of variations which are closely related to Tartar's earlier work and his far reaching programme on oscillations in nonlinear partial differential equations and compensated compactness [Ta79, Ta83, Ta90]. At the heart of the analysis of microstructures in variational problems lie the notions of quasiconvexity and of quasiconvex hulls which remain fifty years after Morrey's seminal work [Mo52, Mo66] one of the fundamental challenges in the calculus of variations with ramifications to a number of other problems [Sv95, A98, B98].

7.1 Introduction

The fundamental contribution in the recent work by Ball&James [BJ87, BJ92] and Chipot&Kinderlehrer [CK88] is to start from a variational approach and not from kinematic theories as in [BMK54, Er80, Er86, WLR53]. The degeneracies of the theory if one assumes invariance under all bijections of the lattice were analyzed in [Er77, Er89, Fo87, Pa77, Pa81, Pi84, Za92]. Applications of the shape memory effect can be found in the conference proceedings [SMST97]. A beautiful account of the relations between the quasiconvex hull of a set and properties of sequences converging to the set is given in [Sv95]. The definition of the Young measure goes back to L. C. Young and was introduced to the analysis of oscillations in partial differential equations by Tartar, see, e.g., [Ta79, Ta83]. The version of the fundamental theorem on Young measures given in Section A.1 follows [B89]. Young measures generated by sequences of gradients were characterized in [KP91], see Theorem A.1.6 for a statement of the result, and the averaging technique is one of their technical tools. A detailed discussion of the different definitions for the semiconvex hulls and the proofs of their equivalence can be found in [M99b]. Surprising existence results for nonconvex variational principles have been obtained by Müller&Šverák by an adaption of Gromov's convex integration method [MS99a] and by Dacorogna&Marcellini [DcM99] using the Baire category argument. However, the rigidity results in [DM95] show that geometry of these solutions is very different from laminates of finite order. The Lavrientiev phenomenon has been studied in [BM85, BK87, NM90] and in [Li95] in connection with the approximation of singular minimizers. Approximations of Young measures can be found in [NW93, NW95, CR00] for scalar problems and in [AP00] for two-dimensional vector valued problems.

7.2 Semiconvex Hulls of Compact Sets

The fact that gradient Young measures supported on three matrices without rank-one connections are trivial, i.e., a Dirac mass placed at one of the three matrices, was first proven in [Sv91]. The paper [Sv95] contains an overview of the different semiconvex hulls of sets and their relation to compactness questions. The example of the discrete set with eight points in Section 2.1 is taken from [DcT00], see Remark 2.1.4 for an account of the partial results in this paper. Generalizations of the notion of quasiconvexity in the sense of Morrey to integrands that depend on higher derivatives were studied by Meyers [M65]. New examples of quasiconvex functions on symmetric matrices were obtained by Šverák [Sv92b] and this paper contains a concise summary of the restrictions they impose on gradient Young measures. It is an open problem whether these functions can be extended to the space of all matrices, see [MS99b] for partial results. The observation that one can find on a one-sheeted hyperboloid at every point two straight lines contained in the hyperboloid that correspond to rank-one directions is due to Šverák [Sv93] where he used this fact to derive the formula of the quasiconvex hull of two martensitic wells, see the statement of his result in Theorem 2.2.2. This paper also contains the proof that the semiconvex hulls of multi-well sets in two-dimensions without rank-one connections are trivial. The locality property of the rank-one convex hull in Proposition 2.1.5 was first mentioned in [Pe93] and later in the context of $\mathcal{D}$-convex sets in [MP98]. Elegant proofs can be found in [Kir00, Mt00]. The paper [BhD01] presents the semiconvex hulls for k-well problems invariant under SO(2); O(2) and O(2,3) in Sections 2.2

and 2.3 as well as the dimension reduction in Section 2.5. The proofs given here simplify the original arguments. The theory of martensitic thin films has been described in [BJ99]. The results about the two-well problem in three dimensions can be found in [DKMS00]. Note that Theorem 2.6.4 establishes the existence of a neighborhood of order one about the identity such that the polyconvex hull of $SO(3) \cup SO(3)H$ is trivial for all H in this neighborhood. A first result in this direction was obtained in [KL00] where the authors prove that the quasiconvex hull is trivial if H is sufficiently close to the identity. The proof relies on a linearization argument, John's estimates for deformations with bounded strains [Jo72a, Jo72b], and the quasiconvexity of the quadratic function $q : \mathbb{M}^{3\times 3} \mapsto \mathbb{R}$, $q(X) = -(\mathrm{cof}(X + X^T))_{33}$, see Theorem A.1.3. The first example of two incompatible wells with nontrivial polyconvex hull was constructed in [DKMS00], and the parameters h_i in this example violate the second set of inequalities in condition ii) in Theorem 2.6.4 with $h_1 = h^{-2} \approx 193.995$ and $h_2 = h_3 = h$. Here h is the larger of the two solutions of the equation $h + 1/h = 14$. The example given in Proposition 2.6.5 slightly improves the original one. It is not known what the optimal value for h is for which the polyconvex hull is trivial and how the topology of the hulls changes as soon as they become larger. The formula for the polyconvex hull of K in (2.53) has first been proven in [DSD00]. This paper contains also the result for the three-dimensional situation in the special case of $\gamma_1 = \gamma_2$, and the same proof yields the result in the general case. It follows also from the characterization of the semiconvex hulls without the constraint that the determinant be positive in [DcT98], see Remark 11 in [DcT00]. Our proof for the representation of K^{pc} follows [DSD00]. The case of arbitrary sets in 2×2 matrices that are given by conditions on the singular values (with and without the restriction that the determinant be positive) has been solved in [CT00]. The paper [B77] contains a characterization of convex and polyconvex functions that depend on singular values. The function $F \to \lambda_1(F) + \lambda_2(F)$ which we use is just an example of a rich family of convex functions depending on the singular values of F. A more general version of Theorem 2.7.5 can be found in [CDDMO01]. The proof presented here is taken from [DSD01]. The problem to maximize the expression $F : R$ for $R \in O(n)$ was also considered in [Ja86], Appendix 3.

7.3 Macroscopic Energy for Nematic Elastomers

The motivation behind the successful efforts to synthesize nematic elastomers [FKR81, KpF91] was the attempt to reproduce the mesophases typical of a liquid crystal within an amorphous, polymeric solid. The resulting physical system has the translational order of a solid phase coupled to the orientational order of a nematic phase. Soft deformation modes whose occurrence had been predicted by theory [GL89], were observed experimentally [KnF], in association with the appearance of domain patterns with a characteristically layered texture. The formation of these microstructures was explained by energy minimization in the framework of continuum models in [WB96, WT96]. The model considered here is due to Bladon, Terentjev&Warner [BTW93]. The mathematical analysis of this model started in [DSD00], where an explicit formula for the relaxation of the energy in a two-dimensional model was obtained. The results presented here are based on [D01, DSD01]. The crucial construction in the proof of the quasiconvexity of the envelope is due to Müller&Šverák [MS99a]. The relaxation result for nematic elastomers constitutes one of the few explicitly know relaxation results and seems to be the first for an SO(3) invariant energy related to phase transformations, see e.g. [Kh67, Pi91, K91, LDR95] for related relaxation questions. In particular it allows one to explore the features of the relaxed energy in numerical simulations, and first results in this direction have been reported in [CDD01]. A different approach has been pursued in [ACF99], and an extension of the results to compressible models was proposed in [Sy01]. The maximal eigenvalue of the cofactor matrix was also used in [BC97, BC99] to separate points from polyconvex hulls of sets.

7.4 Uniqueness and Stability

The uniqueness and stability results are strongly influenced by Luskin's work on uniqueness and stability of microstructure [L96a, L96b, LL98a, LL98b, L98, BLL99, LL99a, LL99b, BL00, L00, EL01, BBL01]. Our approach to a formulation of stability via the the distance d is inspired by the construction of a distance in [CKL91] for the one-dimensional, scalar case. In our presentation we stress the point of view that the stability estimates should be formulated for Young measures and not for functions. The idea to use the cofactor in the proof of uniqueness of simple laminates is inspired by [BJ92] and generalizes results in [BhD01]. The fact that polyconvex measures supported on two incompatible matrices have to be trivial (i.e. a single Dirac mass placed at one of the two matrices) was shown in [BJ87]. The proofs in Section 4.1 are an adaption of the ideas in [BLL99] to the framework with polyconvex measures in the n-dimensional setting. In particular, the proof of the estimates for the excess rotation becomes conceptually much clearer and shorter than in [BLL99]. Possible extensions of this theory described in this chapter are strongly limited by Iqbal's results [Iq99] that the volume fractions of laminates within laminates are typically not uniquely determined from the center of mass of the Young measure. The rates for the finite element minimizer in Section 4.5 are based on a construction that has been known for a long time. A more formal argument can be found in [C91, CC92, CCK95]. Based on ideas from [AcF84, AcF88, Liu77], Zhang constructed a truncation method for functions generating gradient Young measures that was sharpened by Müller (see [Z92, M99a]). The idea to formulate error estimates for finite element minimizers in a distance $d(\cdot,\cdot)$ on the space of all probability measures originates in [CKL91]. Our results cover all triangulations that are regular in the sense of Ciarlet [Ci78].

7.5 Applications to Martensitic Transformations

The transformation mechanisms are well-understood in many materials and we only mention a few typical examples that relate to the analytical results. Indium rich InTl alloys typically undergo a transformation from a face-centered cubic phase into a face-centered tetragonal phase [Gt50]. The transformation strains given in the text for the cubic to trigonal phase transformation correspond to the case that the lengths of the sides of the unit cell in the cubic and the trigonal phase are the same, see [JH00]. This is very nearly the case in AuCd [HS00] and in $TbDyFe_2$, a material with giant magnetostriction [JK93]. The first type of the cubic to orthorhombic phase has not been observed in any material. The second type of the cubic to orthorhombic transformation occurs in materials with an fcc parent phase and is for example found in CuAlNi alloys [DR64].

The uniqueness results follow also from Luskin's work already quoted in Section 7.4. However, the proofs become more transparent and shorter in our general framework developed in Chapter 4. The quasiconvex hull of the three wells describing the cubic to tetragonal transformation has attracted a lot of attentions and is still unknown. A formula for the polyconvex hull of the three wells was recently announced in [Fr00]. The paper [DK02] contains a construction that shows that the interior of the lamination convex hull of three wells with determinant equal to one contains the identity matrix as an interior point. This answers a question raised in [M98] and predicts an ideally soft or 'liquid-like' behavior of shape memory materials. Proposition 5.1.5 shows that it is also difficult to construct explicitly laminates that are supported on all three wells. In fact, any laminate supported on three wells must contain at least four Dirac masses. This was shown in [Fr00] based on a different proof for the polyconvex hull. However, it is worthwhile to note that there exists a short proof for the quasiconvex hull which relies on Theorem 9 in [Sv91] (see also [Sv92b]). This theorem states that any gradient Young measure supported on three incompatible matrices must be a single Dirac mass. The proof of the proposition given in the text then uses crucially the identity (5.6) which is identity (5.7) in [JK89]. The uniqueness part in the cubic to orthorhombic transformation in Theorem 5.3.1 follows also from Theorem 3.1 in [BLL99]. For completeness of our presentation we finally discuss the three tetragonal to monoclinic transformations described in [BL00]. Theorem 5.4.1 summarizes the results in [BL00] and presents a short proof in the general framework described in this text. The new ingredient is the precise characterization of the sets $\mathcal{M}^{pc}(K;F)$ in the exceptional cases for the lattice parameters.

7.6 Algorithmic Aspects

The algorithm for the computation of rank-one convex (or more generally, $\mathcal{D}$-convex) envelopes of functions was defined in [D99] and its convergence (in dependence on the discretization parameter h in the space of all matrices) was established in [DW00]. Theorems 6.1.4 and 6.1.5 and Lemma 6.1.7 can be found there. The algorithm is based on ideas in [KS86] how to find the rank-one convex envelope of a function by performing convexifications along rank-one lines. The book [Dc89] contains a lot of information about nonconvex variational problems, their relaxation, and integral representations for the relaxed energy. Explicit relaxation formulae can be found in [Kh67, Pi91, K91, LDR95, DSD01]. The rigidity results in [DM95] show that every function $\boldsymbol{u} : \Omega \to \mathbb{R}^3$ with $D\boldsymbol{u} \in \mathrm{SO}(3)U_1 \cup \mathrm{SO}(3)U_2$ is (locally) a function of one variable if the set $E = \{\boldsymbol{x} \in \Omega : D\boldsymbol{u}(\boldsymbol{x}) \in \mathrm{SO}(3)U_1\}$ is a set of finite perimeter. Here U_1 and U_2 are two of the thee wells in the cubic to tetragonal phase transformation. This result was extended to the three-well problem in [Kir98]. Various characterizations of semiconvex hulls of sets in terms of convexifications of distance functions can be found in [Z98]. The Kohn-Strang example was first discussed in [KS86] and generalizations and questions concerning the existence of minimizers have been analyzed in [DcM95, AF98]. The construction of the rank-one convex envelope of the Kohn-Strang function in Section 6.1 follows Section 5D in [KS86]. An algorithm for the computation of separately convex envelopes of functions and hulls was presented in [MP98]. See also [AP00] for related ideas. Approximation properties for laminates by laminates supported on finitely many points have been obtained in [Pe93, MS99a]. The four point configuration with nontrivial rank-one convex hull was discovered independently by several authors in different contexts (see e.g. [AH86, BFJK, CT93, Sch74, Ta93]).

A. Convexity Conditions and Rank-one Connections

A.1 Convexity Conditions

Throughout the text we have seen the fundamental importance of various convexity conditions, in particular of rank-one convexity, quasiconvexity and polyconvexity. In this appendix, we give a precise definition of these notions of convexity and we summarize some important results.

Let $M(F)$ denote the vector of all minors (i.e. all subdeterminants) of an $m \times n$ matrix $F \in \mathbb{M}^{m\times n}$ and $d(m,n)$ its length. For $m = n = 2$ and $n = m = 3$ we have $M(F) = \det F$ and $M(F) = (F, \operatorname{cof} F, \det F)$, respectively.

Definition A.1.1. *A function $f : \mathbb{M}^{m\times n} \to \overline{\mathbb{R}} = (-\infty, \infty]$ is said to be*

i) convex *if*

$$f(\lambda A + (1-\lambda)B) \leq \lambda f(A) + (1-\lambda) f(B) \quad \forall A, B \in \mathbb{M}^{m\times n}, \forall \lambda \in [0,1];$$

ii) polyconvex *if there exists a convex function $g : \mathbb{R}^{d(m,n)} \to \overline{\mathbb{R}}$ such that*

$$f(F) = g(M(F));$$

iii) quasiconvex *if for every open and bounded set $\Omega \subset \mathbb{R}^n$ with $|\partial\Omega| = 0$ one has*

$$\int_\Omega f(F + D\varphi)\mathrm{d}x \geq \int_\Omega f(F)\mathrm{d}x \quad \forall F \in \mathbb{M}^{m\times n}, \forall \varphi \in W_0^{1,\infty}(\Omega; \mathbb{R}^m),$$

whenever the integral on the left hand side exists;

iv) rank-one convex *if for all $A, B \in \mathbb{M}^{m\times n}$ with* $\operatorname{rank}(A-B) = 1$ *and for all $\lambda \in [0,1]$*

$$f(\lambda A + (1-\lambda)B) \leq \lambda f(A) + (1-\lambda) f(B).$$

While all these notions coincide for $m = 1$ or $n = 1$, they are fundamentally different for $m \geq 2$ and $n \geq 2$.

Proposition A.1.2. *Assume that $m, n \geq 2$ and that $f : \mathbb{M}^{m\times n} \to \mathbb{R}$ is real valued. Then we have the following implications:*

$$
\begin{array}{ccccc}
f \textit{ convex} & & & & f \textit{ convex} \\
\Downarrow & & & & \not\Uparrow \\
f \textit{ polyconvex} & & & & f \textit{ polyconvex} \\
\Downarrow & & \textit{and} & & \not\Uparrow \\
f \textit{ quasiconvex} & & & & f \textit{ quasiconvex} \\
\Downarrow & & & & \not\Uparrow \qquad (m \geq 3) \\
f \textit{ rank}-\textit{one convex} & & & & f \textit{ rank}-\textit{one convex}
\end{array}
$$

If $f : \mathbb{M}^{m\times n} \to \overline{\mathbb{R}}$, *then*

$$f \textit{ convex} \quad \Rightarrow \quad f \textit{ polyconvex} \quad \Rightarrow \quad f \textit{ rank-one convex} \tag{A.1}$$

and

$$f \textit{ polyconvex} \quad \Rightarrow \quad f \textit{ quasiconvex.}$$

A detailed discussion of these implications and various counterexamples can be found in [Dc89], Theorem 4.1 and [M99b], Lemma 4.3. See also [BM84] for a generalization of the concept of quasiconvexity, called $W^{1,p}$-quasiconvexity. The question of whether rank-one convexity implies quasiconvexity was open for a long time and finally answered by Šverák [Sv92a] for $m \geq 3$. The case $m = n = 2$ is still completely open. For smooth functions, rank-one convexity is equivalent to the Legendre-Hadamard condition that the second gradient be nonnegative on rank-one matrices, i.e. $(D^2 f; R, R) \geq 0$ for all $R \in \mathbb{M}^{m\times n}$ with $\text{rank}(R) = 1$. Based on Šverák's example, Kristensen recently showed that there does not exist an analogous local criterion for quasiconvexity and polyconvexity, see [Kr99, Kr00]. Milton [Mi02] modified the construction of the counterexample and found a set of seven matrices in $\mathbb{M}^{3\times 2}$ for which the quasiconvex hull is bigger than the rank-one convex hull.

A special case are quadratic functions for which rank-one convexity and quasiconvexity are known to be equivalent.

Theorem A.1.3 ([Dc89], Theorem 4.1.7). *Let* A *be a symmetric matrix in* $\mathbb{M}^{mn\times mn}$. *Let*

$$f(X) = \langle AX, X \rangle,$$

where $X \in \mathbb{M}^{m\times n}$. *Then*

i) f *is rank-one convex if and only if* f *is quasiconvex;*

ii) if $m = 2$ *or* $n = 2$ *then*

$$f \textit{ polyconvex} \Leftrightarrow f \textit{ quasiconvex} \Leftrightarrow f \textit{ rank-one convex;}$$

iii) if $m, n \geq 3$ then in general

$$f \text{ rank-one convex} \not\Rightarrow f \text{ polyconvex.}$$

An example for part iii) is due to Terpstra [Tr38] and Serre [Sr83]. A short self-contained proof can be found in [B85].

Rank-one convexity can be viewed as a special case of the more general concept of $\mathcal{D}$-convexity. Let $\mathcal{D} \subset \mathbb{M}^{m\times n}$ be any set. Then a function $f : \mathbb{M}^{m\times n} \to \mathbb{R}$ is said to be $\mathcal{D}$-convex if the functions

$$g : \mathbb{R} \to \mathbb{R}, \quad g(t) = f(F + tD)$$

are convex for all $F \in \mathbb{M}^{m\times n}$ and all $D \in \mathcal{D}$ (see [MP98] for a detailed discussion).

Based on these notions of convexity, we define semiconvex hulls of compact sets $K \subset \mathbb{M}^{m\times n}$ in the following way:

$$K^{\mathrm{pc}} = \{F \in \mathbb{M}^{m\times n} : f(F) \leq \sup_{X\in K} f(X) \ \forall f : \mathbb{M}^{m\times n} \to \mathbb{R} \text{ polyconvex }\}.$$

The quasiconvex hull K^{qc}, the rank-one convex hull K^{rc}, the convex hull $\mathrm{conv}(K)$, and the $\mathcal{D}$-convex hull $K^{\mathcal{D}}$ are defined analogously (in [MP98] the set $K^{\mathcal{D}}$ is called the functionally $\mathcal{D}$-convex hull). In particular for the description of constructions it is convenient to introduce a further hull which has no immediate characterization in terms of convex functions. The lamination convex hull K^{lc} is defined in the following way (see [MS96]): Let $K^{(0)} = K$ and define

$$K^{(i+1)} = \{\lambda A + (1-\lambda)B : A, B \in K^{(i)}, \mathrm{rank}(A - B) = 1, \lambda \in (0,1)\} \cup K^{(i)}.$$

Then

$$K^{\mathrm{lc}} = \bigcup_{i=0}^{\infty} K^{(i)}.$$

The relations in the first part of Proposition A.1.2 between the different notions of convexity imply immediately the following chain of inclusions:

$$K^{\mathrm{lc}} \subseteq K^{\mathrm{rc}} \subseteq K^{\mathrm{qc}} \subseteq K^{\mathrm{pc}} \subseteq \mathrm{conv}(K).$$

Similarly we define semiconvex envelopes of functions $f : \mathbb{M}^{m\times n} \to \mathbb{R}$. The largest polyconvex function less than or equal to f is called the polyconvex envelope of f and denoted by f^{pc}. The quasiconvex, the rank-one convex, the convex, and the $\mathcal{D}$-convex envelopes are defined correspondingly and denoted by f^{qc}, f^{rc}, f^{**}, $f^{\mathcal{D}}$, respectively. In view of Proposition A.1.2 we have for functions $f : \mathbb{M}^{m\times n} \to \mathbb{R}$ the following inequalities,

$$f^{\mathrm{rc}} \geq f^{\mathrm{qc}} \geq f^{\mathrm{pc}}.$$

For extended valued functions $f : \mathbb{M}^{m\times n} \to \overline{\mathbb{R}}$ we define the envelopes analogously, but in this case the inequalities between the envelopes might not hold.

The condition $\mathcal{H}_N$ has been introduced in [Dc85].

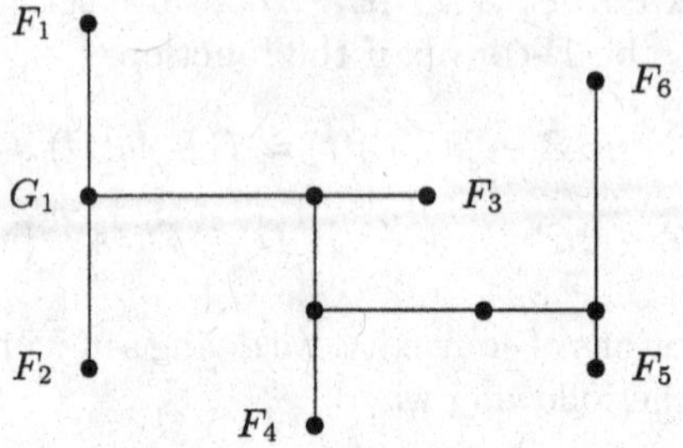

Fig. A.1. *Graphic interpretation of condition $\mathcal{H}_N$: the solid lines in the figure correspond to the rank-one connections, the dots at the end of the lines represent the matrices F_i, and the dots in the middle of the lines indicate the matrices G_i which are recursively constructed.*

Definition A.1.4. *We say that N pairs (λ_i, F_i), $i = 1, \dots, N$, with $\lambda_i \in (0,1)$ and $F_i \in \mathbb{M}^{m\times n}$ satisfy condition $\mathcal{H}_N$ if the following holds:*

- *if $N = 2$ then* $\operatorname{rank}(F_1 - F_2) = 1$ *and* $\lambda_1 + \lambda_2 = 1$;
- *if $N > 2$, then (relabeling the matrices if necessary)* $\operatorname{rank}(F_1 - F_2) = 1$ *and the pairs $(\mu_i, G_i)_{i=1,\dots,N-1}$ defined by*

$$\mu_1 = \lambda_1 + \lambda_2, \quad G_1 = \frac{\lambda_1}{\lambda_1 + \lambda_2} F_1 + \frac{\lambda_2}{\lambda_1 + \lambda_2} F_2,$$

$$\mu_i = \lambda_{i+1}, \qquad G_i = F_{i+1}, \quad i = 2, \dots, N-1,$$

satisfy condition $\mathcal{H}_{N-1}$.

It is an immediate consequence of the foregoing definitions that $F \in K^{(i)}$ implies the existence of at most 2^i pairs $(\lambda_i, F_i)_{i=1,\dots,k}$ with $F_i \in K$ satisfying condition $\mathcal{H}_k$.

We finally describe classes of (homogeneous) probability measures based on these notions of convexity. We denote by $\mathcal{M}(K)$ the set of all Radon measures supported on K equipped with the total variation norm $\|\cdot\|_{\mathcal{M}}$ and by $\mathcal{P}(K)$ the subset of all nonnegative Radon measures with mass one. We write

$$\langle \nu, f \rangle = \int_K f(A)d\nu(A) \quad \forall \nu \in \mathcal{P}(K), \forall f : K \to \mathbb{R},$$

whenever the right hand side exists.

It is convenient to introduce the following notation: if U is invertible and $\nu \in \mathcal{M}(K)$, then we denote by σU the measure given by

$$(\sigma U)(E) = \sigma(EU^{-1}) \text{ where } EU^{-1} = \{F : FU \in E\}.$$

We refer to the Dirac measure placed at a point $F \in \mathbb{M}^{m\times n}$ frequently as the Dirac mass at F.

As pointed out in the introduction, the link between the nonconvex minimization problem and probability measures is the Young measure generated by (subsequences of) minimizing sequences.

Theorem A.1.5 ([B89], Section 2). *Let Ω be Lebesgue measurable, let $K \subset \mathbb{R}^m$ be closed, and let $z_j : \Omega \to \mathbb{R}^m$, $j = 1, 2, \ldots$, be a sequence of Lebesgue measurable functions satisfying $z_j \to K$ in measure as $j \to \infty$, i.e., given any open neighborhood U of K in $\mathbb{R}^m$, then*

$$\lim_{j\to\infty} \left|\{x \in \Omega : z_j(x) \notin U\}\right| = 0.$$

Then there exists a subsequence z_μ of z_j and a family $\{\nu_x\}_{x\in\Omega}$ of positive measures on $\mathbb{R}^m$, depending measurably on x, such that
i) $\|\nu_x\|_{\mathcal{M}} = \int_{\mathbb{R}^m} d\nu_x \le 1$ for a.e. $x \in \Omega$,
ii) $\operatorname{supp} \nu_x \subseteq K$ for a.e. $x \in \Omega$, and
iii) $f(z_\mu) \overset{*}{\rightharpoonup} \langle \nu_x, f\rangle$ in L^∞ for each continuous function $f : \mathbb{R}^m \to \mathbb{R}$ satisfying $f(\lambda) \to 0$ as $|\lambda| \to \infty$.
Suppose further that $\{z_\mu\}$ satisfies the boundedness condition

$$\lim_{k\to\infty} \sup_\mu \left|\{x \in \Omega \cap B_R : |z_\mu(x)| \ge k\}\right| = 0$$

for every $R > 0$, where $B_R = B(\mathbf{0}, R)$. Then $\|\nu_x\|_{\mathcal{M}} = 1$ for a.e. $x \in \Omega$ (i.e., ν_x is a probability measure), and given any measurable subset A of Ω,

$$f(z_\mu) \rightharpoonup \langle \nu_x, f\rangle \quad \text{in } L^1(A)$$

for any continuous function $f : \mathbb{R}^m \to \mathbb{R}$ such that $\{f(z_\mu)\}$ is sequentially weakly relatively compact in $L^1(A)$.

In analogy to Jensen's inequality for convex functions we define the class of polyconvex measures $\mathcal{M}^{\mathrm{pc}}(K) \subseteq \mathcal{P}(K)$ by

$$\mathcal{M}^{\mathrm{pc}}(K) = \{\nu \in \mathcal{P}(K) : f(\langle \nu, id\rangle) \le \langle \nu, f\rangle \ \forall f : K \to \mathbb{R} \text{ polyconvex }\},$$

and we write $\mathcal{M}^{\mathrm{pc}}(K; F)$ for all measures $\mu \in \mathcal{M}^{\mathrm{pc}}(K)$ with center of mass equal to F, i.e., $\langle \mu, id\rangle = F$. We define the class of quasiconvex measures $\mathcal{M}^{\mathrm{qc}}(K)$ (usually called homogeneous gradient Young measures) and rank-one convex measures $\mathcal{M}^{\mathrm{rc}}(K)$ (also called laminates) analogously ([KP91, Pe93, Sv95]). If the pairs (λ_i, F_i), $i = 1, \ldots, N$ satisfy condition $\mathcal{H}_N$, then

$$\nu = \sum_{i=1}^{N} \lambda_i \delta_{F_i}$$

defines a laminate of finite order. If $N = 2$, then ν is called simple laminate. In case that $N = 3$ or $N = 4$ and (up to relabeling the matrices)

$$\operatorname{rank}(F_1 - F_2) = 1, \quad \operatorname{rank}(F_3 - F_4) = 1, \quad \operatorname{rank}(G_1 - G_2) = 1$$

where

$$G_1 = \frac{\lambda_1}{\lambda_1 + \lambda_2} F_1 + \frac{\lambda_2}{\lambda_1 + \lambda_2} F_2, \quad G_2 = \frac{\lambda_3}{\lambda_3 + \lambda_4} F_3 + \frac{\lambda_4}{\lambda_3 + \lambda_4} F_4,$$

the measure ν is called a second order laminate.

A fundamental result by Kinderlehrer&Pedregal states that quasiconvex measures are exactly the Young measures that can be generated by sequences of gradients.

Theorem A.1.6 ([KP91]). *A (weakly* measurable) map $\nu : \Omega \to \mathcal{P}(\mathbb{M}^{m\times n})$ is a gradient Young measure if and only if $\nu_{\boldsymbol{x}} \geq 0$ a.e. and there exist a compact set $K \subset \mathbb{M}^{m\times n}$ and $\boldsymbol{u} \in W^{1,\infty}(\Omega;\mathbb{R}^m)$ such that the following three conditions hold:*

i) $\operatorname{supp} \nu_{\boldsymbol{x}} \subseteq K$ *for a.e.* $\boldsymbol{x}$,
ii) $\langle \nu_{\boldsymbol{x}}, id\rangle = D\boldsymbol{u}(\boldsymbol{x})$ *for a.e.* $\boldsymbol{x}$, *and*
iii) $\langle \nu_{\boldsymbol{x}}, f\rangle \geq f(\langle \nu_{\boldsymbol{x}}, id\rangle)$ *for a.e.* $\boldsymbol{x}$ *and for all quasiconvex functions* $f : \mathbb{M}^{m\times n} \to \mathbb{R}$.

We say that the Young measure $\{\nu_{\boldsymbol{x}}\}_{\boldsymbol{x}\in\Omega}$ is homogeneous if there exists a $\nu \in \mathcal{P}(\mathbb{M}^{m\times n})$ such that $\nu_{\boldsymbol{x}} = \nu$ for a.e. $\boldsymbol{x}$. We identify in this case the family of measures $\{\nu_{\boldsymbol{x}}\}_{\boldsymbol{x}\in\Omega}$ with the single measure ν.

The next theorem describes the close connection between these convex hulls and the corresponding sets of Radon measures.

Theorem A.1.7. *Assume that $K \subset \mathbb{M}^{m\times n}$ is a compact set. Then*

$$K^{\mathrm{qc}} = \big\{ \langle \nu, id\rangle : \nu \in \mathcal{M}^{\mathrm{qc}}(K) \big\}$$

and the analogous statement holds for K^{rc} and K^{pc}.

Since SO(2) is contained in an affine subspace of $\mathbb{M}^{2\times 2}$ it is not difficult to see that

$$\operatorname{conv}(\mathrm{SO}(2)) = \left\{ \begin{pmatrix} a & -b \\ b & a \end{pmatrix}, \, a^2 + b^2 \leq 1 \right\}. \tag{A.2}$$

The convex hull of SO(3) is given by

$$\operatorname{conv}(\mathrm{SO}(3)) = \Big\{ F = QU : Q \in \mathrm{SO}(3),\, U = U^T, \tag{A.3}$$

$$\sum_{i=1}^{3} \varepsilon_i \lambda_i \leq 1 \text{ for } |\varepsilon_i| = 1 \text{ and } \varepsilon_1\varepsilon_2\varepsilon_3 = -1 \Big\}, \tag{A.4}$$

where λ_1, λ_2, λ_3 are the eigenvalues of the symmetric matrix U (see [Ja86]). Moreover,

$$\operatorname{tr}(F) - 2e^T F e \leq 1 \qquad \forall e \in \mathbb{S}^2,\, \forall F \in \operatorname{conv}\mathrm{SO}(3).$$

We will frequently use the important fact that $\mathrm{SO}(n)$ does not support any (nontrivial) gradient Young measure.

Theorem A.1.8 ([Ki88]). *Suppose that* $Du \in \mathrm{SO}(n)$ *a.e. in* Ω. *Then* Du *is constant and* $u(x) = Qx + b$ *with* $Q \in \mathrm{SO}(n)$ *and* $b \in \mathbb{R}^n$. *If* $u_j \in W^{1,\infty}(\Omega; \mathbb{R}^n)$ *satisfies*

$$\operatorname{dist}(Du_j, \mathrm{SO}(n)) \to 0 \text{ in measure,}$$

then $Du_j \to Q$ *in measure where* $Q \in \mathrm{SO}(n)$ *is a constant.*

The rank-one convex hull has a certain locality property which we will use in order to show that certain rank-one convex hulls are trivial.

Proposition A.1.9 ([Pe93, MP98, Mt00, Kir00]). *Assume that* K *is compact and that* K^{rc} *consists of two compact components* C_1 *and* C_2 *with* $C_1 \cap C_2 = \emptyset$. *Then*

$$K^{\mathrm{rc}} = (K \cap C_1)^{\mathrm{rc}} \cup (K \cap C_2)^{\mathrm{rc}}. \tag{A.5}$$

A.2 Existence of Rank-one Connections

The discussion in the introduction shows that formation of microstructure is closely connected to the existence of rank-one connections between energy wells. Here we say that two wells $\mathrm{SO}(m)U_1$ and $\mathrm{SO}(m)U_2$ with $U_1\, U_2 \in \mathbb{M}^{m\times n}$ symmetric and positive definite are *rank-one connected* or *compatible* if there exists a rotation $R \in \mathrm{SO}(m)$ such that

$$RU_2 - U_1 = a \otimes b$$

with $a \in \mathbb{R}^m$, $b \in \mathbb{R}^n$. Otherwise the wells are called *incompatible.*

In this section, we discuss the solvability of the so-called twinning equation $\operatorname{rank}(QA - B) = 1$ in detail for (2×2) and (3×3) matrices. We begin with the case $m = n = 3$. In this case the equation $RU_2 - U_1 = a \otimes m$ is after postmultiplication by U_1^{-1} equivalent to

$$RU_2U_1^{-1} = \mathbb{I} + \boldsymbol{a} \otimes \boldsymbol{n}, \quad \boldsymbol{n} = U_1^{-1}\boldsymbol{m}.$$

By the polar decomposition theorem this equation holds if and only if the matrix $C = U_1^{-1}U_2^2U_1^{-1}$ has the representation

$$C = (\mathbb{I} + \boldsymbol{n} \otimes \boldsymbol{a})(\mathbb{I} + \boldsymbol{a} \otimes \boldsymbol{n}), \quad 1 + \langle \boldsymbol{a}, \boldsymbol{n} \rangle > 0. \tag{A.6}$$

Therefore the existence of rank-one connections is equivalent to finding solutions $\boldsymbol{a}$, $\boldsymbol{n} \in \mathbb{R}^3$ of (A.6) with C given.

Proposition A.2.1 ([BJ87, Kh83]). *Necessary and sufficient conditions for a symmetric* (3×3)*-matrix* $C \neq \mathbb{I}$ *with eigenvalues* $\lambda_1 \leq \lambda_2 \leq \lambda_3$ *to be expressible in the form*

$$C = (\mathbb{I} + \boldsymbol{n} \otimes \boldsymbol{a})(\mathbb{I} + \boldsymbol{a} \otimes \boldsymbol{n})$$

with $1 + \langle \boldsymbol{a}, \boldsymbol{n} \rangle > 0$ *and* $\boldsymbol{a} \neq 0$, $\boldsymbol{n} \neq 0$ *are that* $\lambda_1 > 0$ *(i.e., C is positive definite) and* $\lambda_2 = 1$*. The solutions are given by*

$$\begin{aligned} \boldsymbol{b} =& \varrho \sqrt{\frac{\lambda_3(1-\lambda_1)}{\lambda_3 - \lambda_1}}\, \boldsymbol{e}_1 + \kappa \sqrt{\frac{\lambda_1(\lambda_3 - 1)}{\lambda_3 - \lambda_1}}\, \boldsymbol{e}_3, \\ \boldsymbol{m} =& \frac{1}{\varrho} \frac{\sqrt{\lambda_3} - \sqrt{\lambda_1}}{\sqrt{\lambda_3 - \lambda_1}} \left(-\sqrt{1 - \lambda_1}\boldsymbol{e}_1 + \kappa \sqrt{\lambda_3 - 1}\boldsymbol{e}_3 \right), \end{aligned}$$

where $\varrho \neq 0$ *is a constant and* $\boldsymbol{e}_1$, $\boldsymbol{e}_3$ *are normalized eigenvectors of C corresponding to* λ_1, λ_3*, respectively, and where* κ *can take the values* ± 1.

In the n-dimensional situation we obtain after suitable transformations the condition that at least $n - 2$ eigenvalues have to be equal to one, while the smaller eigenvalue is less than or equal to one, and the larger is larger than or equal to one.

Proposition A.2.2 ([DM95], Proposition 5.2). *Assume that* $A = \mathbb{I}$ *and* $B = \operatorname{diag}(\lambda_1, \ldots, \lambda_n)$ *with* $0 < \lambda_1 \leq \cdots \leq \lambda_n$*. Then the two wells* $\mathrm{SO}(n)$ *and* $\mathrm{SO}(n)B$ *are rank-one connected if and only if* $\lambda_2 = \cdots = \lambda_{n-1} = 1$*. Moreover, the vectors* $\boldsymbol{a}$ *and* $\boldsymbol{b}$ *in the representations* $Q - B = \boldsymbol{a} \otimes \boldsymbol{b}$ *lie in the plane spanned by* $\boldsymbol{e}_1$ *and* $\boldsymbol{e}_n$.

In twinning calculations one often encounters the situation that U_1 and U_2 are related by some rotation $R \in \mathrm{SO}(3)$, through $U_2 = R^T U_1 R$ and therefore

$$C_2 = U_2^T U_2 = R^T U_1^T U_1 R = R^T C_1 R \neq C_1. \tag{A.7}$$

In this situation, the solvability of the twinning equation

$$QU_1 = U_2(\mathbb{I} + \boldsymbol{a} \otimes \boldsymbol{n}), \quad Q \in \mathrm{SO}(3), \tag{A.8}$$

is described by the following proposition.

Proposition A.2.3 ([Er91]). *Assume that the symmetric and positive definite (3×3)-matrices C_1 and C_2 are related by a rotation in the sense of identity (A.7). Then the following assertions hold:*

1. *If there exists a solution of the twinning equation (A.8), then at least one of the following conditions must hold:*
   a) *R represents a $180°$ rotation.*
   b) *$\boldsymbol{e}$ is an eigenvector of C_1.*
   c) *$\boldsymbol{e}$ is perpendicular to an eigenvector of C_1, which can be taken as $(C_1\boldsymbol{e}) \wedge \boldsymbol{e}$, if (b) does not hold.*
2. *If any of the above conditions holds, and $C_2 \neq C_1$, then the twinning equation (A.8) can be solved.*
3. *If the twinning equation (A.8) can be solved, and (a) does not hold, then (A.7) must hold with R replaced by $\bar{R}$ with $\bar{R}$ a $180°$ rotation.*

Condition (1a) is particularly important, since in this case the two solutions can easily be obtained, as discussed in [Er81, Er85, Gr83]. The following version can be found in [Bh92].

Proposition A.2.4. *Let $R = -\mathbb{I}+2\boldsymbol{e}\otimes\boldsymbol{e}$ be a $180°$ rotation about $\boldsymbol{e} \in \mathbb{S}^2$ and assume that U_1 and $U_2 \in \mathbb{M}^{3\times 3}$ are symmetric matrices with $U_1^T U_1 \neq U_2^T U_2$ and $U_2 = R^T U_1 R$. Then there are two solutions of the equation*

$$QU_1 - U_2 = \boldsymbol{a} \otimes \boldsymbol{n}, \quad Q \in \mathrm{SO}(3), \boldsymbol{a}, \boldsymbol{n} \in \mathbb{R}^3, \tag{A.9}$$

and they are given by

$$\boldsymbol{n}_1 = \boldsymbol{e}, \quad \boldsymbol{a}_1 = 2\Big(\frac{U_2^{-T}\boldsymbol{e}}{|U_2^{-T}\boldsymbol{e}|^2} - U_2\boldsymbol{e}\Big)$$

and

$$\boldsymbol{n}_2 = \frac{2}{\varrho}\Big(\boldsymbol{e} - \frac{U_2^T U_2\boldsymbol{e}}{|U_2\boldsymbol{e}|^2}\Big), \quad \boldsymbol{a}_2 = \varrho U_2 \boldsymbol{e},$$

where ϱ is chosen in such a way that $|\boldsymbol{n}_2| = 1$. The corresponding rotations can be found by solving the equation (A.9) for Q.

The two-dimensional situation allows some simplifications.

Proposition A.2.5. *Assume that C_1, $C_2 \in \mathbb{M}^{2\times 2}$ are positive semidefinite, $C_1 = F_1^T F_1$, $C_2 = F_2^T F_2$. Let $\boldsymbol{e} \in \mathbb{S}^1$. Then the following four statements are equivalent:*

i) *there exist $Q \in \mathrm{SO}(2)$ and $\boldsymbol{a} \in \mathbb{R}^2$ such that $QF_1 - F_2 = \boldsymbol{a} \otimes \boldsymbol{e}^\perp$;*
ii) *we have $|F_1\boldsymbol{e}|^2 = |F_2\boldsymbol{e}|^2$;*
iii) *there exists a $\boldsymbol{v} \in \mathbb{R}^2$ such that $C_1 = C_2 + \boldsymbol{v} \otimes \boldsymbol{e}^\perp + \boldsymbol{e}^\perp \otimes \boldsymbol{v}$;*
iv) $\det(C_1 - C_2) \leq 0$.

Moreover, the vector $\boldsymbol{a}$ in statement i) and the vector $\boldsymbol{v}$ in statement iii) are related by $\boldsymbol{v} = F_2^T\boldsymbol{a} + \frac{1}{2}|\boldsymbol{a}|^2\boldsymbol{e}^\perp$. Finally, if $\det F_1 = \det F_2$, then $\boldsymbol{a} = \alpha F_2\boldsymbol{e}$ with $\alpha \in \mathbb{R}$ and the two wells are always rank-one connected.

Remark A.2.6. The observation that two wells with equal determinant in two dimensions are always rank-one connected is at the heart of the characterization of the semiconvex hulls in two dimensions. While this can be proven by a direct calculation, it is worthwhile noting that this fact is also a consequence the following decomposition of matrices (see, e.g., [CK88], Proposition 3.4).

Proposition A.2.7. *Let* $A \in \mathbb{M}^{3\times 3}$ *with* $\det A > 0$. *Then there is a rotation* $Q \in SO(3)$ *and vectors* $\boldsymbol{a}_1, \boldsymbol{n}_1, \boldsymbol{a}_2, \boldsymbol{n}_2 \in \mathbb{R}^3$ *with* $\langle \boldsymbol{a}_i, \boldsymbol{n}_i \rangle = 0$, $i = 1, 2$, *such that*

$$A = (\det A)^{1/3} Q(\mathbb{I} + \boldsymbol{a}_2 \otimes \boldsymbol{n}_2)(\mathbb{I} + \boldsymbol{a}_1 \otimes \boldsymbol{n}_1).$$

If $A \in \mathbb{M}^{2\times 2}$, *then the same result holds with* $\boldsymbol{a}_2 = \boldsymbol{n}_2 = 0$ *and the exponent* $\frac{1}{3}$ *replaced by* $\frac{1}{2}$.

B. Elements of Crystallography

A set of points $\mathcal{L}$ is called a Bravais lattice if and only if there exist three linearly independent vectors $\boldsymbol{g}_1, \boldsymbol{g}_2, \boldsymbol{g}_3 \in \mathbb{R}^3$ such that

$$\mathcal{L} = \mathcal{L}(\boldsymbol{g}_i) = \{x = \sum_{i=1}^{3} n_i \boldsymbol{g}_i,\ n_i \in \mathbb{Z}\}.$$

The vectors $\boldsymbol{g}_i$ are not uniquely determined and a theorem in crystallography (see [Er77]) states that two sets of linearly independent vectors $\boldsymbol{g}_1, \boldsymbol{g}_2, \boldsymbol{g}_3$ and $\widetilde{\boldsymbol{g}}_1, \widetilde{\boldsymbol{g}}_2, \widetilde{\boldsymbol{g}}_3$ generate the same lattice if and only if there exists a matrix $M \in \mathbb{Z}^{3\times3}$ with $\det M = \pm1$ such that

$$\widetilde{\boldsymbol{g}}_i = \sum_{i=1}^{3} M_{ij} \boldsymbol{g}_j.$$

It turns out that there are fourteen different three-dimensional or *Bravais lattices*[1] which are conventionally grouped into seven *crystal systems* sharing some symmetry, see Table B.1: triclinic[2,3], monoclinic[4], orthorhombic[5,6], tetragonal[7,8], hexagonal[9], cubic[10].

The symmetry is described in terms of *crystallographic point groups*, i.e., groups of symmetry operations which map the lattice back to itself and leave at least one point in the lattice fixed. The symbols for the different groups in Hermann-Mauguin notation can be found in Table B.1. Here n denotes a rotation by $2\pi/n$, m a reflection about a mirror plane, $\bar{1}$ an inversion through

[1] Auguste Bravais (1811-63), professor at the École Polytechnique, classified the fourteen space lattices in 1848. The German physicist M. L. Frankenheim found erroneously 15 types in 1835.

[2] *gr.* *τϱί-* three, thrice (in compound words)

[3] *gr.* *κλίνειν* come to lean, make to slope or slant

[4] *gr.* *μόνος* alone, solitary; only

[5] *gr.* *'oϱθόσ* straight

[6] *gr.* *'o 'ϱόμβoσ* rhombus, i.e., a four sided figure with all the sides, but only the opposite angles, equal

[7] *gr.* *τετϱά-* four (in compound words)

[8] *gr.* *'η γωνία* corner, angle

[9] *gr.* *'έξ* (indecl.) six

[10] *gr.* *κύβος*, *lat.* cubus marked on all six sides; anything of cubic shape

a point and $\bar{n}$ an improper rotation, i.e., a rotation by $2\pi/n$ followed by an inversion through a point on the axis. The position of the operation within the symbol refers to a different direction in the lattice related to the operation (axis of rotation, normal to the plane of reflection), see [S69], Chapter 3.5 for further details. For a given lattice $\mathcal{L}(\boldsymbol{g}_i)$, we define the point group $\mathcal{P}(\boldsymbol{g}_i)$ to be the maximal group under which the lattice is invariant. For example, the point group of the cube is the group of 48 orthogonal transformations that map the cube to the cube[11].

There are several ways to choose unit cells for the fourteen different lattices, and the standard cells (highest degree of symmetry) are shown in Figure B.1. The conventional symbols are P for primitive (the unit cell contains just one point), C for C-centered (a lattice point in the center of the C side of the unit cell), F or fc for face centered (lattice points in the center of each face) and I or bc for body centered (a lattice point in the center of the unit cell; German: *innenzentriert*) (see e.g. [S69] for a discussion of the notation). Note that the face centered and the body centered tetragonal systems are equivalent.

The different crystal systems are in particular of importance in diffusionless solid-solid phase transformations which are characterized by a break of symmetry of the underlying Bravais lattice. We call *martensite*[12] the phase that forms as the result of such a transformation. The high temperature phase is frequently called *austenite*[13]. A given system can undergo several transformations at several critical temperatures. For example, immediately after solidification iron forms a body centered cubic (bcc) structure, called δ-ferrite. Upon further cooling, iron transforms to a face centered cubic structure (fcc), called γ or austenite. Finally, iron transforms back to the bcc structure at lower temperatures; this structure is called α, or ferrite. In higher carbon steels, thc fcc austenite transforms to a body centered tetragonal martensite, thus showing the characteristic break of symmetry during the transformation.

The following examples describe typical phase transformations and the corresponding transformation matrices where we always assume that the reference configuration is given in the undistorted austenitic phase. The number N of variants in the low temperature phase is given by

$$N = \frac{\text{order of high temperature point group}}{\text{order of low temperature point group}}$$

see e.g. [VTA74, BJ92] for a discussion of the mathematical concepts. A detailed analysis of the point groups associated with the different variants is contained in [PZ00]. A nice summary of the matrices describing the energy

[11] In the context of continuum models in elasticity the point group is frequently defined as a subgroup of SO(3). This is also our convention in this text.

[12] Adolf Martens (1850-1914), German metallurgist, professor at the TH Berlin.

[13] Sir William Chandler Roberts-Austen (1843-1902), FRS, 19th-century pioneer in metallurgy and alloy phase diagrams. He presented in 1897 his first temperature-composition diagram for the Fe-C system [KPt98].

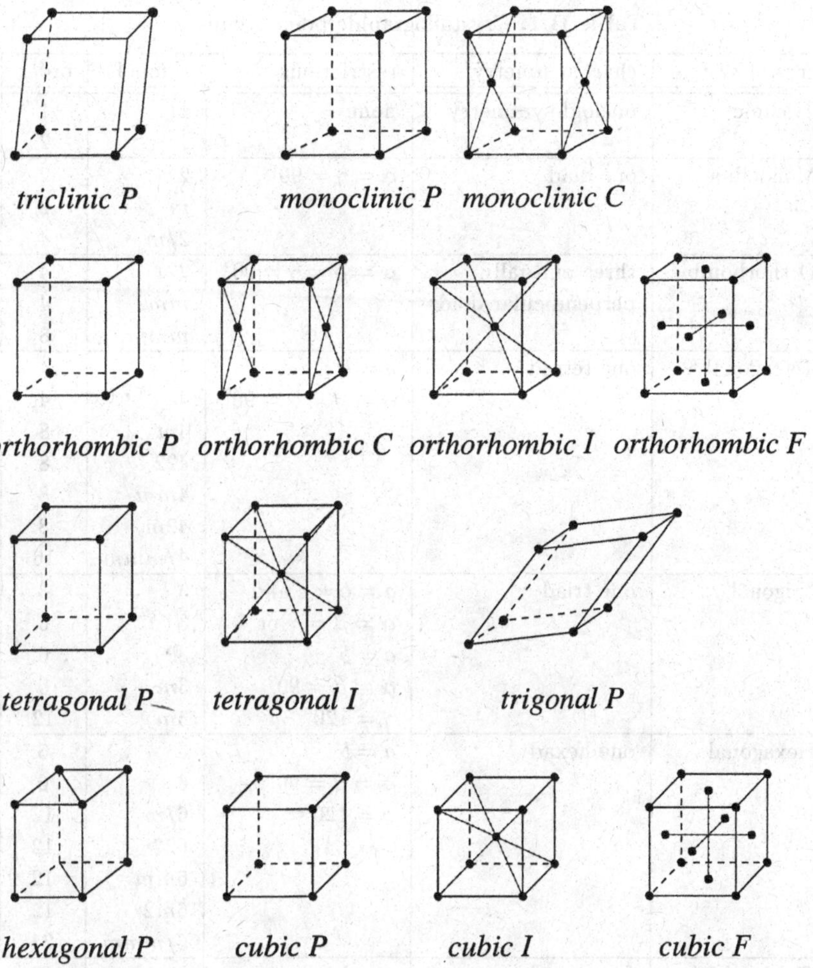

Fig. B.1. *The fourteen Bravais lattices.*

wells (under the assumption that the austenitic phase is the reference configuration) and the relation of their elements to the transformation strains can be found in [H97, JH00]. We include some examples and collect the relevant information about the *twinning systems.* We define twins to be continuous and piecewise homogeneous deformations with constant deformation gradients F_1 and F_2 in layers separated by a hyperplane. Necessarily the two deformation gradients must be rank-one connected and the normal $\boldsymbol{n}$ on the hyperplane is determined from $F_1 - F_2 = \boldsymbol{a} \otimes \boldsymbol{n}$. Since we are interested in zero-energy deformations, we focus on rank-one connections between energy wells $\mathrm{SO}(3)U_i$ and $\mathrm{SO}(3)U_j$, $i \neq j$. By Propositions A.2.3 and A.2.4 the situation $U_i = R^T U_j R$ with R a 180° degree rotation is particularly important.

Table B.1. Crystallographic point groups.

crystal system	char, symmetry	restrictions	symbol	order
Triclinic	onefold symmetry	none	1 $\bar{1}$	1 2
Monoclinic	one diad	$\alpha = \gamma = 90°$	2 m $2/m$	2 2 4
Orthorhombic	three mutually perpendicular diads	$\alpha = \beta = \gamma = 90°$	222 $mm2$ mmm	4 4 8
Tetragonal	one tetrad	$a = b$ $\alpha = \beta = \gamma = 90°$	4 $\bar{4}$ $4m$ 422 $4mm$ $\bar{4}2m$ $4/mmm$	4 4 8 8 8 8 16
Trigonal	one triad	$a = b = c$ and $\alpha = \beta = \gamma$ or $a = b$ $\alpha = \beta = 90°$ $\gamma = 120°$	3 $\bar{3}$ 32 $3m$ $\bar{3}m$	3 6 6 6 12
Hexagonal	one hexad	$a = b$ $\alpha = \beta = 90°$ $\gamma = 120°$	6 $\bar{6}$ $6/m$ 622 $6mm$ $\bar{6}m2$ $6/mmm$	6 6 12 12 12 12 24
Cubic	four triads	$a = b = c$ $\alpha = \beta = \gamma = 90°$	23 $m3$ 432 $\bar{4}3m$ $m3m$	12 24 24 24 48

We call the two solutions given in Proposition A.2.4 a *type-I* and a *type-II twin*, respectively. A twin is called a *compound twin* if it is at the same time a type-I and a type-II twin, i.e., if there exist two distinct 180° degree rotations R_1, R_2 with $U_i = R_1^T U_j R_1$ and $U_i = R_2^T U_j R_2$.

C. Notation

We write $\mathbb{R}^n$ for the n-dimensional real vector space with scalar product $\langle \boldsymbol{u}, \boldsymbol{v} \rangle = \sum_i u_i v_i$, $\mathbb{R}^n_+$ for the positive octant $\{\boldsymbol{x} : x_i \geq 0 \text{ for } i = 1, \ldots, n\}$ in $\mathbb{R}^n$, and $\mathbb{S}^n$ for the unit sphere in $\mathbb{R}^{n+1}$. We use $\mathbb{M}^{m \times n}$ for the space of all (real) $(m \times n)$-matrices with the norm $|F|^2 = \operatorname{tr}(F^T F)$, where $F^T \in \mathbb{M}^{n \times m}$ is the transposed matrix, and the scalar product $F : G = \sum_{i,j} F_{ij} G_{ij}$. If $\boldsymbol{e} \in \mathbb{R}^2$, then $\boldsymbol{e}^\perp = J\boldsymbol{e}$, where J is the counter-clockwise rotation by $\frac{\pi}{2}$.

If $m = n$, then the cofactor matrix $\operatorname{cof} F \in \mathbb{M}^{n \times n}$ of F is defined to be the matrix of all $(n-1) \times (n-1)$ minors of F which satisfies

$$\operatorname{cof} F = (\det F) F^{-T} \quad \text{i.e.} \quad (\operatorname{cof} F) F^T = (\det F)\mathbb{I} \quad \forall F \in \mathbb{M}^{n \times n}, \qquad \text{(C.1)}$$

where for $\boldsymbol{a} \in \mathbb{R}^m$ and $\boldsymbol{b} \in \mathbb{R}^n$ the $(m \times n)$-matrix $\boldsymbol{a} \otimes \boldsymbol{b}$ is given by $(\boldsymbol{a} \otimes \boldsymbol{b})_{i,j} = a_i b_j$, and $\mathbb{I} = \operatorname{diag}(1, \ldots, 1)$ is the identity matrix in $\mathbb{M}^{n \times n}$. We will frequently use the following formula for the inversion of a matrix,

$$(\mathbb{I} + \boldsymbol{a} \otimes \boldsymbol{b})^{-1} = \mathbb{I} - \boldsymbol{a} \otimes \boldsymbol{b} \quad \forall \boldsymbol{a}, \boldsymbol{b} \in \mathbb{R}^n, \ \langle \boldsymbol{a}, \boldsymbol{b} \rangle = 0.$$

In particular, suppose that $F - G = \boldsymbol{a} \otimes \boldsymbol{b}$ and let

$$H_\lambda = \lambda F + (1 - \lambda) G = G + \lambda \boldsymbol{a} \otimes \boldsymbol{b}.$$

If G and H_λ are invertible and if $\langle G^{-1}\boldsymbol{a}, \boldsymbol{b} \rangle = 0$, then

$$\begin{aligned} H_\lambda^{-1} &= G^{-1} - \lambda G^{-1}\boldsymbol{a} \otimes G^{-T}\boldsymbol{b}, \\ \operatorname{cof} H_\lambda &= \operatorname{cof} G - \lambda \operatorname{cof} G\boldsymbol{b} \otimes G^{-1}\boldsymbol{a}. \end{aligned} \qquad \text{(C.2)}$$

The assumption $\langle G^{-1}\boldsymbol{a}, \boldsymbol{b} \rangle = 0$ is for example satisfied if $\det F = \det G$ since the formula

$$\det(\mathbb{I} + \boldsymbol{c} \otimes \boldsymbol{d}) = 1 + \langle \boldsymbol{c}, \boldsymbol{d} \rangle, \quad \boldsymbol{c}, \boldsymbol{d} \in \mathbb{R}^n$$

implies that

$$\det F = \det(G + \boldsymbol{a} \otimes \boldsymbol{b}) = (\det G)(1 + \langle G^{-1}\boldsymbol{a}, \boldsymbol{b} \rangle).$$

We have for $F \in \mathbb{M}^{2 \times 2}$ the expansion

$$\det(A + B) = \det A + A : (\operatorname{cof} B) + \det B, \tag{C.3}$$

while for $F \in \mathbb{M}^{3\times 3}$

$$\operatorname{cof}(F - \mathbb{I}) = \operatorname{cof} F - (\operatorname{tr} F)\mathbb{I} + F^T + \mathbb{I}, \tag{C.4}$$
$$\det(F - \mathbb{I}) = \det F - \operatorname{tr}\operatorname{cof} F + \operatorname{tr} F - \det \mathbb{I}. \tag{C.5}$$

We denote the group of all orthogonal matrices by $\mathrm{O}(n)$, the group of all proper rotations by $\mathrm{SO}(n)$ or $\mathrm{O}^+(n)$ and we define $\mathrm{O}^-(n) = \mathrm{O}(n) \setminus \mathrm{O}^+(n)$ to be the set of all orthogonal matrices with determinant minus one. Finally $\mathrm{O}(n,m) = \{Q \in \mathbb{M}^{m\times n} : Q^T Q = \mathbb{I}\}$.

We will frequently use the following polar decomposition result (see e.g. Theorem 9 in Chapter IX, §14 in [Gnt58]).

Proposition C.1. *Any matrix $F \in \mathbb{M}^{n\times n}$ can be represented as $F = Q_1 S_1 = S_2 Q_2$ with Q_1, $Q_2 \in \mathrm{O}(n)$ and uniquely determined symmetric and positive semidefinite matrices S_1 and S_2. Moreover, if $\det F \neq 0$, then also Q_1 and Q_2 are uniquely determined.*

Definition C.2. *Assume that $F \in \mathbb{M}^{n\times n}$ and that the polar decomposition of F is given by $F = QS$ with $Q \in \mathrm{O}(n)$ and S symmetric and positive definite. We call the (nonnegative) eigenvalues $0 \le \lambda_1 \le \ldots \le \lambda_n$ of S the* singular values *of F. Moreover, if $\det F < 0$, then we call the values $\sigma_i = \lambda_i$, $i = 2, \ldots, n$, and $\sigma_1 = -\lambda_1$ the* signed singular values *of F. By definition, $|\sigma_1| \le \sigma_2 \le \ldots \le \sigma_n$, $\sigma_1 < 0$.*

Remark C.3. Assume $\det F < 0$ and that $F = QS$ is the polar decomposition of F. If $\Lambda^+ = \operatorname{diag}(\lambda_1, \ldots, \lambda_n)$ denotes the diagonal matrix of the singular values of F then $S = Q_1^T \Lambda^+ Q_1$ with $Q_1 \in \mathrm{SO}(n)$. Similarly, if $\Lambda^- = \operatorname{diag}(\sigma_1, \ldots, \sigma_n)$ is the diagonal matrix of the signed singular values of F, then there exists a proper orthogonal matrix $Q_2 \in \mathrm{SO}(n)$ such that $F = Q_2 \Lambda^- Q_1$. The convention to put the minus sign in front of the smallest singular value of F is convenient in the statements of Proposition 2.7.8 and Proposition 2.7.9.

Vectors and matrices

$\mathbb{R}^n$	n-dimensional Euclidean space
$\langle \boldsymbol{u}, \boldsymbol{v} \rangle$	inner product in $\mathbb{R}^n$
$\mathbb{S}^n$	unit sphere in $\mathbb{R}^{n+1}$
$\mathbb{M}^{m\times n}$	space of all real $m \times n$ matrices
$\operatorname{tr} F$	trace of an $n \times n$ matrix F
$F : G$	inner product in $\mathbb{M}^{n\times n}$ defined by $F : G = \operatorname{tr}(F^T G)$
$\lvert F \rvert$	Euclidean norm in $\mathbb{R}^n$ and $\mathbb{M}^{m\times n}$, $\lvert F\rvert^2 = F : F$
$\lambda_i(F)$	singular values of F, $\lambda_1 \leq \cdots \leq \lambda_n$
$\sigma_i(F)$	signed singular values of F, $\lvert\sigma_1\rvert \leq \sigma_2 \leq \ldots \leq \sigma_n$
$\mathbb{I}$	identity matrix in $\mathbb{M}^{n\times n}$
F^T	transposed matrix
$\operatorname{cof} F$	cofactor matrix, $(\operatorname{cof} F)F^T = (\det F)\mathbb{I}$
$M(F)$	vector of all minors of a matrix F
$\mathrm{SO}(n)$	orthogonal matrices in $\mathbb{M}^{n\times n}$ with $\det F = 1$
$\mathrm{O}(n)$	orthogonal matrices in $\mathbb{M}^{n\times n}$ with $\det F = \pm 1$
$\mathrm{O}^{\pm}(n)$	orthogonal matrices with determinant ± 1
$\mathrm{O}(m, n)$	isometries from $\mathbb{R}^m$ into $\mathbb{R}^n$
$R^{\pi}_{i\pm j}$, R^{π}_i	180° rotations with axes $\boldsymbol{e}_i \pm \boldsymbol{e}_j$ and $\boldsymbol{e}_i$
$\boldsymbol{a} \otimes \boldsymbol{b}$	rank-one matrix $(\boldsymbol{a} \otimes \boldsymbol{b})_{ij} = a_i b_j$
J	counterclockwise rotation by $\pi/2$ in the plane
$v^{\perp}$	Jv, v rotated counterclockwise by $\pi/2$ in the plane
$\widehat{\boldsymbol{e}}_1$, $\widehat{\boldsymbol{e}}_2$	standard basis in $\mathbb{R}^2$
$\boldsymbol{e}_1$, $\boldsymbol{e}_2$, $\boldsymbol{e}_3$	standard basis in $\mathbb{R}^3$

Hulls of matrices

$K^{(\mathrm{i})}$	ith lamination hull
K^{lc}	lamination convex hull of K
K^{rc}	rank-one convex hull of K
K^{qc}	quasiconvex hull of K
K^{pc}	polyconvex hull of K
$\operatorname{conv} K$	convex hull of K

Sets of measures

$\operatorname{supp} \mu$	support of the measure μ
$\langle \nu, f \rangle$	integral of f with respect to ν
σU	measure defined by $(\sigma U)(E) = \sigma(EU^{-1})$ for U invertible and $EU^{-1} = \{F : FU \in E\}$
$\mathcal{P}(K)$	set of all probability measures supported on K
$\mathcal{M}^{\mathrm{rc}}(K)$	laminates supported on K
$\mathcal{M}^{\mathrm{rc}}(K; F)$	laminates supported on K with center of mass F
$\mathcal{M}^{\mathrm{qc}}(K)$	gradient Young measures supported on K
$\mathcal{M}^{\mathrm{qc}}(K; F)$	gradient Young measures supported on K with center of mass F
$\mathcal{M}^{\mathrm{pc}}(K)$	polyconvex measures supported on K
$\mathcal{M}^{\mathrm{pc}}(K; F)$	polyconvex measures supported on K with center of mass F

Functions and function spaces

L^p	Lebesgue space
$W^{k,p}$	Sobolev space
$C^{k,\alpha}$	Hölder Spaces
ℓ^p	space of p-summable sequences
$\lvert x \rvert_{\ell^1}$	$\lvert x_1 \rvert + \ldots + \lvert x_n \rvert$ for $x \in \mathbb{R}^n$
$\lvert x \rvert_{\ell^\infty}$	$\max_{i=1,\ldots,n} \lvert x_i \rvert$ for $x \in \mathbb{R}^n$
$(t)^+$	$\max\{t, 0\}$
f^{rc}	rank-one convex envelope of f
f^{qc}	quasiconvex envelope of f
f^{pc}	polyconvex envelope of f
$\widehat{\pi}$	embedding $\mathbb{M}^{2\times 2} \to \mathbb{M}^{3\times 2}$
π	projection from $\mathbb{M}^{m\times n}$ onto a multi-well set K
Π	projection from $\mathbb{M}^{m\times n}$ onto two wells

References

[AcF84] E. ACERBI, N. FUSCO, *Semicontinuity problems in the calculus of variations*, Arch. Rational Mech. Anal. **86** (1984), 125-145

[AcF88] E. ACERBI, N. FUSCO, *An approximation lemma for $W^{1,p}$ functions*, in: *Material instabilities in continuum mechanics and related mathematical problems* (J. M. Ball, ed.), Oxford University Press, 1988, 217-242

[AF98] G. ALLAIRE, G. FRANCFORT, *Existence of minimizers for non-quasiconvex functionals arising in optimal design*, Ann. Inst. H. Poincaré Anal. Non Linéaire **15** (1998), 301-339

[ACF99] D. R. ANDERSON, D. E. CARLSON, E. FRIED, *A continuum-mechanical theory for nematic elastomers*, J. Elasticity **56** (1999), 33-58

[AP00] E. ARANDA, P. PEDREGAL, *On the computation of the rank-one convex hull of a function*, SIAM J. Sci. Comput. **22** (2000), 1772-1790

[A98] K. ASTALA, *Analytic aspects of quasiconformality*, Proceedings of the International Congress of Mathematicians, Vol. II (Berlin, 1998), Doc. Math. 1998, Extra Vol. II, 617-626

[AH86] R. J. AUMANN, S. HART, *Bi-convexity and bi-martingales*, Israel J. Math. **54** (1986), 159-180

[B77] J. M. BALL, *Convexity conditions and existence theorems in nonlinear elasticity*, Arch. Rational Mech. Anal. **63** (1976/77), 337–403

[B84] J. M. BALL, *Differentiability properties of symmetric and isotropic functions*, Duke Math. J. **51** (1984), 699-728

[B85] J. M. BALL, *On the paper "Basic calculus of variations"*, Pacific J. Math. **116** (1985), 7-10

[B89] J. M. BALL, *A version of the fundamental theorem for Young measures*, in: *PDEs and continuum models of phase transitions*, Lecture Notes in Phys. **344**, Springer, 1989, 207-215

[B98] J. M. BALL, *The calculus of variations and materials science*. Current and future challenges in the applications of mathematics (Providence, RI, 1997), Quart. Appl. Math. **56** (1998), 719-740

[BC97] J. M. BALL, C. CARSTENSEN, *Nonclassical austenite-martensite interfaces*, J. Phys. IV France **7** (1997), 35-40

[BC99] J. M. BALL, C. CARSTENSEN, *Compatibility conditions for microstructures and the austenite martensite transition*, Mater. Sc. Engin. A **273-275** (1999), 231-236

[BJ87] J. M. BALL, R. D. JAMES, *Fine phase mixtures as minimizers of energy*, Arch. Rational Mech. Anal. **100** (1987), 13-52

[BJ92] J. M. BALL, R. D. JAMES, *Proposed experimental tests of a theory of fine microstructure and the two-well problem*, Phil. Trans. Roy. Soc. London A **338** (1992), 389-450

[BK87] J. M. BALL, G. KNOWLES, *A numerical method for detecting singular minimizers*, Numer. Math. **51** (1987), 181-197

[BM85] J. M. Ball, V. J. Mizel, *One-dimensional variational problems whose minimizers do not satisfy the Euler-Lagrange equation*, Arch. Rational Mech. Anal. **90** (1985), 325-388

[BM84] J. M. Ball, F. Murat, *$W^{1,p}$-quasiconvexity and variational problems for multiple integrals*, J. Funct. Anal. **58** (1984), 225-253

[BBL01] P. Belik, T. Brule, M. Luskin, *On the numerical modeling of deformations of pressurized martensitic thin films*, M^2AN Math. Model. Numer. Anal. **35** (2001), 525-548

[BL00] P. Belik, M. Luskin, *Stability of microstructure for tetragonal to monoclinic martensitic transformations*, M^2AN Math. Model. Numer. Anal. **34** (2000), 663-685

[BCDM00] H. Ben Belgacem, S. Conti, A. DeSimone, S. Mller, *Rigorous bounds for the Föppl-von Kármán theory of isotropically compressed plates*, J. Nonlinear Sci. **10** (2000), 661-683

[Bh92] K. Bhattacharya, *Self-Accommodation in Martensite*, Arch. Rational Mech. Anal. **120** (1992), 201-244

[BhD01] K. Bhattacharya, G. Dolzmann, *Relaxation of some multi-well problems*, Proc. Roy. Soc. Edinburgh Sect. A **131** (2001), 279-320

[BFJK] K. Bhattacharya, N. B. Firoozye, R. D. James, R. V. Kohn, *Restrictions on microstructure*, Proc. Roy. Soc. Edinburgh Sect. A **124** (1994), 843-878

[BJ99] K. Bhattacharya, R. D. James, *A theory of thin films of martensitic materials with applications to microactuators*, J. Mech. Phys. Solids **47** (1999), 531-576

[BhK96] K. Bhattacharya, R. V. Kohn, *Symmetry, texture and the recoverable strain of shape memory materials*, Acta Materialia **25** (1996), 529-542

[BhK97] K. Bhattacharya, R. V. Kohn, *Elastic energy minimization and the recoverable strains of polycrystalline shape-memory materials*, Arch. Rational Mech. Anal. **139** (1997), 99-180

[BLL99] K. Bhattacharya, B. Li, M. Luskin, *The simply laminated microstructure in martensitic crystals that undergo a cubic-to-orthorhombic phase transformation*, Arch. Ration. Mech. Anal. **149** (1999), 123-154

[BTW93] P. Bladon, E. M. Terentjev, M. Warner., *Transitions and instabilities in liquid-crystal elastomers*, Phys. Rev. E **47** (1993), R3838-R3840

[BMK54] J. S. Bowles, J. K. MacKenzie, *The crystallography of martensitic transformations I and II*, Acta Met. **2** (1954), 129-137, 138-147

[CT00] P. Cardaliaguet, R. Tahraoui, *Equivalence between rank-one convexity and polyconvexity for isotropic sets of $\mathbb{R}^{2\times 2}$ (part I)*, Cahiers de Mathématiques du CEREMADE 0028 (2000)

[CP97] C. Carstensen, P. Plecháč, *Numerical solution of the scalar double-well problem allowing microstructure*, Math. Comp. **66** (1997), 997-1026

[CP00] C. Carstensen, P. Plecháč, *Numerical analysis of compatible phase transitions in elastic solids*, SIAM J. Numer. Anal. **37** (2000), 2061-2081

[CR00] C. Carstensen, T. Roubíček, *Numerical approximation of Young measures in non-convex variational problems*, Numer. Math. **84** (2000), 395-415

[CT93] E. Casadio Tarabusi, *An algebraic characterization of quasiconvex functions*, Ricerche Mat. **42** (1993), 11-24

[C91] M. Chipot, *Numerical analysis of oscillations in nonconvex problems*, Numer. Math. **59** (1991), 747-767

[CC92] M. Chipot, C. Collins, *Numerical approximations in variational problems with potential wells*, SIAM J. Numer. Anal. **29** (1992), 1002-1019

[CCK95] M. Chipot, C. Collins, D. Kinderlehrer, *Numerical analysis of oscillations in multiple well problems*, Numer. Math. **70** (1995), 259-282

[CK88] M. Chipot, D. Kinderlehrer, *Equilibrium configurations of crystals*, Arch. Rational Mech. Anal. **103** (1988), 237-277

[CKO99] R. Choksi, R. V. Kohn, F. Otto, *Domain branching in uniaxial ferromagnets: a scaling law for the minimum energy*, Comm. Math. Phys. **201** (1999), 61-79

[Ch95] J. W. Christian, S. Mahajan, *Deformation Twinning*, Progress in Materials Science **39** (1995), 1-157

[Ci78] P. G. Ciarlet, *The finite element method for elliptic problems*, North-Holland, Amsterdam, 1978

[CKL91] C. Collins, D. Kinderlehrer, M. Luskin, *Numerical approximation of the solution of a variational problem with a double well potential*, SIAM J. Numer. Anal. **28** (1991), 321-332

[CL89] C. Collins, M. Luskin, *The computation of the austenitic-martensitic phase transition*, in: *PDEs and continuum models of phase transitions* (Nice, 1988), Lecture Notes in Phys. **344**, Springer, Berlin-New York, 1989, 34-50

[CLR93] C. Collins, M. Luskin, J. Riordan, *Computational results for a two-dimensional model of crystalline microstructure*, in: *Microstructure and phase transition*, (D. Kinderlehrer, R. D. James, M. Luskin and J. Ericksen, eds.), IMA Vol. Math. Appl. **54**, Springer, New York, 1993, 51-56

[CDD01] S. Conti, A. DeSimone, G. Dolzmann, *Soft elastic response of stretched sheets of nematic elastomers: a numerical study*, J. Mech. Phys. Solids, **50** (2002), 1431-1451

[CDDMO01] S. Conti, A. DeSimone, G. Dolzmann, S. Müller, F. Otto, *Multiscale problems in materials*, in: *Trends in Nonlinear Analysis*, (M. Kirkilionis S. Krömker, R. Rannacher, F. Tomi, eds.), Springer (to appear)

[Dc85] B. Dacorogna, *Remarques sur les notions de polyconvexité, quasi-convexité et convexité de rang* 1, J. Math. Pures Appl. **64** (1985), 403-438

[Dc89] B. Dacorogna, *Direct methods in the calculus of variations*, Springer, 1989

[DcM95] B. Dacorogna, P. Marcellini, *Existence of minimizers for non-quasiconvex integrals*, Arch. Rational Mech. Anal. **131** (1995), 359-399

[DcM99] B. Dacorogna, P. Marcellini, *Implicit partial differential equations*, Birkhäuser, 1999

[DcT98] B. Dacorogna, C. Tanteri, *On the different convex hulls of sets involving singular values*, Proc. Roy. Soc. Edinburgh Sect. A **128** (1998), 1261-1280.

[DcT00] B. Dacorogna, C. Tanteri, *Implicit partial differential equations and the constraints of nonlinear elasticity*, Journal des Mathématiques Pures et Appliquées **81** (2002), 311-341

[DS93] A. DeSimone, *Energy minimizers for large ferromagnetic bodies*, Arch. Rational Mech. Anal. **125** (1993), 99-143

[DS99] A. DeSimone, *Energetics of fine domain structures*, Ferroelectrics **222** (1999), 275-284

[DSD00] A. DeSimone, G. Dolzmann, *Material instabilities in nematic elastomers*, Physica D **136** (2000), 175-191

[DSD01] A. DeSimone, G. Dolzmann, *Macroscopic response of nematic elastomers via relaxation of a class of* SO(3)*-invariant energies*, Arch. Rational Mech. Anal. **161** (2002), 181 - 204

[DSJ97] A. DeSimone, R. D. James, *A theory of magnetostriction oriented towards applications*, J. Appl. Phys **81** (1997), 5706-5708

[D99] G. Dolzmann, *Numerical computation of rank-one convex envelopes*, SIAM J. Numer. Anal. **36** (1999), 1621-1635

[D01] G. Dolzmann, *Variational methods for crystalline microstructure – theory and computation*, Habilitationsschrift, Universität Leipzig, 2001

[DK02] G. DOLZMANN, B. KIRCHHEIM, *Liquid-like response of shape-memory materials*, MPI MIS Preprint 53, 2002

[DKMS00] G. DOLZMANN, B. KIRCHHEIM, S. MÜLLER, V. ŠVERÁK, *The two-well problem in three dimensions*, Calc. Var. Partial Differential Equations **10** (2000), 21-40

[DM95] G. DOLZMANN, S. MÜLLER, *Microstructures with finite surface energy: the two-well problem*, Arch. Rational Mech. Anal. **132** (1995), 101-141

[DW00] G. DOLZMANN, N. J. WALKINGTON, *Estimates for numerical approximations of rank-one convex envelopes*, Numer. Math. **85** (2000), 647-663

[DR64] M. J. DUGGIN, W. A. RACHINGER, *The nature of the martensite transformation in a copper-nickel-aluminum alloy*, Acta Metall. **12** (1964), 529-535

[EL01] Y. EFENDIEV, M LUSKIN, *Stability of microstructures for some martensitic transformations*, Math. Comput. Modelling **34** (2001), 1289-1305

[Er77] J. L. ERICKSEN, *Special topics in nonlinear elastostatics*, in: *Advances in applied mechanics* **17** (C. S. Yih, ed.), Academic Press, 1977, 189-244

[Er80] J. L. ERICKSEN, *Some phase transitions in crystals*, Arch. Rational Mech. Anal. **73** (1980), 99-124

[Er81] J. L. ERICKSEN, *Continuous martensitic transitions on thermoelastic solids*, J. Thermal Stresses **4** (1981), 107-119

[Er85] J. L. ERICKSEN, *Some surface defects in unstressed thermoelastic solids*, Arch. Rational Mech. Anal. **88** (1985), 337-345

[Er86] J. L. ERICKSEN, *Constitutive theory for some constrained elastic crystals*, Int. J. Solids and Structures **22**, (1986), 951-964

[Er89] J. L. ERICKSEN, *Weak martensitic transformations in Bravais lattices*, Arch. Rational Mech. Anal. **107** (1989), 23-36

[Er91] J. L. ERICKSEN, *On kinematic conditions of compatibility*, J. Elasticity **26** (1991), 65-74

[FKR81] H. FINKELMANN, H. J. KOCH, G. REHAGE, *Liquid crystalline elastomers – A new type of liquid crystalline materials*, Macromol. Chem. Rapid Commun. **2** (1981), 317-325

[Fo87] I. FONSECA, *Variational methods for elastic crystals*, Arch. Rational Mech. Anal. **97** (1987), 189-220

[FH90] R. L. FOSDICK, B. HERTOG, *Material symmetry and crystals*, Arch. Rational Mech. Anal. **110** (1990), 43-72

[Fr90] D. A. FRENCH, *On the convergence of finite-element approximations of a relaxed variational problem*, SIAM J. Numer. Anal. **27** (1990), 419-436

[Fr00] G. FRIESECKE, *Quasiconvex hulls and recoverable strains in shape-memory alloys*, TMR meeting 'Phase transitions in crystalline solids', Max Planck Institute for Mathematics in the Sciences, Leipzig, 2000

[Gnt58] F. R. GANTMACHER, Matrizenrechnung, Teil I, VEB Deutscher Verlag der Wissenschaften, Berlin, 1958

[GP99] M. K. GOBBERT, A. PROHL, *A discontinuous finite element method for solving a multiwell problem*, SIAM J. Numer. Anal. **37** (1999), 246-268

[GP00] M. K. GOBBERT, A. PROHL, *A comparison of classical and new finite element methods for the computation of laminate microstructure*, Appl. Num. Math. **36** (2001), 155-178

[GL89] L. GOLUBOVIC, T. C. LUBENSKY, *Nonlinear elasticity of amorphous solids*, Phys. Rev. Lett. **63** (1989), 1082-1085

[Gd94] P. A. GREMAUD, *Numerical analysis of a nonconvex variational problem related to solid-solid phase transitions*, SIAM J. Numer. Anal. **31** (1994), 111-127

[Gr83] M. E. GURTIN, *Two-phase deformations of elastic solids*, Arch. Rational Mech. Anal. **84** (1984), 1-29

[Gt50] L. GUTTMAN, *Crystal structures and transformations in Indium-Thallium solid solutions*, Transactions AIME, J. Metals **188** (1950), 1472-1477

[H97] K. F. HANE, *Microstructures in Thermoelastic Materials*, Dissertation, Department of Aerospace Engineering and Mechanics, University of Minnesota, 1997

[HS00] K. F. HANE, T. W. SHIELD, *Microstructure in the Cubic to Trigonal Transition*, Materials Science and Engineering A **291** (2000), 147-159

[Iq99] Z. IQBAL, *Variational methods in solid mechanics*, Ph.D. Thesis, University of Oxford, 1999

[Ja86] R. D. JAMES, *Displacive phase transformations in solids*, J. Mech. Phys. Solids **34** (1986), 359-394

[JH00] R. D. JAMES, K. F. HANE, *Martensitic transformations and shape-memory materials*, Acta Mater. **48** (2000), 197-222

[JK89] R. D. JAMES, D. KINDERLEHRER, *Theory of diffusionless phase transitions*, in: *PDEs and continuum models of phase transitions* (Nice, 1988), Lecture Notes in Phys. **344**, Springer, Berlin, 1989, 51-84

[JK93] R. D. JAMES, D. KINDERLEHRER, *Theory of magnetostriction with applications to* $Tb_xDy_1 - xFe_2$, Phil. Mag. B **68** (1993), 237-274

[Jo72a] F. JOHN, *Bounds for deformations in terms of average strains*, Inequalities, III, Proc. Third Sympos., Univ. California, Los Angeles, Calif., 1969, Academic Press, New York, 1972, 129-145

[Jo72b] F. JOHN, *Uniqueness of non-linear elastic equilibrium for prescribed boundary displacements and sufficiently small strains*, Comm. Pure Appl. Math. **25** (1972), 617-635

[KPt98] F. X. KAYSER, J. W. PATTERSON, *Sir William Chandler Roberts-Austen – His rôle in the development of binary diagrams and modern physical metallurgy*, J. Phase Equilibria **19** (1998), 11-18

[Kh67] A. KHACHATURYAN, *Some questions concerning the theory of phase transformations in solids*, Soviet Physics - Solid State **8** (1967), 2163-2168

[Kh83] A. KHACHATURYAN, *Theory of structural transformations in solids*, Wiley, 1983

[Ki88] D. KINDERLEHRER, *Remarks about equilibrium configurations of crystals*, in: *Material instabilities in continuum mechanics and related mathematical problems* (J. M. Ball, ed.), Oxford University Press, 1988, 217-242

[KP91] D. KINDERLEHRER, P. PEDREGAL, *Characterization of Young measures generated by gradients*, Arch. Rational Mech. Anal. **115** (1991), 329-365

[KP95] D. KINDERLEHRER, P. PEDREGAL, *Remarks on the two-well problem* (unpublished manuscript)

[Kir98] B. KIRCHHEIM, *Lipschitz minimizers of the 3-well problem having gradients of bounded variation*, MPI MIS Preprint #12, 1998

[Kir00] B. KIRCHHEIM, *Geometry and rigidity of microstructures*, Habilitationsschrift, Universität Leipzig, 2001

[K91] R. V. KOHN, *The relaxation of a double-well energy*, Contin. Mech. Thermodyn. **3** (1991), 193-236

[KL00] R. V. KOHN, V. LODS, *Some results about two incompatible elastic strains*, manuscript

[KM92] R. V. KOHN, S. MÜLLER, *Branching of twins near an austenite-twinned-martensite interface*, Phil. Mag. A **66** (1992), 697-715

[KM94] R. V. KOHN, S. MÜLLER, *Surface energy and microstructure in coherent phase transitions*, Comm. Pure Appl. Math. **47** (1994), 405-435

[KN00] R. V. KOHN, B. NIETHAMMER, *Geometrically nonlinear shape-memory polycrystals made from a two-variant material*, M^2AN Math. Model. Numer. Anal. **34** (2000), 377-398

[KS86] R. V. KOHN, G. STRANG, *Optimal design and relaxation of variational problems II*, Comm. Pure Appl. Math. **39** (1986), 139-182

[Kr99] J. KRISTENSEN, *On the non-locality of quasiconvexity*, Ann. Inst. H. Poincaré Anal. Non Linéaire **16** (1999), 1-13

[Kr00] J. KRISTENSEN, *On conditions for polyconvexity*, Proc. Amer. Math. Soc. **128** (2000), 1793-1797

[KnF] I. KUNDLER, H. FINKELMANN, *Strain-induced director reorientation in nematic liquid single crystal elastomers*, Macromol. Chem. Rapid Communications **16** (1995), 679-686

[KpF91] J. KÜPFER, H. FINKELMANN, *Nematic liquid single crystal elastomers*, Macromol. Chem. Rapid Communications **12** (1991), 717-726

[LDR95] H. LEDRET, A. RAOULT, *The quasiconvex envelope of the Saint Venant-Kirchhoff stored energy function*, Proc. Roy. Soc. Edinburgh **125** (1995), 1179-1192

[LL98a] B. LI, M. LUSKIN, *Nonconforming finite element approximation of crystalline microstructure*, Math. Comp. **67** (1998), 917-946

[LL98b] B. LI, M. LUSKIN, *Finite element analysis of microstructure for the cubic to tetragonal transformation*, SIAM J. Numer. Anal. **35** (1998), 376-392

[LL99a] B. LI, M. LUSKIN, *Approximation of a martensitic laminate with varying volume fractions*, M^2AN Math. Model. Numer. Anal. **33** (1999), 67-87.

[LL99b] B. LI, M. LUSKIN, *Theory and computation for the microstructure near the interface between twinned layers and a pure variant of martensite*, Materials Science & Engineering A **273** (1999), 237-240

[LL00] B. LI, M. LUSKIN, *Approximation of a martensitic laminate with varying volume fractions*, M^2AN Math. Model. Numer. Anal. **33** (1999), 67-87.

[Li95] Z. LI, *A numerical method for computing singular minimizers*, Numer. Math. **71** (1995), 317-330

[Li97] Z. LI, *Simultaneous numerical approximation of microstructures and relaxed minimizers*, Numer. Math. **78** (1997), 21-38

[Li98a] Z. LI, *Laminated microstructure in a variational problem with a non-rank-one connected double well potential*, J. Math. Anal. Appl. **217** (1998), 490-500

[Li98b] Z. LI, *Rotational transformation method and some numerical techniques for computing microstructures*, Math. Models Methods Appl. Sci. **8** (1998), 985-1002

[Liu77] F. C. LIU, *A Luzin type property of Sobolev functions*, Indiana Univ. Math. J. 26 (1977), 645-651

[L96a] M. LUSKIN, *Approximation of a laminated microstructure for a rotationally invariant, double well energy density*, Numer. Math. **75** (1996), 205-221

[L96b] M. LUSKIN, *On the computation of crystalline microstructure*, Acta Numer. **5** (1996), 191-257

[L98] M. LUSKIN, *The stability and numerical analysis of martensitic microstructure*, in ENUMATH 97, H. Bock, F. Brezzi, R. Glowinski, G. Kanschat, Y. Kuznetsov, J. Periaux and R. Rannacher, editors, 1998, 54-69

[L00] M. LUSKIN, *On the stability of microstructure for general martensitic transformations*, in: Lectures on Applied Mathematics, H.-J. Bungartz and R. W. Hoppe and C. Zenger, editors, Springer-Verlag, 2000, 31-44.

[Mt00] J. MATOUŠEK, *On directional convexity*, Discrete Comput. Geom. **25** (2001), 389-403

[MP98] J. MATOUŠEK, P. PLECHÁČ, *On functional separately convex hulls*, Discrete Comput. Geom. **19** (1998), 105-130

[M65] N. G. MEYERS, *Quasi-convexity and lower semi-continuity of multiple variational integrals of any order*, Trans. Amer. Math. Soc. **119** (1965) 125-149

[Mi02] G. W. MILTON, *The theory of composites*, Cambridge monographs on applied and computational mathematics, Cambridge University Press, 2002

[Mo52] C. B. MORREY, *Quasi-convexity and the lower semicontinuity of multiple integrals*, Pacific J. Math. **2** (1952), 25-53

[Mo66] C. B. MORREY, *Multiple integrals in the calculus of variations*, Springer, 1966

[M98] S. MÜLLER, *Microstructures, phase transitions and geometry*, in: Proceedings European Congress of Mathematics, Budapest, July 22-26, 1996 (A. Balog et al., eds), Progress in Mathematics, Birkhäuser 1998

[M99a] S. MÜLLER, *A sharp version of Zhang's theorem on truncating sequences of gradients*, Trans. Amer. Math. Soc. **351** (1999), 4585-4597

[M99b] S. MÜLLER, *Variational methods for microstructure and phase transitions*, in: *Proc. C.I.M.E. summer school 'Calculus of variations and geometric evolution problems'*, Cetraro, 1996, (F. Bethuel, G. Huisken, S. Müller, K. Steffen, S. Hildebrandt, M. Struwe, eds.), Springer LNM 1713, 1999

[MS96] S. MÜLLER, V. ŠVERÁK, *Attainment results for the two-well problem by convex integration*, in: *Geometric analysis and the calculus of variations* (J. Jost, ed.), International Press, 1996, 239-251

[MS99a] S. MÜLLER, V. ŠVERÁK, *Convex integration with constraints and applications to phase transitions and partial differential equations*, J. Eur. Math. Soc. (JEMS) **1** (1999), 393-442

[MS99b] S. MÜLLER, V. ŠVERÁK, *Convex integration for Lipschitz mappings and counterexamples to regularity*, Ann. of Math. (2), to appear

[NM90] P. V. NEGRÓN-MARRERO, *A numerical method for detecting singular minimizers of multidimensional problems in nonlinear elasticity*, Numer. Math. **58** (1990), 135-144

[NW93] R. A. NICOLAIDES, N. J. WALKINGTON, *Computation of microstructure utilizing Young measure representations*, Transactions of the Tenth Army Conference on Applied Mathematics and Computing (West Point, NY, 1992), ARO Rep., 93-1, 1993, 57-68

[NW95] R. A. NICOLAIDES, N. J. WALKINGTON, *Strong convergence of numerical solutions to degenerate variational problems*, Math. Comp. **64** (1995), 117-127

[OG94] M. ORTIZ, G. GIOIA, *The morphology and folding patterns of buckling-driven thin-film blisters*, J. Mech. Phys. Solids **42** (1994), 531-559

[OR99] M. ORTIZ, E. REPETTO, *Nonconvex energy minimization and dislocation structures in ductile single crystals*, J. Mech. Phys. Solids **47** (1999), 397-462

[Pa77] G. P. PARRY, *On the crystallographic point groups and on Cauchy symmetry*, Math. Proc. Cambridge Philos. Soc. **82** (1977), 165-175

[Pa81] G. P. PARRY, *On the relative strengths of stability criteria in anisotropic materials*, Mech. Res. Comm. **7** (1980), 93-98

[Pe93] P. PEDREGAL, *Laminates and microstructure*, Europ. J. Appl. Math. **4** (1993), 121-149

[SMST97] SMST-97-Proceedings, Proceedings of the Second International Conference on Shape Memory and Superelastic Technology Conference held in March 1997, Asilomar, California, USA, (A. Pelton, D. Hodgson, S. Russell, and T. Duerig, eds.), Shape Memory Applications, Inc., 2380 Owen Street, Santa Clara, CA

[Pi91] A. C. PIPKIN, *Elastic materials with two preferred sates*, Q. J. Mech. Appl. Math. **44** (1991), 1-15

[Pi84] M. PITTERI, *Reconciliation of local and global symmetries of crystals*, J. Elasticity **14** (1984), 175-190

[PZ98] M. PITTERI, G. ZANZOTTO, *Generic and non-generic cubic-to-monoclinic transitions and their twins*, Acta. Mater. **46** (1998), 225-237

[PZ00] M. Pitteri, G. Zanzotto, *Continuum models for phase transitions and twinning in crystals*, Chapman & Hall, London, 2002

[Pr00] A. Prohl, *An adaptive finite element method for solving a double well problem describing crystalline microstructure*, M^2AN Math. Model. Numer. Anal. **33** (1999), 781-796

[S69] D. Sands, *Introduction to crystallography*, W. A. Benjamin, Inc., 1969, unabridged republication by Dover, 1993

[Sch74] V. Scheffer, *Regularity and irregularity of solutions to nonlinear second order elliptic systems of partial differential equations and inequalities*, Dissertation, Princeton University, 1974

[Sr83] D. Serre, *Formes quadratiques et calcul des variations*, J. Math. Pures Appl. **62** (1983), 177-196

[Sy01] M. Šilhavý, *Relaxation of a class of* SO(n)*-invariant energies related to nematic elastomers*, manuscript

[Si89] S. A. Silling, *Phase changes induced by deformation in isothermal elastic crystals*, J. Mech. Phys. Solids **37** (1989), 293-316

[Sv91] V. Šverák, *On regularity of the Monge-Ampère equation without convexity assumptions*, preprint, Heriot-Watt University, 1991

[Sv92a] V. Šverák, *Rank-one convexity does not imply quasiconvexity*, Proc. Roy. Soc. Edinburgh Sect. A **120** (1992), 293-300

[Sv92b] V. Šverák, *New examples of quasiconvex functions*, Arch. Rational Mech. Anal. **119** (1992), 293-300

[Sv93] V. Šverák, *On the problem of two wells*, in: *Microstructure and phase transitions*, IMA Vol. Appl Math. **54**, (D. Kinderlehrer, R. D. James, M. Luskin and J. Ericksen, eds.), Springer, 1993, 183-189

[Sv95] V. Šverák, *Lower semicontinuity of variational integrals and compensated compactness*, in: Proc. ICM 1994 (S. D. Chatterji, ed.), vol. 2, Birkhäuser, 1995, 1153-1158

[Ta79] L. Tartar, *Estimations de coefficients homogénéises*, in: *Computing methods in applied sciences and engineering*, Lecture Notes in Math. **704**, Springer, 1979, 364-373

[Ta83] L. Tartar, *The compensated compactness method applied to systems of conservation laws*, in: *Systems of nonlinear partial differential equations*, NATO Adv. Sci. Inst. Ser. C: Math. Phys. Sci. **111**, Reidel, 1983, 263-285

[Ta90] L. Tartar, *H-measures, a new approach for studying homogenisation, oscillations and concentration effects in partial differential equations*, Proc. Roy. Soc. Edinburgh Sect. A **115** (1990), 193-230

[Ta93] L. Tartar, *Some remarks on separately convex functions*, in: *Microstructure and phase transitions*, IMA Vol. Appl Math. **54**, (D. Kinderlehrer, R. D. James, M. Luskin and J. Ericksen, eds.), Springer, 1993, 191-204

[Ta95] L. Tartar, *Beyond Young measures*, Meccanica **30** (1995), 505-526

[Tr38] F. J. Terpstra, *Die Darstellung biquadratischer Formen als Summen von Quadraten mit Anwendungen auf die Variationsrechnung*, Math. Ann. **116** (1938), 166-180

[VTA74] G. Van Tendeloo, S. Amelinckx, *Group-theoretical considerations concerning domain formation in ordered alloys*, Acta Cryst. **A30** (1974), 431-440

[WT96] M. Warner, E. M. Terentjev, *Nematic elastomers – a new state of matter?*, Prog. Polym. Sci. **21** (1996), 835-891

[WLR53] M. S. Wechsler, D. S. Lieberman, T. A. Read, *On the theory of the formation of martensite*, Trans. AIME J. Metals **197** (1953), 1503-1515

[WB96] J. Weilepp, H. R. Brand, *Director reorientation in nematic-liquid-single-crystal elastomers by external mechanical stress*, Europhys. Lett. **34** (1996), 495-500

[Yo37] L. C. YOUNG, *Generalized curves and the existence of an attained absolute minimum in the calculus of variations*, Comptes Rendues de la Société des Sciences et des Lettres de Varsovie, classe III **30** (1937), 212-234

[Za92] G. ZANZOTTO, *On the material symmetry group of elastic crystals and the Born rule*, Arch. Rational Mech. Anal. **121** (1992), 1-36

[Z92] K. ZHANG, *A construction of quasiconvex functions with linear growth at infinity*, Ann. Scuola Norm. Sup. Pisa Cl. Sci. (4) **19** (1992), 313-326

[Z98] K. ZHANG, *On various semiconvex hulls in the calculus of variations*, Calc. Var. Partial Differential Equations **6** (1998), 143-16

Index

Lecture Notes in Mathematics

For information about Vols. 1–1619 please contact your bookseller or Springer-Verlag

Vol. 1620: R. Donagi, B. Dubrovin, E. Frenkel, E. Previato, Integrable Systems and Quantum Groups. Montecatini Terme, 1993. Editors:M. Francaviglia, S. Greco. VIII, 488 pages. 1996.

Vol. 1621: H. Bass, M. V. Otero-Espinar, D. N. Rockmore, C. P. L. Tresser, Cyclic Renormalization and Automorphism Groups of Rooted Trees. XXI, 136 pages. 1996.

Vol. 1622: E. D. Farjoun, Cellular Spaces, Null Spaces and Homotopy Localization. XIV, 199 pages. 1996.

Vol. 1623: H.P. Yap, Total Colourings of Graphs. VIII, 131 pages. 1996.

Vol. 1624: V. Brýnzanescu, Holomorphic Vector Bundles over Compact Complex Surfaces. X, 170 pages. 1996.

Vol. 1625: S. Lang, Topics in Cohomology of Groups. VII, 226 pages. 1996.

Vol. 1626: J. Azéma, M. Emery, M. Yor (Eds.), Séminaire de Probabilités XXX. VIII, 382 pages. 1996.

Vol. 1627: C. Graham, Th. G. Kurtz, S. Méléard, Ph. E. Protter, M. Pulvirenti, D. Talay, Probabilistic Models for Nonlinear Partial Differential Equations. Montecatini Terme, 1995. Editors: D. Talay, L. Tubaro. X, 301 pages. 1996.

Vol. 1628: P.-H. Zieschang, An Algebraic Approach to Association Schemes. XII, 189 pages. 1996.

Vol. 1629: J. D. Moore, Lectures on Seiberg-Witten Invariants. VII, 105 pages. 1996.

Vol. 1630: D. Neuenschwander, Probabilities on the Heisenberg Group: Limit Theorems and Brownian Motion. VIII, 139 pages. 1996.

Vol. 1631: K. Nishioka, Mahler Functions and Transcendence. VIII, 185 pages. 1996.

Vol. 1632: A. Kushkuley, Z. Balanov, Geometric Methods in Degree Theory for Equivariant Maps. VII, 136 pages. 1996.

Vol. 1633: H. Aikawa, M. Essén, Potential Theory – Selected Topics. IX, 200 pages. 1996.

Vol. 1634: J. Xu, Flat Covers of Modules. IX, 161 pages. 1996.

Vol. 1635: E. Hebey, Sobolev Spaces on Riemannian Manifolds. X, 116 pages. 1996.

Vol. 1636: M. A. Marshall, Spaces of Orderings and Abstract Real Spectra. VI, 190 pages. 1996.

Vol. 1637: B. Hunt, The Geometry of some special Arithmetic Quotients. XIII, 332 pages. 1996.

Vol. 1638: P. Vanhaecke, Integrable Systems in the realm of Algebraic Geometry. VIII, 218 pages. 1996.

Vol. 1639: K. Dekimpe, Almost-Bieberbach Groups: Affine and Polynomial Structures. X, 259 pages. 1996.

Vol. 1640: G. Boillat, C. M. Dafermos, P. D. Lax, T. P. Liu, Recent Mathematical Methods in Nonlinear Wave Propagation. Montecatini Terme, 1994. Editor: T. Ruggeri. VII, 142 pages. 1996.

Vol. 1641: P. Abramenko, Twin Buildings and Applications to S-Arithmetic Groups. IX, 123 pages. 1996.

Vol. 1642: M. Puschnigg, Asymptotic Cyclic Cohomology. XXII, 138 pages. 1996.

Vol. 1643: J. Richter-Gebert, Realization Spaces of Polytopes. XI, 187 pages. 1996.

Vol. 1644: A. Adler, S. Ramanan, Moduli of Abelian Varieties. VI, 196 pages. 1996.

Vol. 1645: H. W. Broer, G. B. Huitema, M. B. Sevryuk, Quasi-Periodic Motions in Families of Dynamical Systems. XI, 195 pages. 1996.

Vol. 1646: J.-P. Demailly, T. Peternell, G. Tian, A. N. Tyurin, Transcendental Methods in Algebraic Geometry. Cetraro, 1994. Editors: F. Catanese, C. Ciliberto. VII, 257 pages. 1996.

Vol. 1647: D. Dias, P. Le Barz, Configuration Spaces over Hilbert Schemes and Applications. VII. 143 pages. 1996.

Vol. 1648: R. Dobrushin, P. Groeneboom, M. Ledoux, Lectures on Probability Theory and Statistics. Editor: P. Bernard. VIII, 300 pages. 1996.

Vol. 1649: S. Kumar, G. Laumon, U. Stuhler, Vector Bundles on Curves – New Directions. Cetraro, 1995. Editor: M. S. Narasimhan. VII, 193 pages. 1997.

Vol. 1650: J. Wildeshaus, Realizations of Polylogarithms. XI, 343 pages. 1997.

Vol. 1651: M. Drmota, R. F. Tichy, Sequences, Discrepancies and Applications. XIII, 503 pages. 1997.

Vol. 1652: S. Todorcevic, Topics in Topology. VIII, 153 pages. 1997.

Vol. 1653: R. Benedetti, C. Petronio, Branched Standard Spines of 3-manifolds. VIII, 132 pages. 1997.

Vol. 1654: R. W. Ghrist, P. J. Holmes, M. C. Sullivan, Knots and Links in Three-Dimensional Flows. X, 208 pages. 1997.

Vol. 1655: J. Azéma, M. Emery, M. Yor (Eds.), Séminaire de Probabilités XXXI. VIII, 329 pages. 1997.

Vol. 1656: B. Biais, T. Björk, J. Cvitanic, N. El Karoui, E. Jouini, J. C. Rochet, Financial Mathematics. Bressanone, 1996. Editor: W. J. Runggaldier. VII, 316 pages. 1997.

Vol. 1657: H. Reimann, The semi-simple zeta function of quaternionic Shimura varieties. IX, 143 pages. 1997.

Vol. 1658: A. Pumarino, J. A. Rodrýguez, Coexistence and Persistence of Strange Attractors. VIII, 195 pages. 1997.

Vol. 1659: V, Kozlov, V. Maz'ya, Theory of a Higher-Order Sturm-Liouville Equation. XI, 140 pages. 1997.

Vol. 1660: M. Bardi, M. G. Crandall, L. C. Evans, H. M. Soner, P. E. Souganidis, Viscosity Solutions and Applications. Montecatini Terme, 1995. Editors: I. Capuzzo Dolcetta, P. L. Lions. IX, 259 pages. 1997.

Vol. 1661: A. Tralle, J. Oprea, Symplectic Manifolds with no Kähler Structure. VIII, 207 pages. 1997.

Vol. 1662: J. W. Rutter, Spaces of Homotopy Self-Equivalences – A Survey. IX, 170 pages. 1997.

Vol. 1663: Y. E. Karpeshina; Perturbation Theory for the Schrödinger Operator with a Periodic Potential. VII, 352 pages. 1997.

Vol. 1664: M. Väth, Ideal Spaces. V, 146 pages. 1997.

Vol. 1665: E. Giné, G. R. Grimmett, L. Saloff-Coste, Lectures on Probability Theory and Statistics 1996. Editor: P. Bernard. X, 424 pages, 1997.

Vol. 1666: M. van der Put, M. F. Singer, Galois Theory of Difference Equations. VII, 179 pages. 1997.

Vol. 1667: J. M. F. Castillo, M. González, Three-space Problems in Banach Space Theory. XII, 267 pages. 1997.

Vol. 1668: D. B. Dix, Large-Time Behavior of Solutions of Linear Dispersive Equations. XIV, 203 pages. 1997.

Vol. 1669: U. Kaiser, Link Theory in Manifolds. XIV, 167 pages. 1997.

Vol. 1670: J. W. Neuberger, Sobolev Gradients and Differential Equations. VIII, 150 pages. 1997.

Vol. 1671: S. Bouc, Green Functors and *G*-sets. VII, 342 pages. 1997.

Vol. 1672: S. Mandal, Projective Modules and Complete Intersections. VIII, 114 pages. 1997.

Vol. 1673: F. D. Grosshans, Algebraic Homogeneous Spaces and Invariant Theory. VI, 148 pages. 1997.

Vol. 1674: G. Klaas, C. R. Leedham-Green, W. Plesken, Linear Pro-*p*-Groups of Finite Width. VIII, 115 pages. 1997.

Vol. 1675: J. E. Yukich, Probability Theory of Classical Euclidean Optimization Problems. X, 152 pages. 1998.

Vol. 1676: P. Cembranos, J. Mendoza, Banach Spaces of Vector-Valued Functions. VIII, 118 pages. 1997.

Vol. 1677: N. Proskurin, Cubic Metaplectic Forms and Theta Functions. VIII, 196 pages. 1998.

Vol. 1678: O. Krupková, The Geometry of Ordinary Variational Equations. X, 251 pages. 1997.

Vol. 1679: K.-G. Grosse-Erdmann, The Blocking Technique. Weighted Mean Operators and Hardy's Inequality. IX, 114 pages. 1998.

Vol. 1680: K.-Z. Li, F. Oort, Moduli of Supersingular Abelian Varieties. V, 116 pages. 1998.

Vol. 1681: G. J. Wirsching, The Dynamical System Generated by the 3n+1 Function. VII, 158 pages. 1998.

Vol. 1682: H.-D. Alber, Materials with Memory. X, 166 pages. 1998.

Vol. 1683: A. Pomp, The Boundary-Domain Integral Method for Elliptic Systems. XVI, 163 pages. 1998.

Vol. 1684: C. A. Berenstein, P. F. Ebenfelt, S. G. Gindikin, S. Helgason, A. E. Tumanov, Integral Geometry, Radon Transforms and Complex Analysis. Firenze, 1996. Editors: E. Casadio Tarabusi, M. A. Picardello, G. Zampieri. VII, 160 pages. 1998.

Vol. 1685: S. König, A. Zimmermann, Derived Equivalences for Group Rings. X, 146 pages. 1998.

Vol. 1686: J. Azéma, M. Émery, M. Ledoux, M. Yor (Eds.), Séminaire de Probabilités XXXII. VI, 440 pages. 1998.

Vol. 1687: F. Bornemann, Homogenization in Time of Singularly Perturbed Mechanical Systems. XII, 156 pages. 1998.

Vol. 1688: S. Assing, W. Schmidt, Continuous Strong Markov Processes in Dimension One. XII, 137 page. 1998.

Vol. 1689: W. Fulton, P. Pragacz, Schubert Varieties and Degeneracy Loci. XI, 148 pages. 1998.

Vol. 1690: M. T. Barlow, D. Nualart, Lectures on Probability Theory and Statistics. Editor: P. Bernard. VIII, 237 pages. 1998.

Vol. 1691: R. Bezrukavnikov, M. Finkelberg, V. Schechtman, Factorizable Sheaves and Quantum Groups. X, 282 pages. 1998.

Vol. 1692: T. M. W. Eyre, Quantum Stochastic Calculus and Representations of Lie Superalgebras. IX, 138 pages. 1998.

Vol. 1694: A. Braides, Approximation of Free-Discontinuity Problems. XI, 149 pages. 1998.

Vol. 1695: D. J. Hartfiel, Markov Set-Chains. VIII, 131 pages. 1998.

Vol. 1696: E. Bouscaren (Ed.): Model Theory and Algebraic Geometry. XV, 211 pages. 1998.

Vol. 1697: B. Cockburn, C. Johnson, C.-W. Shu, E. Tadmor, Advanced Numerical Approximation of Nonlinear Hyperbolic Equations. Cetraro, Italy, 1997. Editor: A. Quarteroni. VII, 390 pages. 1998.

Vol. 1698: M. Bhattacharjee, D. Macpherson, R. G. Möller, P. Neumann, Notes on Infinite Permutation Groups. XI, 202 pages. 1998.

Vol. 1699: A. Inoue,Tomita-Takesaki Theory in Algebras of Unbounded Operators. VIII, 241 pages. 1998.

Vol. 1700: W. A. Woyczyński, Burgers-KPZ Turbulence, XI, 318 pages. 1998.

Vol. 1701: Ti-Jun Xiao, J. Liang, The Cauchy Problem of Higher Order Abstract Differential Equations, XII, 302 pages. 1998.

Vol. 1702: J. Ma, J. Yong, Forward-Backward Stochastic Differential Equations and Their Applications. XIII, 270 pages. 1999.

Vol. 1703: R. M. Dudley, R. Norvaiša, Differentiability of Six Operators on Nonsmooth Functions and p-Variation. VIII, 272 pages. 1999.

Vol. 1704: H. Tamanoi, Elliptic Genera and Vertex Operator Super-Algebras. VI, 390 pages. 1999.

Vol. 1705: I. Nikolaev, E. Zhuzhoma, Flows in 2-dimensional Manifolds. XIX, 294 pages. 1999.

Vol. 1706: S. Yu. Pilyugin, Shadowing in Dynamical Systems. XVII, 271 pages. 1999.

Vol. 1707: R. Pytlak, Numerical Methods for Optimal Control Problems with State Constraints. XV, 215 pages. 1999.

Vol. 1708: K. Zuo, Representations of Fundamental Groups of Algebraic Varieties. VII, 139 pages. 1999.

Vol. 1709: J. Azéma, M. Émery, M. Ledoux, M. Yor (Eds), Séminaire de Probabilités XXXIII. VIII, 418 pages. 1999.

Vol. 1710: M. Koecher, The Minnesota Notes on Jordan Algebras and Their Applications. IX, 173 pages. 1999.

Vol. 1711: W. Ricker, Operator Algebras Generated by Commuting Projećtions: A Vector Measure Approach. XVII, 159 pages. 1999.

Vol. 1712: N. Schwartz, J. J. Madden, Semi-algebraic Function Rings and Reflectors of Partially Ordered Rings. XI, 279 pages. 1999.

Vol. 1713: F. Bethuel, G. Huisken, S. Müller, K. Steffen, Calculus of Variations and Geometric Evolution Problems. Cetraro, 1996. Editors: S. Hildebrandt, M. Struwe. VII, 293 pages. 1999.

Vol. 1714: O. Diekmann, R. Durrett, K. P. Hadeler, P. K. Maini, H. L. Smith, Mathematics Inspired by Biology. Martina Franca, 1997. Editors: V. Capasso, O. Diekmann. VII, 268 pages. 1999.

Vol. 1715: N. V. Krylov, M. Röckner, J. Zabczyk, Stochastic PDE's and Kolmogorov Equations in Infinite Dimensions. Cetraro, 1998. Editor: G. Da Prato. VIII, 239 pages. 1999.

Vol. 1716: J. Coates, R. Greenberg, K. A. Ribet, K. Rubin, Arithmetic Theory of Elliptic Curves. Cetraro, 1997. Editor: C. Viola. VIII, 260 pages. 1999.
Vol. 1717: J. Bertoin, F. Martinelli, Y. Peres, Lectures on Probability Theory and Statistics. Saint-Flour, 1997. Editor: P. Bernard. IX, 291 pages. 1999.
Vol. 1718: A. Eberle, Uniqueness and Non-Uniqueness of Semigroups Generated by Singular Diffusion Operators. VIII, 262 pages. 1999.
Vol. 1719: K. R. Meyer, Periodic Solutions of the N-Body Problem. IX, 144 pages. 1999.
Vol. 1720: D. Elworthy, Y. Le Jan, X-M. Li, On the Geometry of Diffusion Operators and Stochastic Flows. IV, 118 pages. 1999.
Vol. 1721: A. Iarrobino, V. Kanev, Power Sums, Gorenstein Algebras, and Determinantal Loci. XXVII, 345 pages. 1999.
Vol. 1722: R. McCutcheon, Elemental Methods in Ergodic Ramsey Theory. VI, 160 pages. 1999.
Vol. 1723: J. P. Croisille, C. Lebeau, Diffraction by an Immersed Elastic Wedge. VI, 134 pages. 1999.
Vol. 1724: V. N. Kolokoltsov, Semiclassical Analysis for Diffusions and Stochastic Processes. VIII, 347 pages. 2000.
Vol. 1725: D. A. Wolf-Gladrow, Lattice-Gas Cellular Automata and Lattice Boltzmann Models. IX, 308 pages. 2000.
Vol. 1726: V. Marić, Regular Variation and Differential Equations. X, 127 pages. 2000.
Vol. 1727: P. Kravanja M. Van Barel, Computing the Zeros of Analytic Functions. VII, 111 pages. 2000.
Vol. 1728: K. Gatermann Computer Algebra Methods for Equivariant Dynamical Systems. XV, 153 pages. 2000.
Vol. 1729: J. Azéma, M. Émery, M. Ledoux, M. Yor Séminaire de Probabilités XXXIV. VI, 431 pages. 2000.
Vol. 1730: S. Graf, H. Luschgy, Foundations of Quantization for Probability Distributions. X, 230 pages. 2000.
Vol. 1731: T. Hsu, Quilts: Central Extensions, Braid Actions, and Finite Groups. XII, 185 pages. 2000.
Vol. 1732: K. Keller, Invariant Factors, Julia Equivalences and the (Abstract) Mandelbrot Set. X, 206 pages. 2000.
Vol. 1733: K. Ritter, Average-Case Analysis of Numerical Problems. IX, 254 pages. 2000.
Vol. 1734: M. Espedal, A. Fasano, A. Mikelić, Filtration in Porous Media and Industrial Applications. Cetraro 1998. Editor: A. Fasano. 2000.
Vol. 1735: D. Yafaev, Scattering Theory: Some Old and New Problems. XVI, 169 pages. 2000.
Vol. 1736: B. O. Turesson, Nonlinear Potential Theory and Weighted Sobolev Spaces. XIV, 173 pages. 2000.
Vol. 1737: S. Wakabayashi, Classical Microlocal Analysis in the Space of Hyperfunctions. VIII, 367 pages. 2000.
Vol. 1738: M. Émery, A. Nemirovski, D. Voiculescu, Lectures on Probability Theory and Statistics. XI, 356 pages. 2000.
Vol. 1739: R. Burkard, P. Deuflhard, A. Jameson, J.-L. Lions, G. Strang, Computational Mathematics Driven by Industrial Problems. Martina Franca, 1999. Editors: V. Capasso, H. Engl, J. Periaux. VII, 418 pages. 2000.
Vol. 1740: B. Kawohl, O. Pironneau, L. Tartar, J.-P. Zolesio, Optimal Shape Design. Tróia, Portugal 1999. Editors: A. Cellina, A. Ornelas. IX, 388 pages. 2000.
Vol. 1741: E. Lombardi, Oscillatory Integrals and Phenomena Beyond all Algebraic Orders. XV, 413 pages. 2000.
Vol. 1742: A. Unterberger, Quantization and Non-holomorphic Modular Forms. VIII, 253 pages. 2000.
Vol. 1743: L. Habermann, Riemannian Metrics of Constant Mass and Moduli Spaces of Conformal Structures. XII, 116 pages. 2000.
Vol. 1744: M. Kunze, Non-Smooth Dynamical Systems. X, 228 pages. 2000.
Vol. 1745: V. D. Milman, G. Schechtman, Geometric Aspects of Functional Analysis. Israel Seminar 1999-2000. VIII, 289 pages. 2000.
Vol. 1746: A. Degtyarev, I. Itenberg, V. Kharlamov, Real Enriques Surfaces. XVI, 259 pages. 2000.
Vol. 1747: L. W. Christensen, Gorenstein Dimensions. VIII, 204 pages. 2000.
Vol. 1748: M. Ruzicka, Electrorheological Fluids: Modeling and Mathematical Theory. XV, 176 pages. 2001.
Vol. 1749: M. Fuchs, G. Seregin, Variational Methods for Problems from Plasticity Theory and for Generalized Newtonian Fluids. VI, 269 pages. 2001.
Vol. 1750: B. Conrad, Grothendieck Duality and Base Change. X, 296 pages. 2001.
Vol. 1751: N. J. Cutland, Loeb Measures in Practice: Recent Advances. XI, 111 pages. 2001.
Vol. 1752: Y. V. Nesterenko, P. Philippon, Introduction to Algebraic Independence Theory. XIII, 256 pages. 2001.
Vol. 1753: A. I. Bobenko, U. Eitner, Painlevé Equations in the Differential Geometry of Surfaces. VI, 120 pages. 2001.
Vol. 1754: W. Bertram, The Geometry of Jordan and Lie Structures. XVI, 269 pages. 2001.
Vol. 1755: J. Azéma, M. Émery, M. Ledoux, M. Yor, Séminaire de Probabilités XXXV. VI, 427 pages. 2001.
Vol. 1756: P. E. Zhidkov, Korteweg de Vries and Nonlinear Schrödinger Equations: Qualitative Theory. VII, 147 pages. 2001.
Vol. 1757: R. R. Phelps, Lectures on Choquet's Theorem. VII, 124 pages. 2001.
Vol. 1758: N. Monod, Continuous Bounded Cohomology of Locally Compact Groups. X, 214 pages. 2001.
Vol. 1759: Y. Abe, K. Kopfermann, Toroidal Groups. VIII, 133 pages. 2001.
Vol. 1760: D. Filipović, Consistency Problems for Heath-Jarrow-Morton Interest Rate Models. VIII, 134 pages. 2001.
Vol. 1761: C. Adelmann, The Decomposition of Primes in Torsion Point Fields. VI, 142 pages. 2001.
Vol. 1762: S. Cerrai, Second Order PDE's in Finite and Infinite Dimension. IX, 330 pages. 2001.
Vol. 1763: J.-L. Loday, A. Frabetti, F. Chapoton, F. Goichot, Dialgebras and Related Operads. IV, 132 pages. 2001.
Vol. 1764: A. Cannas da Silva, Lectures on Symplectic Geometry. XII, 217 pages. 2001.
Vol. 1765: T. Kerler, V. V. Lyubashenko, Non-Semisimple Topological Quantum Field Theories for 3-Manifolds with Corners. VI, 379 pages. 2001.
Vol. 1766: H. Hennion, L. Hervé, Limit Theorems for Markov Chains and Stochastic Properties of Dynamical Systems by Quasi-Compactness. VIII, 145 pages. 2001.
Vol. 1767: J. Xiao, Holomorphic Q Classes. VIII, 112 pages. 2001.
Vol. 1768: M.J. Pflaum, Analytic and Geometric Study of Stratified Spaces. VIII, 230 pages. 2001.
Vol. 1769: M. Alberich-Carramiñana, Geometry of the Plane Cremona Maps. XVI, 257 pages. 2002.

Vol. 1770: H. Gluesing-Luerssen, Linear Delay-Differential Systems with Commensurate Delays: An Algebraic Approach. VIII, 176 pages. 2002.
Vol. 1771: M. Émery, M. Yor, Séminaire de Probabilités 1967-1980. A Selection in Martingale Theory. IX, 553 pages. 2002.
Vol. 1772: F. Burstall, D. Ferus, K. Leschke, F. Pedit, U. Pinkall, Conformal Geometry of Surfaces in S^4. VII, 89 pages. 2002.
Vol. 1773: Z. Arad, M. Muzychuk, Standard Integral Table Algebras Generated by a Non-real Element of Small Degree. X, 126 pages. 2002.
Vol. 1774: V. Runde, Lectures on Amenability. XIV, 296 pages. 2002.
Vol. 1775: W. H. Meeks, A. Ros, H. Rosenberg, The Global Theory of Minimal Surfaces in Flat Spaces. Martina Franca 1999. Editor: G. P. Pirola. X, 117 pages. 2002.
Vol. 1776: K. Behrend, C. Gomez, V. Tarasov, G. Tian, Quantum Comohology. Cetraro 1997. Editors: P. de Bartolomeis, B. Dubrovin, C. Reina. VIII, 319 pages. 2002.
Vol. 1777: E. García-Río, D. N. Kupeli, R. Vázquez-Lorenzo, Osserman Manifolds in Semi-Riemannian Geometry. XII, 166 pages. 2002.
Vol. 1778: H. Kiechle, Theory of K-Loops. X, 186 pages. 2002.
Vol. 1779: I. Chueshov, Monotone Random Systems. VIII, 234 pages. 2002.
Vol. 1780: J. H. Bruinier, Borcherds Products on O(2,1) and Chern Classes of Heegner Divisors. VIII, 152 pages. 2002.
Vol. 1781: E. Bolthausen, E. Perkins, A. van der Vaart, Lectures on Probability Theory and Statistics. Ecole d' Eté de Probabilités de Saint-Flour XXIX-1999. Editor: P. Bernard. VIII, 466 pages. 2002.
Vol. 1782: C.-H. Chu, A. T.-M. Lau, Harmonic Functions on Groups and Fourier Algebras. VII, 100 pages. 2002.
Vol. 1783: L. Grüne, Asymptotic Behavior of Dynamical and Control Systems under Perturbation and Discretization. IX, 231 pages. 2002.
Vol. 1784: L.H. Eliasson, S. B. Kuksin, S. Marmi, J.-C. Yoccoz, Dynamical Systems and Small Divisors. Cetraro, Italy 1998. Editors: S. Marmi, J.-C. Yoccoz. VIII, 199 pages. 2002.
Vol. 1785: J. Arias de Reyna, Pointwise Convergence of Fourier Series. XVIII, 175 pages. 2002.
Vol. 1786: S. D. Cutkosky, Monomialization of Morphisms from 3-Folds to Surfaces. V, 235 pages. 2002.
Vol. 1787: S. Caenepeel, G. Militaru, S. Zhu, Frobenius and Separable Functors for Generalized Module Categories and Nonlinear Equations. XIV, 354 pages. 2002.
Vol. 1788: A. Vasil'ev, Moduli of Families of Curves for Conformal and Quasiconformal Mappings.IX, 211 pages. 2002.
Vol. 1789: Y. Sommerhäuser, Yetter-Drinfel'd Hopf algebras over groups of prime order. V, 157 pages. 2002.
Vol. 1790: X. Zhan, Matrix Inequalities. VII, 116 pages. 2002.
Vol. 1791: M. Knebusch, D. Zhang, Manis Valuations and Prüfer Extensions I: A new Chapter in Commutative Algebra. VI, 267 pages. 2002.
Vol. 1792: D. D. Ang, R. Gorenflo, V. K. Le, D. D. Trong, Moment Theory and Some Inverse Problems in Potential Theory and Heat Conduction. VIII, 183 pages. 2002.
Vol. 1793: J. Cortés Monforte, Geometric, Control and Numerical Aspects of Nonholonomic Systems. XV, 219 pages. 2002.
Vol. 1794: N. Pytheas Fogg, Substitution in Dynamics, Arithmetics and Combinatorics. Editors: V. Berthé, S. Ferenczi, C. Mauduit, A. Siegel. XVII, 402 pages. 2002.
Vol. 1795: H. Li, Filtered-Graded Transfer in Using Noncommutative Gröbner Bases. IX, 197 pages. 2002.
Vol. 1796: J.M. Melenk, hp-Finite Element Methods for Singular Perturbations. XIV, 318 pages. 2002.
Vol. 1797: B. Schmidt, Characters and Cyclotomic Fields in Finite Geometry. VIII, 100 pages. 2002.
Vol. 1798: W.M. Oliva, Geometric Mechanics. XI, 270 pages. 2002.
Vol. 1799: H. Pajot, Notes on Analytic Capacity Notes on Analytic Capacity, Rectifiability, Menger Curvature and the Cauchy Integral. VIII, 119 pages. 2002.
Vol. 1801: J. Azéma, M. Émery, M. Ledoux, M. Yor, Séminaire de Probabilités XXXVI. VIII, 499 pages. 2003.
Vol. 1802: V. Capasso, E. Merzbach, B.G. Ivanoff, M. Dozzi, R. Dalang, T. Mountford, Topics in Spatial Stochastic Processes. Martina Franca, Italy 2001. Editor: E. Merzbach. VIII, 241 pages. 2003.
Vol. 1803: G. Dolzmann, Variational Methods for Crystalline Microstructure - Analysis and Computation. VIII, 212 pages. 2003.
Vol. 1804: I. Cherednik, Ya. Markov, R. Howe, G. Lusztig, Iwahori-Hecke Algebras and their Representation Theory. Martina Franca, Italy 1999. Editors: V. Baldoni, D. Barbasch. X, 103 pages. 2003.
Vol. 1805: F. Cao, Geometric Curve Evolution and Image Processing. VIII, 200 pages. 2003.
Vol. 1806: H. Broer, I. Hoveijn. G. Lunther, G. Vegter, Bifurcations in Hamiltonian Systems. Computing Singularities by Gröbner Bases. XIV, 169 pages. 2003.
Vol. 1807: V. D. Milman, G. Schechtman, Geometric Aspects of Functional Analysis. Israel Seminar 2000-2002. VIII, 429 pages. 2003.

Recent Reprints and New Editions

Vol. 1200: V. D. Milman, G. Schechtman, Asymptotic Theory of Finite Dimensional Normed Spaces. 1986. – Corrected Second Printing. X, 156 pages. 2001.
Vol. 1618: G. Pisier, Similarity Problems and Completely Bounded Maps. 1995 – Second, Expanded Edition VII, 198 pages. 2001.
Vol. 1629: J. D. Moore, Lectures on Seiberg-Witten Invariants. 1997 – Second Edition. VIII, 121 pages. 2001.
Vol. 1638: P. Vanhaecke, Integrable Systems in the realm of Algebraic Geometry. 1996 – Second Edition. X, 256 pages. 2001.
Vol. 1702: J. Ma, J. Yong, Forward-Backward Stochastic Differential Equations and Their Applications. 1999. – Corrected Second Printing. XIII, 270 pages. 2000.

4. Lecture Notes are printed by photo-offset from the master-copy delivered in camera-ready form by the authors. Springer-Verlag provides technical instructions for the preparation of manuscripts. Macro packages in TEX, LATEX2e, LATEX2.09 are available from Springer's web-pages at

http://www.springer.de/math/authors/b-tex.html.

Careful preparation of the manuscripts will help keep production time short and ensure satisfactory appearance of the finished book.

The actual production of a Lecture Notes volume takes approximately 12 weeks.

5. Authors receive a total of 50 free copies of their volume, but no royalties. They are entitled to a discount of 33.3% on the price of Springer books purchase for their personal use, if ordering directly from Springer-Verlag.

Commitment to publish is made by letter of intent rather than by signing a formal contract. Springer-Verlag secures the copyright for each volume. Authors are free to reuse material contained in their LNM volumes in later publications: A brief written (or e-mail) request for formal permission is sufficient.

Addresses:

Professor Jean-Michel Morel
CMLA, École Normale Supérieure de Cachan
61 Avenue du Président Wilson
94235 Cachan Cedex France
e-mail: Jean-Michel.Morel@cmla.ens-cachan.fr

Professor Bernard Teissier
Institut de Mathématiques de Jussieu
Equipe "Géométrie et Dynamique"
175 rue du Chevaleret
75013 PARIS
e-mail: Teissier@ens.fr

Professor F. Takens, Mathematisch Instituut
Rijksuniversiteit Groningen, Postbus 800
9700 AV Groningen, The Netherlands
e-mail: F.Takens@math.rug.nl

Springer-Verlag, Mathematics Editorial, Tiergartenstr. 17
D-69121 Heidelberg, Germany
Tel.: +49 (6221) 487-701
Fax: +49 (6221) 487-355
e-mail: lnm@Springer.de